Probability and Mathematical Statistics (Continued)

WILLIAMS • Diffusions, Markov Processes, and Mart
Foundations
ZACKS • Theory of Statistical Inference

Applied Probability and Statistics

ANDERSON, AUQUIER, HAUCK, OAKES, VANDAELE, and WEISBERG • Statistical Methods for Comparative Studies
ARTHANARI and DODGE • Mathematical Programming in Statistics
BAILEY • The Elements of Stochastic Processes with Applications to the Natural Sciences
BAILEY • Mathematics, Statistics and Systems for Health
BARNETT and LEWIS • Outliers in Statistical Data
BARTHOLEMEW • Stochastic Models for Social Processes, *Second Edition*
BARTHOLOMEW and FORBES • Statistical Techniques for Manpower Planning
BECK and ARNOLD • Parameter Estimation in Engineering and Science
BELSLEY, KUH, and WELSCH • Regression Diagnostics: Identifying Influential Data and Sources of Collinearity
BENNETT and FRANKLIN • Statistical Analysis in Chemistry and the Chemical Industry
BHAT • Elements of Applied Stochastic Processes
BLOOMFIELD • Fourier Analysis of Time Series: An Introduction
BOX • R. A. Fisher, The Life of a Scientist
BOX and DRAPER • Evolutionary Operation: A Statistical Method for Process Improvement
BOX, HUNTER, and HUNTER • Statistics for Experimenters: An Introduction to Design, Data Analysis, and Model Building
BROWN and HOLLANDER • Statistics: A Biomedical Introduction
BROWNLEE • Statistical Theory and Methodology in Science and Engineering, *Second Edition*
BURY • Statistical Models in Applied Science
CHAMBERS • Computational Methods for Data Analysis
CHATTERJEE and PRICE • Regression Analysis by Example
CHERNOFF and MOSES • Elementary Decision Theory
CHOW • Analysis and Control of Dynamic Economic Systems
CLELLAND, BROWN, and deCANI • Basic Statistics with Business Applications, *Second Edition*
COCHRAN • Sampling Techniques, *Third Edition*
COCHRAN and COX • Experimental Designs, *Second Edition*
CONOVER • Practical Nonparametric Statistics, *Second Edition*
CORNELL • Experiments with Mixtures: Designs, Models and The Analysis of Mixture Data
COX • Planning of Experiments
DANIEL • Biostatistics: A Foundation for Analysis in the Health Sciences, *Second Edition*
DANIEL • Applications of Statistics to Industrial Experimentation
DANIEL and WOOD • Fitting Equations to Data: Computer Analysis of Multifactor Data, *Second Edition*
DAVID • Order Statistics, *Second Editiion*
DEMING • Sample Design in Business Research
DODGE and ROMIG • Sampling Inspection Tables, *Second Edition*
DRAPER and SMITH • Applied Regression Analysis
DUNN • Basic Statistics: A Primer for the Biomedical Sciences, *Second Edition*
DUNN and CLARK • Applied Statistics: Analysis of Variance and Regression
ELANDT-JOHNSON • Probability Models and Statistical Methods in Genetics
ELANDT-JOHNSON and JOHNSON • Survival Models and Data Analysis

continued on back

Theory and Applications of Stochastic Differential Equations

Theory and Applications of Stochastic Differential Equations

ZEEV SCHUSS

Department of Mathematical Sciences
Tel-Aviv University, Ramat-Aviv
Israel

John Wiley & Sons
New York • Chichester • Brisbane • Toronto

Library of Congress Cataloging in Publication Data:

Schuss, Zeev, 1937–

Theory and applications of stochastic differential equations.

(Wiley series in probability and mathematical statistics)

Includes bibliographical references and index.

1. Stochastic differential equations. I. Title.

QA274.23.S38 519.2 80-14767

ISBN 0-471-04394-X

Printed in the United States of America

10 9 8 7 6 5 4 3 2

To my good friend Bernie Matkowsky

Preface

The purpose of this book is to present sources, theory, and applications of stochastic differential equations of Itô's type; that is, differential equations that contain white noise. It gives the basic theory and a wide range of applications. The main theme of the book is the study of first passage problems by modern singular perturbation methods and their role in various fields of science. The material is so presented as to make the theory available to the applied mathematicians, physicists, chemists, and engineers, who are usually well versed in classical analysis but may feel uneasy in the realms of modern probability and measure theory. The prerequisites for this book are therefore a working knowledge of advanced calculus, elementary theory of ordinary and partial differential equations, and of course, elementary probability theory. The professional probabilist will find here some new analytic methods for the computation of first passage times, transition and exit probabilities, and other quantities of interest. Special stress has been put on modeling phenomena in a variety of scientific areas by stochastic differential equations. Thus phenomena in chemical kinetics, solid-state diffusion, genetics, filtering of signals from noise, and more are modeled.

Since Einstein's creation of the mathematical theory of the Brownian motion and molecular diffusion, much scientific work has been done on its applications in such diverse areas as molecular and atomic physics, chemical kinetics, solid-state theory, stability of structures, population genetics, communication, and many other branches of the natural and social sciences and engineering. The most prominent work in the early stages of the theory of stochastic differential equations was done by Einstein, Smoluchowski, Langevin, Ornstein and Uhlenbeck, and Kramers and was summarized in Chandrasekhar's fundamental paper in 1943. The mathematical theory of stochastic differential equations was developed considerably in the last 25 years and several very rigorous mathematical texts on this subject have appeared. Some very important results were discovered by the mathematical researchers in this field; in particular, the equations for first passage times and exit distributions were derived. The Itô and

Stratonovich calculi in particular gave the theory of stochastic differential equations an enormous push forward. Unfortunately, the gap between the mathematical theory and the sources of the problems has widened to such an extent that, generally, physicists, chemists, and engineers remained unaware of the mathematical techniques now available, while the mathematicians remained unaware of the sources and applications of their theories. The complexity of the theory and the mathematical rigor made the mathematical texts virtually unapproachable to the nonspecialist. This book is an attempt to bridge this gap.

Chapters 1 and 2 present the relevant probability theory, construction of the Brownian motion, and the theory of Itô and Stratonovich integration and calculus. The more demanding and mathematically rigorous material has been relegated to separate sections marked by an asterisk. Such sections should be omitted in the first reading of the book. The basic theory of stochastic differential equations is presented in Chapters 3 through 5. Special attention should be given to the exercises because they contain many classical applications of the theory; in particular, Einstein's and Smoluchowski's theories of diffusion and their applications are contained in the exercises. Chapter 4 also establishes the connection between Markov and diffusion processes on the one hand and solutions of stochastic differential equations on the other. In Chapter 5 the relationship between stochastic differential equations and partial differential equations is demonstrated; the basic equations of Fokker–Planck, Kolmogorov, Dynkin, Feynman, and Kac are derived; and boundary behavior is discussed. A method for the treatment of first passage problems by partial differential equations is developed through the Itô calculus. The main contribution of this book are Chapters 6 through 9. Chapter 6 presents the asymptotic theory of stochastic differential equations and its applications in statistical mechanics, transport theory, and mathematical genetics. In Chapter 7 singular perturbation problems that arise from the Smoluchowski–Kramers theory are treated by a new method, developed by B. Matkowsky and myself. Physical applications of the theory are presented in Chapter 8. Mathematical models of chemical reactions, diffusion, and conductivity in ionic crystals are given. Chapter 9 contains elements of filtering theory in state space and the role of first passage times is shown. Finally, Chapter 10 contains Smoluchowski's theory in the context of the kinetic theory of gases, and a short review of some basic notions in the theory of classical mechanics and partial differential equations. The uniform mathematical treatment reduces many of the problems to that of the determination of the expected first exit time by solving singularly perturbed boundary value problems for partial differential equations. The singular perturbation methods presented in this book lead to explicit expressions for probabilistic and consequently physical quantities of interest, such as the steric factor in the Arrhenius law, reaction rates in multistage chemical reactions, the diffusion tensor for atomic migration in crystals, the electric conductivity in ionic crystals, and the "click" frequency in a FM filter. It is my hope that the book will bridge the gap between the mathematical theory of stochastic differential equations and the natural sciences by giving the scientist a new mathematical tool and by giving the mathematicians some insight into the role of stochastic differential equations in the sciences.

My interest in stochastic differential equations was a consequence of a course on this subject given by H. McKean, Jr., at the Weizmann Institute of Science in Rehovot, for which I am grateful. The idea of the book began with a series of lectures in the Applied Mathematics Seminar that I gave at RPI in 1975–1976. It continued with a set of lecture notes from the Applied Mathematics Institute, the University of Delaware in 1976–1977, and completed at Tel-Aviv University and Northwestern University in the summers of 1978 and 1979. I am grateful to these institutions for their hospitality. The book is based on my joint work with B. Matkowsky, E. Larsen, B. Levikson, S. Eliezer and B. Z. Bobrovsky, for whose cooperation I am deeply indebted. The preparation of this manuscript was partially supported by AFOSR grants 77-3372 and 78-3602B in 1977–1980 at the University of Delaware and Northwestern University.

ZEEV SCHUSS

Ramat-Aviv, Israel
September 1980

Contents

CHAPTER 1

Review of Probability Theory

1.1 EVENTS AND SAMPLE SPACES

Consider the experiment of tossing a fair coin three times. The possible outcomes of this experiment are HHH, HHT, HTH, THH, HTT, THT, TTH, and TTT, where H denotes heads and T denotes tails. Each possible outcome of the experiment is called an *elementary event*. Thus there are eight elementary events in the experiment of tossing a coin three times. More complicated events can be expressed as combinations of the elementary events. Thus the event "two or more heads turn up" in the coin-tossing experiment, which we denote by D, consists of the elementary events HHH, HHT, HTH, THH; that is,

$$D = \{\text{HHH}, \text{HHT}, \text{HTH}, \text{THH}\}.$$

The set Ω of all elementary events corresponding to an experiment is called a *sample space* and each elementary event, denoted by ω, is called a *point* in Ω. In the particular example under consideration, we have

$$\Omega = \{\text{HHH}, \text{HHT}, \text{HTH}, \text{THH}, \text{TTH}, \text{THT}, \text{HTT}, \text{TTT}\}.$$

We write $\omega \in \Omega$ to read "ω is a point in (or an element of) Ω." Any event A that consists of elementary events is a subset of Ω. In particular, the impossible event $\varnothing$, that is, an event that contains no elementary events, is called *the empty* set. Let A and B be events in Ω; then we say that *A is a subset of B* if every element of A is also an element of B, and we write $A \subset B$. Obviously, $A \subset \Omega$, $\varnothing \subset A$, and $A \subset A$. For example, the set D in the coin-tossing experiment is a subset of the set (event) E: "at least one head turns up." More specifically,

$$\{\text{HHH},\text{HHT},\text{HTH},\text{THH}\}=D\subset E$$
$$=\{\text{HHH},\text{THH},\text{HTH},\text{HHT},\text{TTH},\text{THT},\text{HTT}\}.$$

We say that two subsets A and B of Ω are *equal* if they consist of the same elements; that is, $A=B$ if $A\subset B$ and $B\subset A$. We say that two subsets of Ω (two events), A and B, are *disjoint* if they have no common elements. Thus the set F: "at least two tails turn up" and the set D in the coin-tossing experiment are disjoint sets. However, the sets E and F are not disjoint.

Very often, the sample space contains infinitely many points (elementary events). Consider, for example, the experiment of sampling the velocities of the molecules of a monatomic gas consisting of n molecules of mass m in thermal equilibrium. Let $\mathbf{v}_i=(v_i^1, v_i^2, v_i^3)^T$ $(i=1,2,\ldots,n)$ be the velocity vectors of the molecules at the moment the experiment is conducted. We assume that the gas is ideal, thus neglecting the potential energy of intermolecular forces. Denoting the total energy of the gas by E, we have

$$\sum_{i=1}^{n} |\mathbf{v}_i|^2=\frac{2E}{m},$$

where $|\mathbf{v}_i|^2=\mathbf{v}_i\cdot\mathbf{v}_i=\sum_{j=1}^{3}v_i^jv_i^j$. Since E is assumed constant, we see that any outcome of the experiment is a point on the surface of the $3n$-dimensional sphere S of radius $(2E/m)^{1/2}$. We may identify the sample space for this experiment with the set of all points on the surface S. A typical event is G: "the first component of $\mathbf{v}_i$ satisfies the inequality $a<\mathbf{v}_i^1<b$." The set G is therefore a spherical zone on S.

Given two subsets A and B of a sample space Ω, we denote by $A\cup B$ the subset of Ω whose elements are those which belong to A or to B. The set $A\cup B$ is called the *union* of A and B. More generally, given a finite or infinite sequence $\{A_j\}, j=1,2,\ldots,$ of subsets of Ω, we denote by

$$A=A_1\cup A_2\cup\cdots\equiv\bigcup_j A_j$$

the set whose elements are those which belong to *at least* one of the sets A_j. Thus the event $\cup_j A_j$ occurs if at least one of the events A_j $(j=1,2,\ldots)$ does. The set $A=\cup_j A_j$ is called the union of the sets A_j. It is easy to see that $A\cup A=A$, $A\cup\Omega=\Omega$ and $A\cup\varnothing=A$. If $A\subset\Omega$ and $B\subset\Omega$, then the set $A-B$ consists of those elements of A which are not elements of B. The set $A-B$ is called the *difference* of A and B. Thus in the coin-tossing example,

we have

$$E-D=\{\text{HTT},\text{THT},\text{TTH}\},$$

that is, $E-D$: "exactly one head turns up," so that $E-D$ occurs if E occurs but not D. We also have $D-E=\varnothing$. Obviously, $A-\varnothing=A$, $A-A=\varnothing$, and $A-\Omega=\varnothing$. If A and B are disjoint events, then $A-B=A$. Given a finite or infinite sequence $\{A_j\}$, $j=1,2,\ldots,$ of subsets of Ω, we denote by

$$A\equiv A_1\cap A_2\cap\cdots\equiv\bigcap_j A_j$$

the set whose elements belong to all the sets A_j, $j=1,2,\ldots$. The set $A=\cap_j A_j$ is called the *intersection* of the sets A_j. The event $\cap_j A_j$ occurs if *all* the events A_j, $j=1,2,\ldots,$ occur. Obviously, $A\cap A=A=A\cap\Omega$, $A\cap\varnothing=\varnothing$; if $A\subset B$, then $A\cap B=A$. Thus in the coin-tossing example, $E\cap D=D$. Clearly, two sets A and B are disjoint if and only if $A\cap B=\varnothing$. If $A\subset\Omega$, we denote by A^c the difference $\Omega-A$. The set A^c is called the *complement* of A in Ω. It contains those elements of Ω that do not belong to A. Thus the set D^c in the coin-tossing example is the event "at most one head turns up in three tosses of a coin," and we have

$$D^c=\{\text{TTT},\text{TTH},\text{THT},\text{HTT}\}.$$

The set G^c in the monatomic gas example consists of all the points on S outside the spherical zone G.

EXERCISE 1.1.1

(i) Construct a sample space Ω corresponding to the experiment of throwing a die.
(ii) How many elementary events are there in Ω?
(iii) How many events consisting of two elementary events are there in Ω?

EXERCISE 1.1.2

Let the sample space consist of n elementary events.

(i) What is the number of events that consist of exactly k elementary events in Ω?
(ii) What is the number of all events in Ω?

EXERCISE 1.1.3

In random sampling of families the event A occurs if the sampled family has only one child and the event B occurs if the family has at least one child. It is known that there are no families having more than n children. Describe the elementary events and express A and B in terms of the elementary events. Describe the events $A \cup B$, A^c, B^c, $B-A$, and $A-B$.

EXERCISE 1.1.4

In the monatomic gas model, let the events A and B be given by $A=\{a<v_i^1<b\}$ and $B=\{c<v_i^2<d\}$. Describe geometrically the events $A \cup B$, $A \cap B$, and $(A-B) \cup (B-A)$.

EXERCISE 1.1.5

Prove de Morgan's law, $(A \cup B)^c = A^c \cap B^c$; $(A \cap B)^c = A^c \cup B^c$.

1.2 PROBABILITY MEASURE

In the experiment of tossing a fair coin three times, we intuitively assign the probability $\frac{1}{8}$ to each of the possible eight outcomes, as all seem to us equally likely. To the event D we assign the probability $\frac{1}{2}$ because it contains one-half of all possible events in Ω. Obviously, we assign the probability zero to the impossible event $\varnothing$ and the probability 1 to the sure event Ω. To make the intuitive notion of probability mathematically precise, we introduce a system of axioms that formalize the basic properties we expect probability to have. We describe first the set $\mathfrak{B}$ of *random events* on which the probability measure is defined. The elements of $\mathfrak{B}$ are subsets of Ω and $\mathfrak{B}$ has the following properties:

(i) $\Omega \in \mathfrak{B}$.
(ii) If $A \in \mathfrak{B}$ and $B \in \mathfrak{B}$, then $A-B \in \mathfrak{B}$.
(iii) If $A_j \in \mathfrak{B}, j=1,2,\ldots,$ is a sequence of elements of $\mathfrak{B}$, then $\cup_{j=1}^{\infty} A_j \in \mathfrak{B}$.

In particular, complements and intersections of random events are random events. A *probability measure* $P(\cdot)$ is a function defined on the set $\mathfrak{B}$ of random events, which satisfies the following axioms:

Axiom 1. To every element A of $\mathfrak{B}$ there corresponds a number $P(A)$

which satisfies the inequality

$$0 \leqslant P(A) \leqslant 1.$$

Axiom 2. $P(\Omega)=1$.

Axiom 3. If $A_j \in \mathfrak{B}$, $j=1,2,\ldots$, is a finite or infinite sequence of disjoint events, that is, $A_i \cap A_j = \varnothing$ if $i \neq j$, then

$$P\left(\bigcup_j A_j\right) = \sum_j P(A_j).$$

To illustrate the axioms, we again consider the experiment of tossing a fair coin three times. The sample space consists of eight elementary events, and $\mathfrak{B}$ consists of *all* subsets of Ω. It is easy to see that properties (i)–(iii) are satisfied. To each random event in $\mathfrak{B}$ we assign the probability

$$P(A) = \frac{\text{number of elements in } A}{8}.$$

It is easy to verify that Axioms (1)–(3) are satisfied in this case. In the case of the monatomic gas the elementary events have been identified with points of the sphere S of radius $(2E/m)^{1/2}$. We expect the probability of the event G in S to be proportional to the *area* of G; for example, for $G = \{a < v_i^1 < b\}$, elementary calculus shows that

$$P(G) = c\int_a^b \left(1 - \frac{x^2 m}{2E}\right)^{(3n-3)/2} dx \tag{1.2.1}$$

where c is a proportionality constant. Since by (2) $P(S)=1$, we must have

$$c = \frac{1}{\int_{-(2E/m)^{1/2}}^{(2E/m)^{1/2}} (1 - x^2 m/2E)^{(3n-3)/2}\, dx}.$$

The set $\mathfrak{B}$ of random events in S cannot be taken to be the set of *all* subsets of S, since it can be shown that there is no function $P(A)$ defined on the set of all subsets A of S such that (1.2.1) is satisfied and (1)–(3) hold. This is a consequence of the existence of nonmeasurable sets in S (Halmos 1959). The set $\mathfrak{B}$ is defined as the smallest set with properties (i)–(iii) that contains all the events $\{a < v_i^j < b\}$, where $i = 1,2,\ldots,n$; $j = 1,2,3$; and $-2E/m \leqslant a \leqslant b \leqslant 2E/m$. Thus whenever the area of a subset A

of S is defined, we set

$$P(A)=\frac{\text{area of } A}{\text{area of } S}.$$

The computation of $\lim_{n\to\infty} P(G)$ leads to the well-known result of Maxwell: Assuming that the energy is proportional to the number of particles in the gas, we set $E=\gamma n$, where γ is a constant independent of n; hence

$$P\{a<v_i^1<b\}=\frac{\int_a^b\left(1-\frac{x^2m}{2\gamma n}\right)^{(3n-3)/2}dx}{\int_{-(2\gamma n/m)^{1/2}}^{(2\gamma n/m)^{1/2}}(1-x^2m/2\gamma n)^{(3n-3)/2}\,dx}$$

$$\to\left(\frac{3m}{4\pi\gamma}\right)^{1/2}\int_a^b e^{-3mx^2/4\gamma}\,dx.$$

Setting $\gamma=3kT/2$, we obtain Maxwell's result:

$$\lim_{n\to\infty} P\{a<v_i^1<b\}=\left(\frac{m}{2\pi kT}\right)^{1/2}\int_a^b e^{-mx^2/2kT}\,dx.$$

We call T absolute temperature and k is called Boltzmann's constant. A little more sophisticated example is that of the theory of the game of heads and tails. The possible outcomes of this game are all the infinite sequences of H and T. Thus the sample space Ω is the set of all sequences $A_1, A_2,\ldots,$ where each A_j is either the symbol H or the symbol T. There are infinitely many distinct sequences of this kind, and in fact, the elements of Ω cannot even be arranged in a sequence; that is, the set Ω is uncountable (Kamke 1950). If we are to assign a probability measure to each outcome of the game, it would necessarily be $P(\{A_1, A_2,\ldots\})=0$. For, obviously all sequences must have the same probability, and if $P(\{A_1, A_2,\ldots\})=c>0$, then for any sequence of distinct outcomes $A^i=\{A_1^i, A_2^i,\ldots\}$, $i=1,2,\ldots,$ we have $A^i\cap A^j=\varnothing$; therefore, by (3),

$$P\left(\bigcup_j A^j\right)=\sum_j P(A^j)=\sum_j c=\infty,$$

which contradicts (1). We therefore take the elementary events to be the sets of sequences, k of whose places ($k=1,2,\ldots$) are fixed. Clearly, the probability of an elementary event, in which k is the number of fixed places, is the probability of an outcome of the experiment of tossing a fair coin k times. Thus the probability measure assigned to such an elementary event is $1/2^k$. To construct a probability measure on Ω that assigns the elementary events the probability $1/2^k$ if k places are fixed, we map Ω onto the unit interval by assigning to each sequence $A_1, A_2, \ldots$ the number

$$t = \sum_{i=1}^{\infty} 2^{-n}\varepsilon_n, \tag{1.2.2}$$

where $\varepsilon_n = 1$ if $A_n = \mathrm{H}$ and $\varepsilon_n = 0$ if $A_n = \mathrm{T}$. Obviously, $0 \leqslant t \leqslant 1$. This correspondence is not one-to-one since the sequence H, T, T,... and T, H, H,... are mapped into the same number, namely into the number $\frac{1}{2}$, as

$$2^{-1} = \sum_{n=2}^{\infty} 2^{-n}.$$

The sequences that are mapped into the same numbers correspond to the "dyadic rationals," that is, to numbers of the form $r/2^s$, where r and s are positive integers. It is easy to see that the set of all such numbers can be arranged in a sequence (how?) $A = [A^1, A^2, \ldots]$, say (i.e., the set is countable). Since $P(A^j) = 0$ and $A^i \cap A^j = \varnothing$ if $i \neq j$, we must have, by (3), $P(A) = 0$. Identifying all dyadic rationals with sequences that end with T, T,..., we obtain a one-to-one correspondence between Ω and the interval $[0, 1]$. Thus the elementary set $B_1 = \{H, A_1, A_2, \ldots\}$ is mapped onto the interval $\left[\frac{1}{2}, 1\right]$, which is the set of all numbers in the interval $[0, 1]$ whose first digit is 1 in their binary expansion. The event B_1 corresponds to the outcome H in a single toss of a fair coin; thus we assign to B_1 the probability $\frac{1}{2}$. This probability is the length of the interval $\left[\frac{1}{2}, 1\right]$ corresponding to B_1 in the one-to-one correspondence between Ω and the interval $[0, 1]$. It is now easy to see that any elementary event is mapped onto a finite union of intervals whose end points are dyadic rationals, and the probability of such an elementary event equals the sum of the lengths of the dyadic intervals onto which it is mapped. It is also easy now to construct a probability measure on Ω which is consistent with the probability measure assigned to the elementary events. We simply assign to any interval $[a, b] \subset [0, 1]$ its length $b - a$ and extend the definition by properties (1)–(3) to the set $\mathfrak{B}$ of random events (Halmos 1959). The set $\mathfrak{B}$ of random events in Ω can be described by its image in the interval $[0, 1]$ under the one-to-one mapping described above. The image of $\mathfrak{B}_\Omega$ in the interval $[0, 1]$ is the so-called Borel set $\mathfrak{B}$, which is the smallest set containing all the subintervals of $[0, 1]$ such that $\mathfrak{B}$ has the properties (i)–(iii).

Using property (3) of the probability measure, we can derive the formula

$$P(A \cup B) = P(A) + P(B) - P(A \cap B). \tag{1.2.3}$$

Indeed,

$$A \cup B = A \cup [B - (A \cap B)]$$

$$B = [A \cap B] \cup [B - (A \cap B)]$$

and obviously

$$\varnothing = A \cap [B - (A \cap B)]$$

and

$$\varnothing = [A \cap B] \cap [B - (A \cap B)].$$

Hence, by (3),

$$P(A \cup B) = P(A) + P[B - (A \cap B)]$$

and

$$P(B) = P(A \cap B) + P[B - (A \cap B)].$$

It follows that (1.2.3) holds.

EXERCISE 1.2.1 (Kac 1959)

Let $H_n(\omega)$ be the number of H's in the sequence $\omega = \{A_1, A_2, \dots\} \in \Omega$ in the game of heads and tails. Using the identity

$$\text{(1.2.4)} \qquad \lim_{n\to\infty} 2^{-n} \sum_{|k-n/2|<\alpha\sqrt{n}} \binom{n}{k} = (2\pi)^{-1/2} \int_{-2\alpha}^{2\alpha} e^{-x^2/2}\, dx,$$

show that

$$\lim_{n\to\infty} P\left\{ \left| H_n - \frac{n}{2} \right| < \alpha\sqrt{n} \right\} = (2\pi)^{-1/2} \int_{-2\alpha}^{2\alpha} e^{-x^2/2}\, dx.$$

Use this result to devise a test for the fairness of a coin.

EXERCISE 1.2.2

Prove (1.2.4) by using Stirling's formula (Feller 1957).

EXERCISE 1.2.3

Let $A_1, A_2, \ldots, A_n$ ($n \geqslant 3$) be random events. Show that

$$P\left(\bigcup_{j=1}^{n} A_j\right) = \sum_{j=1}^{n} P(A_j) - \sum_{j,k=1}^{n} P(A_j \cap A_k) + \sum_{\substack{i,j,k=1 \\ i<j<k}}^{n} P(A_i \cap A_j \cap A_k) + \cdots$$

$$+(-1)^{n+1} P\left(\bigcap_{j=1}^{n} A_j\right) \qquad \text{(Poincaré).}$$

EXERCISE 1.2.4

Show that

$$P(A) + P(A^c) = 1.$$

EXERCISE 1.2.5

Let $\{A_n\}$, $n = 1, 2, \ldots$, be an increasing sequence of random events; that is, $A_n \subset A_{n+1}$. Show that

$$P\left(\bigcup_{n=1}^{\infty} A_n\right) = \lim_{n\to\infty} P(A_n).$$

EXERCISE 1.2.6

Let $\{A_n\}$, $n = 0, 1, \ldots$, be the sequence of all elementary events in Ω and assume that $P(A_n) = c/n!$, $n = 0, 1, \ldots$. Find c.

EXERCISE 1.2.7

Let $\{A_n\}$, $n = 1, 2, \ldots$, be a decreasing sequence of random events; that is, $A_{n+1} \subset A_n$. Show that

$$P\left(\bigcap_{n=1}^{\infty} A_n\right) = \lim_{n\to\infty} P(A_n).$$

EXERCISE 1.2.8

Show that if $A \subset B$, then $P(A) \leqslant P(B)$.

EXERCISE 1.2.9

What is the probability that heads turn up m times in n tosses of a fair coin?

EXERCISE 1.2.10*

Prove the Borel-Cantelli lemma: let $\{A_n\}$ be an infinite sequence of events such that $\Sigma_n P(A_n) < \infty$. Then $P(\cap_{m=1}^{\infty} \cup_{n=m}^{\infty} A_n) = 0$. The set $\cap_{m=1}^{\infty} \cup_{n=m}^{\infty} A_n$ is called "A_n infinitely often" (A_n i.o.). We have $\omega \in (A_n \text{ i.o.})$ if and only if $\omega \in \cap_{n=1}^{\infty} A_{m_n}$ for some sequence $m_n \to \infty$. A partial converse is also true (Fisz 1963).

1.3 CONDITIONAL PROBABILITY AND INDEPENDENCE

Consider an experiment that is repeated n times and suppose that m times ($m \leqslant n$) the event B occurred and in k of the experiments ($k \leqslant n$) the event A occurred. The frequencies of the events are given by $p(A) = k/n$ and $p(B) = m/n$. If the event $A \cap B$ occurred l times ($l \leqslant m$), then the frequency at which A occurred, provided that B occurred, is $p(A|B) = l/m$. Since

$$\frac{l}{m} = \frac{l/n}{m/n} = \frac{p(A \cap B)}{p(B)},$$

we have

$$p(A|B) = \frac{p(A \cap B)}{p(B)}.$$

Using this relation, we define the conditional probability $P(A|B)$ by setting

$$P(A|B) = \frac{P(A \cap B)}{P(B)} \tag{1.3.1}$$

whenever $P(B)>0$. The function

$$P_B(A)=P(A|B) \tag{1.3.2}$$

is a probability measure in B, where B is now considered as a smaller sample space. The measure $P(A|B)$ is the probability of the event A provided that B occurs. To show that $P_B(A)$ in (1.3.2) is a probability measure on the sample space B, we first describe the set of random events in B. If A is a random event in Ω, we say that $A\cap B$ is a random event in B; thus the random events in B are the intersections of random events in Ω with B. It is easily seen that the set of random events in B satisfies (i)–(iii). To verify (1)–(3) for $P_B(\cdot)$, we note first that since $A\cap B\subset B$, we have by Exercise 1.2.8, $P(A\cap B)\leqslant P(B)$; hence

$$0\leqslant P_B(A)=P(A|B)=\frac{P(A\cap B)}{P(B)}\leqslant 1,$$

so that (1) holds. Obviously, $P_B(B)=P(B|B)=P(B\cap B)/P(B)=1$, so (2) holds. Finally, let $A_i\cap A_j=\varnothing$ for $i\neq j$. Then

$$P_B\Big(\bigcup_j A_j\Big)=\frac{P\Big[\Big(\bigcup_j A_j\Big)\cap B\Big]}{P(B)}=\frac{P\Big[\bigcup_j (A_j\cap B)\Big]}{P(B)}=\frac{\sum_j P(A_j\cap B)}{P(B)}$$

$$=\sum_j P(A_j|B)=\sum_j P_B(A_j)$$

as $(A_i\cap B)\cap(A_j\cap B)=\varnothing$, hence (3) follows.

EXERCISE 1.3.1

Show that

(i) $P(A\cap B)=P(B)P(A|B)=P(A)P(B|A)$.
(ii) $P(A\cap B\cap C)=P(A)P(B|A)P(C|A\cap B)$.

EXERCISE 1.3.2

Assume that $A_i\cap A_j=\varnothing$ if $i\neq j$ and $\cup_j A_j=\Omega$, where $\{A_j\}$ is a finite or infinite sequence of random events in Ω.

(i) Prove the *absolute probability formula*

$$P(B)=\sum_j P(A_j)P(B|A_j)$$

given that $P(A_j)>0, j=1,2,\ldots,$ and that B is any random event in Ω.

(ii) Prove *Bayes' formula*: Under the conditions of part (i), if $P(B)>0$, then

$$P(A_i|B)=\frac{P(A_i)P(B|A_i)}{\sum_j P(A_j)P(B|A_j)}.$$

EXERCISE 1.3.3

Find the probability that in the game of heads and tails, the number of heads turned up in the first m tosses of a fair coin is k, given that the number of heads turned up in n tosses is l $(k\leqslant l, n>m, l\leqslant m)$ (Feller 1957).

EXERCISE 1.3.4

Three dice are thrown. What is the probability that one of the dice shows 4, given that the total sum of the points is 10?

If the fact that B has occurred does not influence the probability of occurrence of the event A, we say that A is independent of B. Mathematically, this is expressed by the requirement

$$P(A|B)=P(A).$$

Using (1.3.1), we obtain in this case

$$P(A\cap B)=P(A)P(B). \tag{1.3.3}$$

We say therefore that the random events A and B are *independent* if (1.3.3) holds.

EXERCISE 1.3.5

Show that if $P(A|B)=P(A)$, then $P(B|A)=P(B)$.

We say that the events $A_1, A_2, \ldots, A_n$ are independent if

$$P\left(\bigcap_{j=1}^{s} A_{i_j}\right)=\prod_{j=1}^{s} P(A_{i_j})$$

for all $1 \leqslant s \leqslant n$ and all choices of $1 \leqslant i_1 < i_2 < \cdots < i_s \leqslant n$. For example, let A be the event "heads turns up in the first toss of a fair coin" and let B be the event "tails turns up in the second toss of a fair coin." Obviously, A and B are independent; therefore, the probability

$$P(A \cap B)=\tfrac{1}{2}\cdot\tfrac{1}{2}=\tfrac{1}{4}.$$

EXERCISE 1.3.6

What is the probability of at least k heads in n tosses of a fair coin?

We say that the infinite sequence of events $\{A_j\}, j=1,2,\ldots,$ is independent if each finite subsequence is independent. Following (Kac 1959) we consider, as an example, the game of heads and tails. Let H_n denote the number of heads that turn up in the first n tosses of a fair coin. We shall compute the probability

$$P\left(\lim_{n\to\infty} \frac{H_n}{n}=\beta\right) \tag{1.3.4}$$

where β is any number. The event

$$\lim_{n\to\infty} \frac{H_n}{n}=\beta \tag{1.3.5}$$

is the set of all sequences $\omega=\{A_1, A_2, \ldots\}$ such that the limit (1.3.5) exists and equals β. Using the mapping (1.2.2) we see that the probability (1.3.4) is the measure of the numbers t in the interval $[0,1]$ such that t is given by (1.2.2), where

$$\lim_{n\to\infty} \sum_{m=1}^{n} \frac{\varepsilon_m(t)}{n}=\beta.$$

The event $\{\varepsilon_1(t)=1\}$ is the image of the set of all sequences in Ω whose first place is H, so that it is the set of all numbers t in [0, 1] whose first digit in their binary expansion is 1. Thus the image of the set $\{\varepsilon_1(t)=1\}$ is the interval $\left[\frac{1}{2},1\right]$. Similarly, $\{\varepsilon_j(t)=1\}=\{$all numbers t whose jth digit is 1$\}$. Obviously, the measure μ of such sets in the interval [0, 1] is given by

$$\mu\{\varepsilon_j(t)=1\}=\mu\{\varepsilon_j(t)=0\}=\tfrac{1}{2},$$

where $\mu([\alpha,\beta])=\beta-\alpha$ for any $[\alpha,\beta]\subset[0,1]$ [$\mu(\cdot)$ is called the Lebesgue measure in [0, 1] (Halmos 1959)]. The events $\{\varepsilon_j(t)=\eta_j\}$, $j=1,2,\ldots$, $\eta_j=0$ or 1, are clearly independent, so that

$$\mu\left(\bigcap_{j=1}^{n}\{\varepsilon_j(t)=\eta_j\}\right)=\prod_{j=1}^{n}\mu\{\varepsilon_j(t)=\eta_j\}=2^{-n}.$$

We introduce the auxiliary functions $r_i(t)=1-2\varepsilon_i(t)$ and note that

$$\sum_{j=1}^{n}\frac{r_j(t)}{n}=2\left(\frac{1}{2}-\sum_{j=1}^{n}\frac{\varepsilon_j(t)}{n}\right).$$

The events $\{r_j(t)=\zeta_j\}$, $j=1,2,\ldots$, $\zeta_j=1$ or -1 are also independent. Since $\varepsilon_j(t)$ is a step function, we have

$$\mu\{\varepsilon_j(t)=1\}=\int_0^1\varepsilon_j(t)\,dt=\tfrac{1}{2};$$

hence $\int_0^1 r_j(t)\,dt=0$. It follows that

$$\int_0^1\left(\prod_{j=1}^{k}r_{i_j}(t)\right\}dt=\prod_{j=1}^{k}\left\{\int_0^1 r_{i_j}(t)\,dt\right\}=0 \tag{1.3.6}$$

for all $i_1<i_2<\cdots<i_k$. By (1.3.6), we have

$$\int_0^1\left\{\sum_{j=1}^{n}r_j(t)\right\}^4 dt=\int_0^1\left\{\sum_{j=1}^{n}r_j^4+\sum_{i<j}r_i^2r_j^2\right\}dt=n+6\binom{n}{2}.$$

Hence

$$\sum_{n=1}^{\infty}\int_0^1\left\{\frac{\left[\sum_{j=1}^{n}r_j(t)\right]^4}{n^4}\right\}dt=\sum_{n=1}^{\infty}O(n^{-2})<\infty.$$

It follows that the series $\Sigma_{n=1}^{\infty}[\Sigma_{j=1}^{n} r_j(t)]^4/n^4$ converges for almost all t, that is, except on a set of measure zero [by Lebesgue's monotone convergence theorem (Halmos 1959)]. Hence

$$\sum_{j=1}^{n} \frac{r_j(t)}{n} = 2\left[\tfrac{1}{2} - \sum_{j=1}^{n} \varepsilon_j(t)\right] \to 0 \qquad \text{as } n\to\infty$$

almost everywhere. Thus the probability (1.3.4) is 1 if $\beta=\frac{1}{2}$ and it is 0 otherwise. This example shows the importance of constructing the sample space and the probability measure.

1.4 RANDOM VARIABLES

A one-dimensional random variable is a single-valued function $X(\omega)$ defined on a sample space Ω such that the set $\{\omega | X(\omega) \leqslant x\}$ is a random event for any real number x. Consider, for example, the game of heads and tails. We define a random variable $e_k(\omega)$ by setting $e_k(\omega)=1$ if the outcome of the kth toss in the sequence ω (ω is any sequence of heads and tails in Ω) is H and $e_k(\omega)=-1$ if it is T. The sets $\{\omega | e_k(\omega) \leqslant x\}$ are the sets $\{\omega | e_k(\omega)=-1\}$ if $-1 \leqslant x < 1$, $\varnothing$ if $x<-1$, and Ω if $x \geqslant 1$. The sets $\varnothing$ and Ω are obviously random events, by properties (i)–(iii) in Section 1.2. The event $\{\omega | e_k(\omega)=-1\}$ is an elementary event in Ω, as defined in Section 1.2; thus $e_k(\omega)$ is a random event. It is easy to see that a sum of random variables is a random variable; thus

$$X_n(\omega) = \sum_{j=1}^{n} e_j(\omega) \tag{1.4.1}$$

is a random variable. It represents the "capital" after n tosses. Alternatively, if we start at the origin and move one step to the right if H turns up and one step to the left if T turns up, then $X_n(\omega)$ represents the position of this random walk on the x-axis after n tosses of a coin. The event $\{X_n(\omega)=k\}$ is the set of all possible sequences (all points ω in Ω) such that the random walk X_n is at the point k on the x-axis after n tosses of the coin. We delete the symbol ω and use the notation $\{X_n=k\}$ to denote the set $\{\omega | X_n(\omega)=k\}$.

EXERCISE 1.4.1

Using the fact that $P(e_k=1)=\frac{1}{2}$, show that $P(X_n=k)=\binom{n}{j}2^{-n}=[n!/(n-j)!j!]2^{-n}$, where $j=(n-k)/2$ if n and k have the same parity, and $P(X_n=k)=0$ otherwise.

Let X be a random variable on a sample space Ω. The function $F(x)=P(X\leqslant x)=$[the probability of the set (event) of points ω in Ω such that $X(\omega)\leqslant x$] is called the *distribution function* of the variable X. Obviously, $F(x)$ is a nondecreasing function of x, and $\lim_{x\to-\infty}F(x)=0$, $\lim_{x\to+\infty}F(x)=1$. If there is a nonnegative function $f(s)$ such that

$$F(x)=\int_{-\infty}^{x}f(s)\,ds,$$

then $f(x)$ is called the *probability density function* of the random variable X. We have, of course,

$$P(a\leqslant X\leqslant b)=\int_{a}^{b}f(x)\,dx.$$

A random variable is called *Gaussian or normal* $N(0,1)$ if its density is given by

$$f(x)=\frac{e^{-x^2/2}}{\sqrt{2\pi}}.$$

A random variable X is called normal $N(m,\sigma)$, where σ is a positive number, if the random variable $Y=(X-m)/\sigma$ is normal $N(0,1)$.

EXERCISE 1.4.2

Show that a random variable X is normal $N(m,\sigma)$ if its density is given by

$$f(x)=\frac{e^{-(x-m)^2/2\sigma^2}}{\sigma\sqrt{2\pi}}.$$

EXERCISE 1.4.3

Let $F(x)$ be the distribution function of a random variable X.

(i) Let $Y=-X$. Show that $F_Y(y)=P(Y\leqslant y)=1-F(-y)$ and $f_Y(y)=$

$f(-y)$, given that $F(x)$ is continuous. What is the correct formula if $F(x)$ is discontinuous at a point?

(ii) Let $Y=aX+b$, where a and b are constants. Show that for positive a

$$F_Y(y)=F\left(\frac{y-b}{a}\right)$$

$$f_Y(y)=\frac{f[(y-b)/a]}{a}.$$

What is the formula in the event that a is negative?

(iii) Let $Y=X^2$. Show that

$$F_Y(y)=\begin{cases}0 & \text{for } y\leqslant 0\\ F(\sqrt{y})-F(-\sqrt{y}) & \text{for } y\geqslant 0\end{cases}$$

$$f_Y(y)=\begin{cases}0 & \text{for } y\leqslant 0\\ \dfrac{f(\sqrt{y})+f(-\sqrt{y})}{2\sqrt{y}} & \text{for } y>0.\end{cases}$$

(iv) Let $y=g(x)$ be a strictly monotone increasing, continuously differentiable function whose inverse is given by $x=h(y)$. Let $Y=g(X)$. Show that

$$f_Y(y)=f[h(y)]\cdot|h'(y)|.$$

(v) Let X be a Gaussian variable. Find the densities of the following random variables: (1) X, (2) X^2, (3) X^k, where k is an integer; (4) e^X, (5) $\sin X$. Note that not all functions in this problem satisfy the conditions of (iv).

(vi) Let X be a random variable, let $f(x)$ be its density, and let $F(x)=\int_{-\infty}^{x} f(s)\,ds$ be its distribution function. Find the density and distribution functions of the random variable $F(X)$.

An n-dimensional random variable is a collection $X=(X_1,X_2,\ldots,X_n)^T$ of n random variables on a sample space Ω such that the set

$$\{\omega\,|\,X_1(\omega)\leqslant x_1, X_2(\omega)\leqslant x_2,\ldots, X_n(\omega)\leqslant x_n\}$$

is a random event for all real numbers $x_1, x_2,\ldots, x_n$. For example, the position of a point in phase space is a $6N$-dimensional random variable,

$X_1, \ldots, X_{3N}$ are the $3N$ space coordinates, and $X_{3N+1}, \ldots, X_{6N}$ are the $3N$ velocity components of an N-particle system. In this case the sample space is the $6N$-dimensional Euclidean space R^{6N}, and the set of random events is the smallest set with properties (i)–(iii) of Section 1.2 that contains all the half-spaces $\{x_j \leqslant a\}, j=1,2,\ldots,6N$, where a is any real number.

The (*joint*) *distribution function* $F(\mathbf{x}) = F(x_1, \ldots, x_n)$ is defined by

$$F(\mathbf{x}) = P(X_1 \leqslant x_1, \ldots, X_n \leqslant x_n)$$

(the probability that $X_1 \leqslant x_1$ *and* $X_2 \leqslant x_2$, *and* ..., $X_n \leqslant x_n$). Obviously,

$$F(x_1, \ldots, x_j, -\infty, x_{j+2}, \ldots, x_n) = 0$$

$$F(\infty, \infty, \ldots, \infty) = 1.$$

EXERCISE 1.4.4

Let (X, Y) be a two-dimensional random variable and let $F(x, y)$ be its distribution function. Show that

$$P(a < X \leqslant b, c < Y \leqslant d) = F(b, d) - F(b, c) - F(a, d) + F(a, c).$$

If there is a nonnegative function of n variables $f(\mathbf{x}) = f(x_1, \ldots, x_n)$ such that

$$F(\mathbf{x}) = \int_{-\infty}^{x_1} \cdots \int_{-\infty}^{x_n} f(s_1, \ldots, s_n)\, ds_1 \cdots ds_n,$$

then $f(\mathbf{x})$ is called the (probability) density of $\mathbf{X}$. An n-dimensional random variable is called Gaussian or normal $N(\mathbf{0}, \mathbf{I})$ if its density is given by

$$f(\mathbf{x}) = (2\pi)^{-n/2} e^{-|\mathbf{x}|^2/2},$$

where $\mathbf{x} = (x_1, \ldots, x_n)$ and $|\mathbf{x}|^2 = \sum_{j=1}^n x_j^2$. A random variable $\mathbf{X} = (X_1, \ldots, X_n)$ has an n-dimensional Gaussian or *normal distribution* $N(\mathbf{m}, \boldsymbol{\sigma})$, where $\mathbf{m}$ is a vector and σ is a symmetric matrix with positive eigenvalues, if its density is given by

$$f(\mathbf{x}) = (2\pi)^{-n/2} (\det \sigma)^{-1/2} \exp\left[-\tfrac{1}{2} (\mathbf{x} - \mathbf{m})^T \boldsymbol{\sigma}^{-1} (\mathbf{x} - \mathbf{m}) \right].$$

Here $\boldsymbol{\sigma}^{-1}$ is the inverse of $\boldsymbol{\sigma}$, and

$$(\mathbf{x}-\mathbf{m})^T\boldsymbol{\sigma}^{-1}(\mathbf{x}-\mathbf{m})=\sum_{i,j=1}^{n}\sigma_{ij}^{-1}(x_i-m_i)(x_j-m_j).$$

Let $F(\mathbf{x})$ be the distribution function of an n-dimensional random variable $\mathbf{X}=(X_1,\ldots,X_n)$. The distribution function of X_j can be found from $F(\mathbf{x})$ by

$$F_j(x_j)=F(\infty,\ldots,\infty,x_j,\infty,\ldots,\infty)$$
$$=\int_{-\infty}^{\infty}\cdots\int_{-\infty}^{x_j}\cdots\int_{-\infty}^{\infty}f(x_1,\ldots,s_j,\ldots,x_n)\,dx_1\cdots ds_j\cdots dx_n.$$

The function $f_j(x_j)$ is called the *marginal* density of X_j. The marginal distribution of $(X_1,\ldots,X_k)$ $(k<n)$ is defined by

$$F(x_1,\ldots,x_k)$$
$$=\int_{-\infty}^{x_1}\cdots\int_{-\infty}^{x_k}\int_{-\infty}^{\infty}f(s_1,\ldots,s_k,x_{k+1},\ldots,x_n)\,ds_1\cdots ds_k\,dx_{k+1}\cdots dx_n.$$

EXERCISE 1.4.5

(i) Show that the marginal distributions of the n-dimensional Gaussian variable are Gaussian.

(ii) Let X be an n-dimensional normal variable $N(\mathbf{m},\boldsymbol{\sigma})$. Find the marginal distributions of $\mathbf{X}$.

Diagonalize $\boldsymbol{\sigma}$ by changing variables $\mathbf{x}-\mathbf{m}=\mathbf{A}\mathbf{y}$, where $\mathbf{A}$ is an orthogonal matrix.)

EXERCISE 1.4.6

Let the joint density of X and Y be given by

$$f(x,y)=\frac{1}{2\pi}\exp\left[-\left(\frac{x^2-2xy+2y^2}{2}\right)\right].$$

(i) Is (X,Y) normal? Find $\mathbf{m}$ and $\boldsymbol{\sigma}$ if the answer is positive.

(ii) Find the marginal distributions of X and Y.

Let $U=g(X,Y)$, $V=h(X,Y)$ be a one-to-one transformation of the random variables (X,Y) whose joint density is $f(x,y)$. The joint density of (U,V) can be found as follows. Let $x=G(u,v)$, $y=H(u,v)$ be the inverse transformation and let $J=\partial(x,y)/\partial(u,v)$ be the Jacobian of the transformation. Then

$$\phi(u,v)=f(G(u,v),H(u,v))|J|$$

is the density of (U,V). (Verify!)

EXERCISE 1.4.7

Let $f(x,y)$ be the density of (X,Y). Find the density of:

(i) $X\pm Y$. (*Hint.* Use $U=X$, $V=X\pm Y$.)
(ii) XY.
(iii) X/Y.
(iv) $\begin{pmatrix} U \\ V \end{pmatrix} = \mathbf{A} \begin{pmatrix} X \\ Y \end{pmatrix} + \mathbf{B}$, where $\mathbf{A}=$ matrix, $\mathbf{B}=$vector.

EXERCISE 1.4.8

Let $\mathbf{X}$ be an n-dimensional $N(\mathbf{0},\boldsymbol{\sigma})$ and let $\mathbf{A}$ be an $r\times n$ matrix. Show that $\mathbf{Y}=\mathbf{AX}$ ($\mathbf{X}=$column vector) is $N(\mathbf{0},\boldsymbol{\tau})$. Find $\boldsymbol{\tau}$.

1.5 DISCRETE VARIABLES AND THE δ-FUNCTION

Consider the experiment of tossing a possibly unfair coin once. The sample space consists of two elementary events $\Omega=\{\mathrm{H},\mathrm{T}\}$, the set of random events consists of all the subsets of Ω, and thus $\mathfrak{B}=\{\varnothing,\{\mathrm{H}\},\{\mathrm{T}\},\Omega\}$. The probability function on $\mathfrak{B}$ is defined by

$$P(\{H\})=p,\ P(\{T\})=q=1-p, \qquad 1\geqslant p\geqslant 0.$$

We define a random variable $X(\omega)$ on Ω by setting $X(\omega)=1$ if $\omega=\mathrm{H}$ and $X(\omega)=0$ if $\omega=\mathrm{T}$. It is easy to see that $\{X(\omega)\leqslant x\}\in\mathfrak{B}$ for all real x, so that $X(\omega)$ is a random variable. The distribution function of $X(\omega)$ is given by

$$F(x)=P(X(\omega)\leqslant x)=\begin{cases}1 & \text{if} \quad x\geqslant 1\\ q & \text{if} \quad 0\leqslant x<1\\ 0 & \text{if} \quad x<0.\end{cases}$$

We see from Fig. 1.5.1 that $F(x)$ is a step function. $F(x)$ is right continuous and it has no finite density function $f(x)$, since $F'(x)=0$ at all points where $F(x)$ is differentiable. To define a density for $F(x)$, we need the concept of Dirac's *δ-function*. First we define some auxiliary concepts.

The class $\mathfrak{D}$ of all infinitely differentiable real functions $f(x)$ such that

$$|x|^N\frac{d^nf(x)}{dx^n}\to 0 \qquad \text{as} \quad |x|\to\infty$$

for all $n\geqslant 0$ and $N>0$ is called the class of *test functions*. We say that a sequence $\{f_n(x)\}$ of functions in $\mathfrak{D}$ converges to $f(x)$ in $\mathfrak{D}$ if

$$\max_x\left|\frac{d^k[f_n(x)-f(x)]}{dx^k}\right|\to 0 \qquad \text{as} \quad n\to\infty \tag{1.5.1}$$

for all k. The *δ-function* is the linear functional (a function from $\mathfrak{D}$ to $\mathbb{R}$) that assigns to each test function $f(x)$ the real number $f(0)$. We write

$$\langle\delta(x), f(x)\rangle=f(0).$$

Very often the notation

$$\int_{-\infty}^{\infty}\delta(x)f(x)\,dx=f(0) \tag{1.5.2}$$

is used. It should be noted, however, that there is no finite function $\delta(x)$ that satisfies (1.5.2) for all test functions (Lighthill 1958). It is easy to see

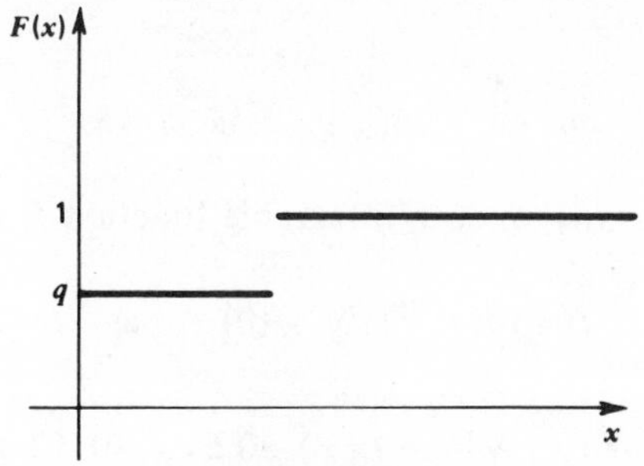

Figure 1.5.1

that if $f_n(x)\to f(x)$ in $\mathfrak{D}$ as $n\to\infty$, then

$$\langle\delta, f_n\rangle\to\langle\delta, f\rangle \quad \text{as} \quad n\to\infty.$$

We express this fact by saying that δ is a *continuous* linear functional. Note that this type of continuity does not imply the continuity of δ as a function. The rules for calculations with the δ-function are the usual rules of the integral calculus. Thus $\delta_y(x)\equiv\delta(x-y)$ is formally defined by

$$\int_{-\infty}^{\infty}\delta(x-y)f(x)\,dx=\int_{-\infty}^{\infty}\delta(z)f(z+y)\,dz=f(y),$$

where the change of variables $z=x-y$ has been used. If $f(x)\equiv 1$, we define, by analogy with (1.5.1),

$$\int_{-\infty}^{\infty}\delta(x)\,dx=1. \tag{1.5.3}$$

Note that 1 is not an element of $\mathfrak{D}$, but if $f_n(x)=1$ for $|x|<n, f_n(x)=0$ for $|x|>n+1$, $0\leqslant f_n(x)\leqslant 1$, and $f_n(x)\in\mathfrak{D}$, then for all y,

$$\int_{-\infty}^{\infty}f_n(x)\,\delta(x-y)\,dx\to 1 \quad \text{as} \quad n\to\infty.$$

Since $f_n(x)\to 1$ as $n\to\infty$ and

$$\frac{d^k}{dx^k}[\,f_n(x)-1]\to 0 \quad \text{as} \quad n\to\infty,$$

we can justify the definition (1.5.3).

The δ-function can be also defined as a *limit* of functions as follows. Let

$$u(x,t)=\frac{1}{\sqrt{2\pi t}}e^{-x^2/2t}, \qquad t>0, \quad -\infty<x<\infty.$$

Then

$$u(x,t)\to\delta(x) \quad \text{as} \quad t\to\infty$$

in the sense that for any smooth integrable function $f(x)$,

$$\int_{-\infty}^{\infty}f(x)u(x,t)\,dx\to f(0) \quad \text{as} \quad t\to 0. \tag{1.5.4}$$

Indeed, $f(x)=f(0)+f_1(x)$, where $f_1(x)\to 0$ as $x\to 0$. Changing variables by setting

$$x = z\sqrt{t}\,,$$

we obtain

$$\int_{-\infty}^{\infty} f(x)u(x,t)\,dx = \frac{1}{\sqrt{2\pi}}\int_{-\infty}^{\infty} f(z\sqrt{t}\,)e^{-z^2/2}\,dz$$

$$= \frac{f(0)}{\sqrt{2\pi}}\int_{-\infty}^{\infty} e^{-z^2/2}\,dz + \frac{1}{\sqrt{2\pi}}\int_{-\infty}^{\infty} f_1(z\sqrt{t}\,)e^{-z^2/2}\,dz.$$

It is well known that

$$\frac{1}{\sqrt{2\pi}}\int_{-\infty}^{\infty} e^{-z^2/2}\,dz = 1;$$

therefore, assuming that $f_1(x)$ is bounded, we have

$$\int_{-\infty}^{\infty} f_1(z\sqrt{t}\,)e^{-z^2/2}\,dz \to 0 \qquad \text{as} \quad t \to 0$$

by standard convergence theorems [Lebesgue's dominated convergence theorem (Halmos 1959)]. It follows that (1.5.4) holds. The δ-function is a particular case of a generalized function. A *generalized function* φ is a linear functional that assigns to each test function $f(x)$ a real number, so that

$$\langle \varphi, \alpha f + \beta g \rangle = \alpha\langle \varphi, f \rangle + \beta\langle \varphi, g \rangle$$

for all real numbers α and β and for all test functions f and g. Generalized functions are *continuous* in the sense that if $f_n(x) \to f(x)$ in the sense of (1.5.1), then $\langle \varphi, f_n(x)\rangle \to \langle \varphi, f(x)\rangle$ as $n \to \infty$. Any integrable function $h(x)$ defines a generalized function acting on test functions $f(x)$ by

$$\langle h, f \rangle \equiv \int h(x)f(x)\,dx.$$

We say that a sequence $\{\varphi_n\}$, $n = 1, 2, \ldots$, of generalized functions converges to a generalized function φ, if for each test function $f(x)$ we have $\langle \varphi_n, f \rangle \to \langle \varphi, f \rangle$ as $n \to \infty$. We also write $\langle \varphi, f \rangle \equiv \int \varphi(x)f(x)\,dx$ even if φ is *not* a function. We denote the set of all generalized functions by $\mathfrak{D}'$. To evaluate the "indefinite integral" of δ,

$$\theta(x)=\int_{-\infty}^{x}\delta(s)\,ds$$

we use the approximation

$$\theta_t(x)=\int_{-\infty}^{x}u(z,t)\,dz=\int_{-\infty}^{\infty}\chi_{(-\infty,x)}(z)u(z,t)\,dz,$$

where $\chi_{(a,b)}(z)=1$ if $z\in(a,b)$ and $\chi_{(a,b)}(z)=0$ otherwise. For each $x>0$, we have the limit $\lim_{t\to 0}\theta_t(x)=1$ as in (1.5.4), and for each $x<0$, we have the limit

$$\lim_{t\to 0}\theta_t(x)=0.$$

Thus

$$\theta(x)=\begin{cases}1 & \text{if} \quad x>0\\ 0 & \text{if} \quad x<0.\end{cases}$$

To make $\theta(x)$ a probability distribution function, we define $\theta(0)=1$ so as to make $\theta(x)$ right continuous. This does not affect $\theta(x)$ as a generalized function, because setting $\theta(0)$ to be *any* number does not change the value of the integral

$$\int_{-\infty}^{\infty}f(x)\theta(x)\,dx$$

if $f(x)$ is any function in $\mathfrak{D}$. We can now define a density for the probability distribution of the variable $X(\omega)$. Since

$$F(x)=P(X(\omega)\leqslant x)=q\theta(x)+p\theta(x-1),$$

we have

$$f(x)=F'(x)=q\,\delta(x)+p\,\delta(x-1). \tag{1.5.5}$$

Since $(1/\sqrt{2\pi t}\,)e^{-x^2/2t}\to 0$ as $t\to 0$ for all $|x|>0$, we must have $\delta(x)=0$ if $x\neq 0$. Thus all the density in (1.5.5) is concentrated at the points $x=0$ and $x=1$. We can define now the density of any probability distribution function $F(x)$ at a jump discontinuity x_0 of $F(x)$ by setting

$$f(x_0)=[\,F(x_0^+)-F(x_0^-)\,]\,\delta(x-x_0),$$

where $F(x_0^{\pm})\equiv\lim_{\varepsilon\downarrow 0}F(x_0\pm\varepsilon)$.

EXERCISE 1.5.1

Construct the distribution and density functions for the random variable

$$X(\omega)=e_1(\omega)+e_2(\omega),$$

where Ω is the sample space for the experiment of tossing a coin twice with the probability measure

$$P(\{\mathrm{H}\})=p, \qquad P(\{\mathrm{T}\})=q,$$

and note that the outcomes of the two tosses are independent.

The variables e_1 and e_2 are defined by $e_i(\omega)=1$ if heads turns up in the ith toss, and $e_i(\omega)=0$ otherwise ($i=1,2$).

EXERCISE 1.5.2

For what values of a is the functional

$$f(x)=a\,\delta(cx+d)$$

a probability density given c and d? (Use a change of variables in the integral.)

EXERCISE 1.5.3

Find the distribution function whose density is given by

$$f(x)=\sum_{n=1}^{\infty} 2^{-n}\,\delta(x-n).$$

EXERCISE 1.5.4

For what value of c is the functional

$$f(x)=c\,\delta(e^{x}-1)$$

a probability density?

EXERCISE 1.5.5

In higher dimensions we define $\delta(\mathbf{x})$ by the relation

$$\int_{R^n} \delta(\mathbf{x}) f(\mathbf{x})\, dx = f(\mathbf{0})$$

for all smooth integrable functions. Show that

$$(2\pi t)^{-n/2} e^{-|\mathbf{x}|^2/2t} \to \delta(\mathbf{x}) \qquad \text{as} \quad t \to 0.$$

A random variable X is called *discrete* if its density is given by

$$f_X(x) = \sum_i p_i\, \delta(x - x_i),$$

$$p_i > 0 \qquad \text{and} \qquad \sum_i p_i = 1.$$

In this case

$$F_X(x) - F_X(x-) = P(X(\omega) = x).$$

At any point of continuity of $F_X(x)$ (i.e., at $x \neq x_i$), we have

$$P(X(\omega) = x) = 0.$$

Thus a discrete variable can assume only a countable set of values with positive probability. We have

$$P(X(\omega) = x_i) = p_i$$

and

$$F_X(x) = \int_{-\infty}^{x} \sum_i p_i\, \delta(s - x_i)\, dx = \sum_{x_i \leqslant x} p_i.$$

1.6 CONDITIONAL DISTRIBUTIONS AND INDEPENDENCE

Let $X(\omega)$ be a random variable on a sample space Ω. For any random event B, we have

$$P(X(\omega) \leqslant x \mid B) = \frac{P((X(\omega) \leqslant x) \cap B)}{P(B)}. \tag{1.6.1}$$

The function $F_{X|B}(x) \equiv P(X(\omega) \leqslant x | B)$ is called the *conditional distribution* of X in B. Let $Y(\omega)$ be another random variable defined on Ω and set

$$B = \{y \leqslant Y(\omega) \leqslant y + \Delta y\}.$$

Then (1.6.1) becomes

$$F_{X|y \leqslant Y \leqslant y+\Delta y}(x) = \frac{P\{(X(\omega) \leqslant x) \cap (y \leqslant Y(\omega) \leqslant y + \Delta y)\}}{P\{y \leqslant Y(\omega) \leqslant y + \Delta y\}}$$

provided that $P(y \leqslant Y(\omega) \leqslant y + \Delta y) > 0$. Let $f(x, y)$ be the joint probability density of X and Y, so that

$$P\{y \leqslant Y(\omega) \leqslant y + \Delta y\} = \int_y^{y+\Delta y} \int_{-\infty}^{\infty} f(x, z)\, dx\, dz.$$

Then

(1.6.2)
$$F_{X|y \leqslant Y \leqslant y+\Delta y}(x) = \frac{\displaystyle\int_y^{y+\Delta y} \int_{-\infty}^{x} f(t, s)\, dt\, ds}{\displaystyle\int_y^{y+\Delta y} \int_{-\infty}^{\infty} f(t, s)\, dt\, ds}.$$

Dividing numerator and denominator by Δy and letting $\Delta y \to 0$, we obtain

$$F_{X|Y}(x|y) \equiv \frac{\displaystyle\int_{-\infty}^{x} f(t, y)\, dt}{\displaystyle\int_{-\infty}^{\infty} f(t, y)\, dt}$$

$$= \frac{\partial F_{X,Y}(x, y)/\partial y}{dF_Y(y)/dy}$$

where $F_{X,Y}(x, y) = \int_{-\infty}^{x} \int_{-\infty}^{y} f(t, s)\, ds\, dt$ is the joint probability distribution of X and Y, and $F_Y(y)$ is the marginal distribution of Y. Differentiating (1.6.2) with respect to x, we get the *conditional density* of X on Y:

(1.6.3)
$$f_{X|Y}(x|y) = \frac{f(x, y)}{\displaystyle\int_{-\infty}^{\infty} f(x, y)\, dx}.$$

We can write Bayes' rule in the form

$$f_{X|Y}(x|y) = f_{Y|X}(y|x) \frac{f_X(x)}{f_Y(y)}.$$

Note that X and Y may be considered to be vectors, so that (1.6.3) takes the form

$$f_{(X_1, X_2, \ldots, X_n)|(Y_1, Y_2, \ldots, Y_m)}[(x_1, x_2, \ldots, x_1)|(y_1, y_2, \ldots, y_m)]$$

$$= \frac{f(x_1, x_2, \ldots, x_n, y_1, y_2, \ldots, y_m)}{\int_{-\infty}^{\infty} \cdots \int_{-\infty}^{\infty} f(x_1, x_2, \ldots, x_n, y_1, y_2, \ldots, y_m)\, dx_1\, dx_2 \cdots dx_n}.$$

If $P(X \in S) > 0$, where S is a set in R, we have

$$P\{X \leqslant x | X \in S\} = \frac{P[(X \leqslant x) \cap (X \in S)]}{P\{X \in S\}}$$

$$= \frac{\int_{-\infty}^{x} \chi_S(x) f(x)\, dx}{\int_S f(x)\, dx,}$$

where $f(x)$ is the density of X. We see that the density of the conditional distribution is given by

$$f_{X|S}(x) = \frac{\chi_s(x) f(x)}{\int_S f(x)\, dx}.$$

If the joint probability distribution of the two-dimensional random variable (X, Y) satisfies

$$F_{(X,Y)}(x, y) = F_X(x) F_Y(y),$$

then X and Y are said to be *independent* variables. In this case the density satisfies

$$f_{(X,Y)}(x, y) = f_X(x) f_Y(y),$$

where $f_X(x) = F'_X(x)$ and $f_Y(y) = F'_Y(y)$.

It is easy to see that if X and Y are independent, then

$$F_{X|Y}(x|y) = F_X(x)$$

and

$$F_{Y|X}(y|x) = F_Y(y).$$

EXERCISE 1.6.1

Let X be a uniform variable on $[a, b]$; that is,

$$f_X(x)=\frac{\chi_{[a,b]}(x)}{b-a}.$$

Find the conditional density

$$f_{X|[\alpha,\beta]}(x),$$

where $[\alpha, \beta]\subset[a, b]$. Find $F_{X|[\alpha,\beta]}(x)$.

EXERCISE 1.6.2

Let X and Y be two independent Gaussian variables. Find the densities of the variables in Exercise 1.4.7. Which of the variables are Gaussian?

EXERCISE 1.6.3

Are the variables X and Y in Exercise 1.4.6 independent? Find $F_{X|Y}(x|y)$, $F_{Y|X}(y|x)$, and the corresponding densities.

The random variables $X_1 \cdots X_n$ are said to be independent if

$$F(x_1,\ldots,x_n)=\prod_{i=1}^{n} F_i(x_i).$$

In this case

$$f(x_1,\ldots,x_n)=\prod_{i=1}^{n} f_i(x_i).$$

EXERCISE 1.6.4

Let $\mathbf{X}=(X_1,\ldots,X_n)$ be an n-dimensional Gaussian variable $N(\mathbf{0}, I)$. Are the variables $X_1,\ldots,X_n$ independent? What is the answer if $\mathbf{X}$ is $N(\mathbf{m}, \boldsymbol{\sigma})$?

EXERCISE 1.6.5

Denote by $\mathbf{v}=(v_1, v_2, v_3)^{\mathrm{T}}$ the velocity vector of a gas particle. Following Maxwell, assume that $\mathbf{v}$ is a three-dimensional random variable whose density $h(\mathbf{v})$ satisfies the following two postulates:

(i) v_j $(j=1,2,3)$ are identically distributed independent random variables.

(ii) The density of $\mathbf{v}$ is a function of the kinetic energy of a particle; that is,

$$h(\mathbf{v})=g\left(\tfrac{1}{2}m|\mathbf{v}|^2\right)\equiv p(|\mathbf{v}|),$$

where m is the mass of a particle. Show that $\mathbf{v}$ must be $N(\mathbf{0}, k\mathbf{I})$, where k is some positive constant and $(\mathbf{I})_{ij}=\delta_{ij}$ (Kronecker's δ), that is, $\mathbf{v}$, is a Gaussian variable. [*Hint.* Set $h(\mathbf{v})=f(v_1)f(v_2)f(v_3)$ and derive the differential equation $p'(s)/sp(s)=\text{constant}$.]

1.7 EXPECTATION, VARIANCE, AND OTHER MOMENTS

Let X be a random variable and let $f(x)$ be its density. For any function $g(x)$ the integral

$$Eg(X)=\int_{-\infty}^{\infty} g(x)f(x)\,dx$$

is called the *expected value* or *mean value or the expectation* of the random variable $g(X)$.

Consider, for example, the random variable X_n, which takes on the values $j=0,1,2,\ldots,n$ with the probability

$$P\{X_n(\omega)=j\}=\binom{n}{j}p^j(1-p)^{n-j}, \qquad 0<p<1.$$

Its density is given by

$$f_{X_n}(x)=\sum_{j=0}^{n}\binom{n}{j}p^j(1-p)^{n-j}\,\delta(x-j). \tag{1.7.1}$$

We have

$$Eg(X_n)=\int_{-\infty}^{\infty} f_{X_n}(x)g(x)\,dx=\sum_{j=0}^{n}\binom{n}{j}p^j(1-p)^{n-j}g(j).$$

The variable X_n, whose density is given by (1.7.1), is called a *binomial* variable $B(n,p)$. We have

$$EX_n = np, \qquad EX_n^2 = np[1+(n-1)p].$$

EXERCISE 1.7.1

Using (1.7.1), calculate $Eg(X_n)$ if:

(i) $g(x)=x$.
(ii) $g(x)=x^2$.
(iii) $g(x)=x^k$ where k is any positive integer.

If X is a Gaussian variable $N(0,1)$, then

$$EX = \frac{1}{\sqrt{2\pi}}\int_{-\infty}^{\infty} xe^{-x^2/2}\,dx = 0$$

$$EX^2 = \frac{1}{\sqrt{2\pi}}\int_{-\infty}^{\infty} x^2e^{-x^2/2}\,dx = 1.$$

EXERCISE 1.7.2

Let X be a Gaussian variable. Show that $EX^{2k+1}=0$, $EX^{2k}=1\cdot 3\cdot \ldots \cdot (2k-1)$ $(k=0,1,2,\ldots)$.

The expectation $m_k = EX^k$ $(k=0,1,2,\ldots)$ is called the *moment of order k* of X.

The expression $E(X-c)^k$, where c is a constant, is called the kth *moment of X about c*. In case $c=EX$, we call $\mu_k = E(X-EX)^k$ the kth *central moment*. The second central moment is called the *variance* of X, and its square root σ is called the *standard deviation* of X,

$$\sigma^2 = \operatorname{Var} X = E(X-EX)^2 = m_2 - m_1^2.$$

It is a measure of dispersion of X.

EXERCISE 1.7.3

Let X_n be $B(n,p)$. Show that $\operatorname{Var} X = np(1-p)$.

EXERCISE 1.7.4

Let $S(\nu)$ be a random variable such that

$$P(S(\nu)=n)=\frac{e^{-\nu}\nu^n}{n!}.$$

$S(\nu)$ is called a *Poisson variable*. Show that $ES(\nu)=\nu$ and $\operatorname{Var} S(\nu)=\nu$.

EXERCISE 1.7.5

Show that if $np_n \to \nu$, then $B(n, p_n) \to S(\nu)$ in distribution as $n \to \infty$; that is, $P(B(n, p_n)=k) \to P(S(\nu)=k)$.

EXERCISE 1.7.6

Prove that

$$\frac{B(n,p)-EB(n,p)}{\sqrt{\operatorname{Var} B(n,p)}} \to N(0,1) \quad \text{as} \quad n\to\infty \quad \text{in distribution}$$

(De Moivre–Laplace).

EXERCISE 1.7.7

Show that

$$\frac{S(\nu)-ES(\nu)}{\sqrt{\operatorname{Var} S(\nu)}} \to N(0,1) \quad \text{as} \quad \nu\to\infty \quad \text{in distribution}$$

(use Stirling's formula).

EXERCISE 1.7.8

Let $EX=m$ and $\operatorname{Var} X=\sigma^2$.

(i) Let $Y=aX+b$; find EY and $\operatorname{Var} Y$.

(ii) The variable $Y=(X-m)/\sigma$ is called a normalized variable. Find EY and $\operatorname{Var} Y$.

EXERCISE 1.7.9

Show that

(i) $E\Sigma_{j=1}^{n} a_j X_j = \Sigma_{j=1}^{n} a_j EX_j$ (a_j = real numbers).

(ii) If $X_1, \ldots, X_n$ are independent, then

$$E\left(\prod_{i=1}^{n} X_i\right) = \prod_{i=1}^{n} (EX_i)$$

and

$$\operatorname{Var}\left(\sum_{j=1}^{n} X_j\right) = \sum_{j=1}^{n} \operatorname{Var} X_j.$$

EXERCISE 1.7.10

Prove Chebyshev's inequality,

$$P(|X-m_1|>k\sigma) \leqslant \frac{1}{k^2} \qquad (k>0)$$

where $m_1 = EX$ and $\sigma^2 = \operatorname{Var} X$. [*Hint.* Use $\int_0^\infty y f(y)\,dy \geqslant k\int_k^\infty f(y)\,dy$.]

In higher dimensions we define

$$Eg(\mathbf{X}) = \int_{-\infty}^{\infty} \cdots \int_{-\infty}^{\infty} g(x_1, \ldots, x_n) f(x_1, \ldots, x_n)\, dx_1 \cdots dx_n$$

where $f(\mathbf{x}) = f(x_1, \ldots, x_n)$ is the density of $\mathbf{X} = (X_1, \ldots, X_n)$. In particular, $\mathbf{m} = E\mathbf{X} = (EX_1, EX_2, \ldots, EX_n) = (m_1, \ldots, m_n)$, where

$$m_j = \int_{-\infty}^{\infty} \cdots \int_{-\infty}^{\infty} x_j f(x_1, \ldots, x_n)\, dx_1 \cdots dx_n.$$

The *moment of order* $k_1+k_2+\cdots+k_n$ is defined by

$$m_{k_1,\ldots,k_n}=EX_1^{k_1}\cdots X_n^{k_n}.$$

The *central moment* of order $k_1+\cdots+k_n$ is defined by

$$m_{k_1,\ldots,k_n}=E(X_1-m_1)^{k_1}\cdots(X_n-m_n)^{k_n}.$$

The *covariance matrix* of $\mathbf{X}$ is defined by

$$(\mathrm{Cov}\,\mathbf{X})_{ij}=E(X_i-m_i)(X_j-m_j).$$

EXERCISE 1.7.11

(i) If X and Y are (one-dimensional) random variables, we define $\mathrm{Cov}(X,Y)$ by

$$\mathrm{Cov}(X,Y)=E[(X-EX)(Y-EY)].$$

Show that $\mathrm{Var}(X+Y)=\mathrm{Var}\,X+\mathrm{Var}\,Y+2\,\mathrm{Cov}(X,Y)-2EXEY$.

(ii) The number

$$\rho=\frac{\mathrm{Cov}(X,Y)}{(\mathrm{Var}\,X\,\mathrm{Var}\,Y)^{1/2}}\equiv\frac{\mu_{11}}{\sigma_1\sigma_2}$$

is called the *correlation coefficient* of X and Y. Show that $-1\leqslant\rho\leqslant 1$ and $\rho^2=1$ if and only if

$$p(Y=aX+b)=1.$$

(iii) If X and Y are independent, show that $\rho=0$ (Fisz 1963).

EXERCISE 1.7.12

Let the two-dimensional random variable (X,Y) have the joint probability density given in Exercise 1.4.6. Calculate EX, EY, $\mathrm{Var}\,X$, $\mathrm{Var}\,Y$, $\mathrm{Cov}(X,Y)$, and ρ.

EXERCISE 1.7.13

Let (X, Y) be a two-dimensional variable $N(\mathbf{m}, \boldsymbol{\sigma})$. Show that

$$f(x,y)=\frac{1}{2\pi\sigma_1\sigma_2\sqrt{1-\rho^2}}\exp\left(-\left\{\frac{1}{2(1-\rho^2)}\left[\frac{(x-m_1)^2}{\sigma_1^2}-\frac{2\rho(x-m_1)(y-m_2)}{\sigma_1\sigma_2}+\frac{(y-m_2)^2}{\sigma_2^2}\right]\right\}\right),$$

where $\sigma_1^2=\operatorname{Var} X$, $\sigma_2^2=\operatorname{Var} Y$, and ρ is the correlation coefficient.

EXERCISE 1.7.14

Let $\mathbf{X}=(X_1,\ldots, X_n)$ be $N(\mathbf{m}, \boldsymbol{\sigma})$. Find $E\mathbf{X}$, $(\operatorname{Cov}\mathbf{X})_{ij}$.

The *characteristic function* of an n-dimensional variable $\mathbf{X}=(X_1,\ldots, X_n)$ is defined by

$$\phi(\mathbf{t})=\phi(t_1,\ldots,t_n)=Ee^{i\mathbf{t}\cdot\mathbf{X}}$$

$$\equiv\int_{-\infty}^{\infty}\cdots\int_{-\infty}^{\infty}\exp\left(i\sum_{j=1}^{n}t_jx_j\right)f(x_1,\ldots,x_n)\,dx_1\cdots dx_n \qquad (i=\sqrt{-1}\,).$$

The characteristic function is therefore the Fourier transform of the density.

For example, let X be a $B(n, p)$ variable. Then

$$\phi_{B(n,p)}(t)=\int_{-\infty}^{\infty}e^{itx}\sum_{j=0}^{n}\binom{n}{j}p^j(1-p)^{n-j}\,\delta(x-j)\,dx$$

$$=\sum_{j=0}^{n}\binom{n}{j}p^j(1-p)^{n-j}e^{ijt}=\left[pe^{it}+1-p\right]^n.$$

If X is a Poisson variable $S(\nu)$, then

$$\phi_{S(\nu)}(t)=\int_{-\infty}^{\infty}e^{itx}\left[\sum_{n=0}^{\infty}\frac{e^{-\nu}\nu^n\,\delta(x-n)}{n!}\right]dx$$

$$=e^{-\nu}\sum_{n=0}^{\alpha}\frac{(\nu e^{it})^n}{n!}=e^{\nu(e^{it}-1)}.$$

Let X be a $B(n,p)$ variable and Y an $S(\nu)$ variable independent of X. We shall compute the density function of $X+Y$ using characteristic functions. Since the variables are independent, the joint density is a product of the densities

$$f_{X,Y}(x,y)=f_X(x)f_Y(y).$$

It follows that

$$\phi_{X+Y}(t)=Ee^{it(X+Y)}=\int_{-\infty}^{\infty}\int_{-\infty}^{\infty}e^{itx}e^{ity}f_X(x)f_Y(y)\,dx\,dy$$

$$=\phi_X(t)\phi_Y(t).$$

Since the product of Fourier transforms is the Fourier transform of the convolution of the two functions, we have

$$f_{X+Y}(x)=f_X(x)*f_Y(x)=\int_{-\infty}^{\infty}f_X(s)f_Y(x-s)\,ds$$

$$=\int_{-\infty}^{\infty}\sum_{j=0}^{n}\binom{n}{j}p^j(1-p)^{n-j}\,\delta(s-j)\sum_{k=0}^{\infty}\frac{e^{-\nu}\nu^k\,\delta(x-s-k)}{k!}\,ds$$

$$=\sum_{j=0}^{n}\sum_{k=0}^{\infty}\frac{\binom{n}{j}p^j(1-p)^{n-j}\nu^k e^{-\nu}\,\delta(x-j-k)}{k!}.$$

Thus

$$P(B(n,p)+S(\nu)=l)=\sum_{j+k=l}\frac{e^{-\nu}\binom{n}{j}p^j(1-p)^{n-j}\nu_k}{k!}.$$

EXERCISE 1.7.15

Compute

$$P(B(n,p)-S(\nu)=l),$$

assuming that $B(n,p)$ and $S(\nu)$ are independent.

EXERCISE 1.7.16

If $X_1, \ldots, X_n$ are independent variables, show that

$$\phi(\mathbf{t}) = \prod_{i=1}^{n} \phi_i(t_i) \qquad \text{where} \quad \phi_j(s) = Ee^{isX_j}.$$

EXERCISE 1.7.17

(i) Show that the characteristic function of a Gaussian variable $N(0,1)$ is $e^{-t^2/2}$.

(ii) If X is $N(m, \sigma)$, show that $\phi(t) = \exp(itm - \frac{1}{2}\sigma^2 t^2)$.

EXERCISE 1.7.18

(i) Let $X_1, \ldots, X_n$ be independent $N(m_j, \sigma_j)$ variables. Show that $z = \sum_{j=1}^{n} X_j$ is $N(m, \sigma)$, where $m = \sum_{j=1}^{n} m_j$ and $\sigma^2 = \sum_{j=1}^{n} \sigma_j^2$. This is the *addition theorem* for normal variables. (*Hint*. Use characteristic functions.) [The converse is also true (Cramér 1936).]

(ii) Show that any linear combination of independent normal variables is a normal variable; that is, if $X_1, \ldots, X_n$ are independent $N(m_j, \sigma_j)$ and $a_1, \ldots, a_n$ are numbers, then $X = \sum_{j=1}^{n} a_j X_j$ is $N(m, \sigma)$. Find m and σ.

EXERCISE 1.7.19

The random variable X is said to have a *log-normal distribution* if its density is of the form

$$f(x) = \begin{cases} 0 & x \leqslant c \\ \dfrac{1}{(x-c)\sigma\sqrt{2\pi}} \exp\left\{ -\dfrac{[\log(x-c)-m]^2}{2\sigma^2} \right\} & x > c \end{cases}$$

where c is constant. Find EX and $\operatorname{Var}(X)$.

The expectation of a random variable can be computed under the condition that another variable takes on a given value. Thus if X and Y are two random variables, then

$$E(Y|X=x)=\frac{\int_{-\infty}^{\infty} yf_{X,Y}(x,y)\,dy}{f_X(x)}.$$

If A is any interval, then

$$E(Y|X\in A)=\frac{\int_{-\infty}^{\infty}\int_A yf_{X,Y}(x,y)\,dx\,dy}{P(X\in A)}$$

$$=\frac{\int_A f_X(x)\int_{-\infty}^{\infty}[yf_{X,Y}(x,y)/f_X(x)]\,dy\,dx}{P(X\in A)}$$

$$=\int_A \frac{f_X(x)}{P(X\in A)}E(Y|X=x)\,dx=E[E(Y|X=x)|X\in A].$$

If, in particular, Y is the characteristic function of a random event B, then

$$P(Y=1)=P(B)$$

and for any random variable X, we have

$$P(B|X\in A)=E[P(B|X=x)|X\in A]$$

and letting $A=(-\infty,\infty)$, we get

$$P(B)=EP(B|X=x).$$

Thus $P(B|X=x)$ is a random variable (Fisz 1963).

CHAPTER 2

The Brownian Motion

2.1 INTRODUCTION, LANGEVIN'S EQUATION, AND THE BROWNIAN MOTION

The perpetual irregular motions of small grains or particles of colloidal size immersed in a fluid were first noticed by the British botanist Brown in 1826. The chaotic motion of such a particle is called *Brownian motion* and a particle performing such a motion is called a Brownian particle. The chaotic perpetual motion of a Brownian particle is the result of its collisions with the molecules of the surrounding fluid. The colloidal particle is much bigger and heavier than the colliding molecules of the fluid, so that each collision has a negligible effect, but the superposition of many small interactions produces an observable effect. The molecular collisions of a Brownian particle occur in very rapid succession and their number is tremendous. Thus a Brownian particle (e.g., colloidal gold particles of radius $50\,\mu\mu$) will suffer about 10^{21} collisions per second if immersed in liquid under normal conditions. This frequency is so high that the small changes in the particle's path caused by each single impact are too fine to be discerned by the observer. Thus the exact path of the particle cannot be followed in any detail but has to be described statistically. The influence of the surrounding medium on the motion of a Brownian particle can be described by a combination of two forces, as follows. Since the particle under consideration is much larger than the particles of the surrounding fluid, the collective effect of the interaction between the Brownian particle and the fluid may be considered to be of hydrodynamical character. More precisely, we may take one force acting on the Brownian particle to be the force of dynamical friction. The frictional force exerted by a fluid on a small sphere immersed in it can be determined from Stokes' law, which states that the drag force per unit mass acting on a spherical particle of radius a is given by $-\beta\mathbf{v}$, where $\beta=6\pi a\eta/m$, η is the coefficient of dynamical viscosity of the surrounding fluid, m is the particle's mass, and $\mathbf{v}$ is the particle's velocity. The second force acting on a Brownian particle is due to the individual collisions with the particles of the surrounding fluid. It produces instantaneous changes in the acceleration of the particle and it is random both in magnitude and direction. The principal assumptions concerning this fluctuating force $\mathbf{f}(t)$ (per unit mass) are the following: (i) $\mathbf{f}(t)$ is statistically independent of $\mathbf{v}(t)$, (ii) the variations of $\mathbf{f}(t)$ are much more frequent than the variations in $\mathbf{v}(t)$, and (iii) the average of $\mathbf{f}(t)$ is zero. Newton's equations of motion are then given by

$$\frac{d\mathbf{v}(t)}{dt} = -\beta\mathbf{v}(t) + \mathbf{f}(t). \tag{2.1.1}$$

Equation (2.1.1) is called Langevin's equation. The statistical properties of $\mathbf{f}(t)$ can be deduced from this equation by matching its solution with known physical laws. The solution of the stochastic (random) differential equation (2.1.1) determines the transition probability density $p(\mathbf{v}, t, \mathbf{v}_0)$ of the random process $\mathbf{v}(t)$, that is, a function $p(\mathbf{v}, t, \mathbf{v}_0)$ such that

$$P(\mathbf{v}(t) \in A \mid \mathbf{v}(0) = \mathbf{v}_0) = \int_A p(\mathbf{v}, t, \mathbf{v}_0)\, d\mathbf{v}.$$

We assume that the initial velocity $\mathbf{v}_0$ is given, so that we must have

$$p(\mathbf{v}, t, \mathbf{v}_0) \to \delta(\mathbf{v} - \mathbf{v}_0) \qquad \text{as} \quad t \to 0.$$

Further, we know from statistical physics that the density $p(\mathbf{v}, t, \mathbf{v}_0)$ must approach the Maxwellian density for the temperature T of the surrounding medium, independently of $\mathbf{v}_0$ as $t \to \infty$. Hence

$$p(\mathbf{v}, t, \mathbf{v}_0) \to \left(\frac{m}{2\pi kT}\right)^{3/2} \exp\left(\frac{-m|\mathbf{v}|^2}{2kT}\right) \qquad \text{as} \quad t \to \infty. \tag{2.1.2}$$

This demand on $p(\mathbf{v}, t, \mathbf{v}_0)$, in turn, requires $\mathbf{f}(t)$ to have certain statistical properties. For, according to (2.1.1), the formal solution is given by

$$\mathbf{v}(t) = \mathbf{v}_0 e^{-\beta t} + \int_0^t e^{-\beta(t-s)} \mathbf{f}(s)\, ds. \tag{2.1.3}$$

Consequently, the statistical properties of the integral must be the same as those of the difference $\mathbf{v}(t) - \mathbf{v}_0 e^{-\beta t}$. Since $\mathbf{v}(t) - \mathbf{v}_0 e^{-\beta t} \approx \mathbf{v}(t)$ for large t, the integral must have in the limit a Gaussian density. Writing the integral as a finite (Riemann) sum

$$\int_0^t e^{-\beta(t-s)}\mathbf{f}(s)\,ds \approx e^{-\beta t}\sum_n e^{\beta n\,\Delta t}\mathbf{f}(n\,\Delta t)\,\Delta t$$

$$\equiv e^{-\beta t}\sum_n e^{\beta n\,\Delta t}\,\Delta\mathbf{b}_n,$$

where $\Delta\mathbf{b}_n = \mathbf{f}(n\,\Delta t)\,\Delta t$, we obtain for large t

(2.1.4) $$\mathbf{v} \approx \sum_n e^{\beta(n\,\Delta t - t)}\,\Delta\mathbf{b}_n.$$

The random variables $\Delta\mathbf{b}_n$ express the random accelerations suffered by a Brownian particle in the time interval $(n\,\Delta t, (n+1)\,\Delta t)$. Thus we may assume that the variables $\Delta\mathbf{b}_n$ are statistically independent of each other, since the successive collisions are completely chaotic. We shall assume that the time intervals Δt are large compared to the average period of a single fluctuation of $\mathbf{f}(t)$. The period of fluctuation of $\mathbf{f}(t)$ is of the order of the time between successive collisions between the Brownian particle and the molecules of the surrounding fluid; in a liquid, this is generally of the order of 10^{-21} sec. Accordingly, each acceleration $\Delta\mathbf{b}_n$ is the result of many collisions, so that we may assume that all $\Delta\mathbf{b}_n$ have the same statistical properties (Chandrasekhar 1954). Therefore, if we choose $\Delta\mathbf{b}_n$ to be zero mean Gaussian variables, then $\mathbf{v}(t)$ will be Gaussian, as required. To compute the variance of $\Delta\mathbf{b}_n$, we set $\operatorname{Var}\Delta\mathbf{b}_n = 2q\,\Delta t$, and using (2.14), we obtain

$$E|\mathbf{v}|^2 = \sum_n 2q\,\Delta t e^{2\beta(n\,\Delta t - t)} \to 2q\int_0^t e^{2\beta(s-t)}\,ds$$

$$= \frac{q}{\beta}(1 - e^{-2\beta t}) \qquad \text{as} \quad \Delta t \to 0.$$

On the other hand, we have

$$E|\mathbf{v}|^2 \to \frac{kT}{m} \qquad \text{as} \quad t \to \infty.$$

Hence

(2.1.5) $$q = \frac{\beta kT}{m}.$$

Let $\mathbf{x}(t)$ be the displacement of the Brownian particle; then

$$\mathbf{x}(t)=\mathbf{x}_0+\int_0^t \mathbf{v}(s)\,ds. \tag{2.1.6}$$

Substituting (2.1.3) in (2.1.6), we obtain

$$\mathbf{x}(t)=\mathbf{x}_0+\int_0^t\left[\mathbf{v}_0e^{-\beta s}+e^{-\beta s}\int_0^s e^{\beta u}\mathbf{f}(u)\,du\right]ds.$$

Changing the order of integration, we obtain

$$\mathbf{x}(t)-\mathbf{x}_0-\frac{\mathbf{v}_0(1-e^{-\beta t})}{\beta}=-e^{-\beta t}\int_0^t\frac{e^{\beta s}\mathbf{f}(s)\,ds}{\beta}+\int_0^t\frac{\mathbf{f}(s)\,ds}{\beta} \tag{2.1.7}$$

$$\equiv\int_0^t g(s)\mathbf{f}(s)\,ds$$

where $g(s)=(1-e^{\beta(s-t)})/\beta$. Using a finite-sum approximation to the integral again, we conclude that the variable

$$\mathbf{x}(t)-\mathbf{x}_0-\frac{\mathbf{v}_0(1-e^{-\beta t})}{\beta}$$

is a zero mean Gaussian variable with variance

$$\sigma^2=2q\int_0^t g^2(s)\,ds=\frac{q}{\beta^3}(2\beta t-3+4e^{-\beta t}-e^{-2\beta t}). \tag{2.1.8}$$

Hence the probability density of $\mathbf{x}(t)$ is given by

$$p(\mathbf{x},t,\mathbf{x}_0,\mathbf{v}_0)=\left\{\frac{m\beta^2}{2kT[2\beta t-3+4e^{-\beta t}-e^{-2\beta t}]}\right\}^{3/2} \tag{2.1.9}$$

$$\cdot\exp\left\{-\frac{m\beta^2|\mathbf{x}-\mathbf{x}_0(1-e^{-\beta t})/\beta|^2}{2kT[2\beta t-3+4e^{-\beta t}-e^{-2\beta t}]}\right\}.$$

For large t, we have

$$p(\mathbf{x},t,\mathbf{x}_0,\mathbf{v}_0)\approx\frac{1}{(4\pi Dt)^{3/2}}\exp\left(\frac{-|\mathbf{x}-\mathbf{x}_0|^2}{4Dt}\right) \tag{2.1.10}$$

where

(2.1.11) $$D = \frac{kT}{m\beta} = \frac{kT}{6\pi a\eta}.$$

It follows that p satisfies the diffusion equation

(2.1.12) $$\frac{\partial p}{\partial t} = D\Delta p.$$

Formula (2.1.11) for the diffusion coefficient is due to Einstein (1956). We can define now the Brownian motion $\mathbf{x}(t)$ by the following properties:

(2.1.13)(i) $$P(\mathbf{x}(t) \in A | \mathbf{x}(0) = \mathbf{x}_0) = (4\pi Dt)^{-3/2} \int_A e^{-|\mathbf{x}-\mathbf{x}_0|^2/4Dt}\, d\mathbf{x}.$$

(ii) The increments $\mathbf{x}(t+s) - \mathbf{x}(t)$ and $\mathbf{x}(t) - \mathbf{x}(t-u)$ are independent and are independent of t for $s \geq 0$ and $u \geq 0$.

(iii) The paths of $\mathbf{x}(t)$ are continuous.

(iv) The joint probability distribution of $(\mathbf{x}(t_1), \mathbf{x}(t_2) \ldots \mathbf{x}(t_n))$ is Gaussian for every finite sequence $t_1 < t_2 < \cdots < t_n$.

Let $w(t)$ be the one-dimensional analog of $\mathbf{x}(t)$ with $D = \frac{1}{2}$ and $w(0) = 0$. The *transition distribution* of $w(t)$ is given by

(2.1.14)(i) $$P(a \leq w(t) \leq b | w(s) = x)$$
$$= [2\pi(t-s)]^{-1/2} \int_a^b e^{-(x-y)^2/2(t-s)}\, dy.$$

(ii) The joint probability distribution of $(w(t_1), \ldots, w(t_n))$ is zero mean Gaussian, so it is determined by the covariances $Ew(t_i)w(t_j)$. It is easy to determine the covariances using the independence of increments and the fact that $Ew^2(t) = t$. Indeed, assuming that $t < s$, we have

(2.1.15) $$Ew(t)w(s) = E[w(s) - w(t)]w(t) + Ew^2(t)$$
$$= E[w(s) - w(t)]Ew(t) + t = t = t \wedge s \equiv \min(t, s).$$

It can be easily seen that the following are Brownian motions.

(2.1.16) (i) $w_1(t)=w(t+s)-w(s)$.

(ii) $w_2(t)=cw\left(\frac{t}{c^2}\right)$, where c is any positive constant.

(iii) $w_3(t)=tw\left(\frac{1}{t}\right)$

Further properties of $w(t)$ are discussed in Section 2.3*. It is shown there i.a. that $w(t)$ is nowhere differentiable and that

$$\limsup_{t\to\infty} w(t)=+\infty, \qquad \liminf_{t\to\infty} w(t)=-\infty.$$

EXERCISE 2.1.1

Carry out the calculations to show (2.1.7)–(2.1.11) and (2.1.16)(i)–(iii).

EXERCISE 2.1.2

Show that the mean-square displacement of the Brownian particle along any given direction (the x-direction, say) is given by the formula

$$\tfrac{1}{3}E|\mathbf{x}(t)-\mathbf{x}_0|^2=2Dt=\left(\frac{kT}{3\pi a\eta}\right)t$$

for large t, but for small t obtain

$$E|\mathbf{x}(t)-\mathbf{x}_0|^2=\left(\frac{3kT}{m}\right)t^2=E|\mathbf{v}_0|^2t^2,$$

assuming that $\mathbf{v}_0$ has the Maxwell distribution (2.1.2).

EXERCISE 2.1.3

Let $\mathbf{x}(t)=\int_0^t g(s)\mathbf{f}(s)\,ds$, where $g(t)$ is a continuous function. Show that $\mathbf{x}(t)$ is a Gaussian variable and show that $E\mathbf{x}(t)=0$,

$$E|\mathbf{x}(t)|^2=2q\int_0^t g^2(s)\,ds.$$

EXERCISE 2.1.4

Let $\mathbf{x}(t)=\int_0^t g(s)\mathbf{f}(s)\,ds$ and let $\mathbf{y}(t)=\int_0^t h(s)\mathbf{f}(s)\,ds$. Show that the variable $\mathbf{z}(t)=(\mathbf{x}(t),\mathbf{y}(t))$ is Gaussian, $E\mathbf{z}(t)=0$, and $\operatorname{Cov}(\mathbf{x}(t),\mathbf{y}(t))=\begin{pmatrix} A & B \\ B & C \end{pmatrix}$, where $A=2q\int_0^t g^2(s)\,ds$, $B=2q\int_0^t h^2(s)\,ds$, and $C=2q\int_0^t g(s)h(s)\,ds$.

EXERCISE 2.1.5

Find the joint transition probability density of $\mathbf{x}(t)$ and $\mathbf{v}(t)$, where $\mathbf{v}(t)$ is given by (2.1.3) and $\mathbf{x}(t)$ is given by (2.1.6).

EXERCISE 2.1.6

The Langevin equation for the Brownian harmonic oscillator is given by

$$\ddot{x}+\beta\dot{x}+\omega^2 x=f(t), \tag{2.1.17}$$

where $x(t)$ is the one-dimensional displacement of the particle and the random force $f(t)$ is the one-dimensional analog of $\mathbf{f}(t)$.

(i) Solve equation (2.1.17).
(ii) Find the velocity $v(t)=\dot{x}(t)$, where $x(t)$ is the solution of (2.1.17).
(iii) Find the transition probability density of $x(t)$.
(iv) Find the transition probability density of $v(t)$.
(v) Find the joint probability density of $(x(t), v(t))$.
(vi) Show that $Ex(t)\to 0$, $Ev(t)\to 0$, $Ex(t)v(t)\to 0$, $Ex^2(t)\to kT/m\omega^2$, and $Ev^2(t)\to kT/m$ as $t\to\infty$ (Chandrasekhar 1954).

EXERCISE 2.1.7

An electron in an ionized gas (plasma) may be considered a Brownian harmonic oscillator. The elastic restoring force is the Coulomb force of attraction and repulsion of the neighboring heavier ions and the other electrons. The collisions are the electric interactions of the particles. A laser beam of frequency ω directed at the plasma may be considered an additional force (electromagnetic field) acting on the electron. The resulting Langevin equation is given by

$$\ddot{x}+\beta\dot{x}+\omega_p^2 x=E_0\sin\omega t+\mathbf{f}(t), \tag{2.1.18}$$

where ω_p is the frequency of plasma oscillations. Here β is the effective electron collision frequency. Solve (2.1.18) and analyze the solution in the following cases: (i) $\beta \ll \omega_p$ and (ii) $\omega=\omega_p$. Carry out the computations as in (i)–(v) in Exercise (2.1.6) (Eliezer and Schuss 1979).

EXERCISE 2.1.8

Write down the Langevin equation for a charged Brownian particle in a constant magnetic field. Carry out the computations indicated in (i)–(v) of Exercise (2.1.6).

2.2 RANDOM WALK AND COIN TOSSING. BACKWARD AND FORWARD EQUATIONS

A motion of a particle of gas under low pressure or a movement of any body that is subject to collisions with other bodies is similar in nature to the Brownian motion. The duration of the collisions is assumed to be very short on the time scale of the observation of the motion, and the frequency of collisions is assumed relatively high on the same time scale. The collisions are also assumed to be random, as in Section 2.1. Such is the case of the random flights of a small star, that is, a star that is small in comparison with the other stars in cosmic space, and this space plays the role of the liquid (e.g., the "milk" in the Milky Way) in Brownian motion. The collisions are to be understood as the appearances of the investigated star within the gravitational field of other stars. The orbit of such a small star is a random path that is rectilinear between short collisions but may be considered rapidly fluctuating on the time scale of cosmic changes. The simplest mathematical model of such a motion in one dimension is the random walk, in which a particle suffers displacements in the form of a series of steps of equal length, each step being taken, either in the forward or backward direction, with equal probability $\frac{1}{2}$ at equal time intervals. A more mathematically formal description is given by the following coin-tossing model. Starting at $x_0=a(a=0,\pm 1,\ldots,))$, flip a coin and define

$$e_n=\begin{cases} 1 & \text{if heads turns up at the } n\text{th toss,}\\ -1 & \text{if tails turns up at the } n\text{th toss;}\end{cases}$$

that is, move one step to the right along the x-axis if heads turns up, and one step to the left if tails turns up. The capital or the location of this process at time n is given by

$$X_n = a + e_1 + e_2 + \cdots + e_n.$$

The variables e_n are independent and $P(e_n = 1) = P(e_n = -1) = \frac{1}{2}$. Obviously, we have

$$P(X_n = k \mid X_0 = 0) \equiv P_0(X_n = k) = 2^{-n} \binom{n}{\frac{n-k}{2}}$$

if n and k have the same parity, and $P_0(X_n = k) = 0$ otherwise.

The sequence X_n of random variables is called a *random process*. It has the following Markov property: "The motion begins afresh at any fixed time m," or, more precisely,

(2.2.1) $$P(X_{n+m} \in B \mid X_m = b, X_0 = a) = P(X_{n+m} \in B \mid X_m = b) = P_b(X_n \in B),$$

where B is any set of integers (Fig. 2.2.1). The *transition probability* of X_n is given by

$$p(n, a, b) = p_a(X_n = b) = P_0(X_n = b - a),$$

and it satisfies the following equation:

(2.2.2) $$p(n+1, a, b) = \tfrac{1}{2}\left[\, p(n, a+1, b) + p(n, a-1, b)\right],$$

for the probability of going from a to b in $n+1$ steps is the probability of taking the first step to the right and then going to b in n steps, plus the probability of going first one step to the left and then reaching b in n steps. We call a the "backward" variable and b the "forward" variable. From

Figure 2.2.1

(2.2.2) we obtain

$$(2.2.3)\quad p(n+1,a,b)-p(n,a,b)=\tfrac{1}{2}[p(n,a+1,b) -2p(n,a,b)+p(n,a-1,b)].$$

Equation (2.2.3) is therefore called the *backward equation* of the process X_n.

EXERCISE 2.2.1

Derive an analogous *forward equation* (with respect to b) for $p(n, a, b)$.

We shall show that the probability distribution of the normalized random walk X_n converges to that of the Brownian motion as $n\to\infty$.

Let $x_n(t)$ be the broken linear interpolation between the points $\left(\frac{X_m}{\sqrt{n}}, \frac{m}{n}\right)$ in the (t, x) plane (Fig. 2.2.2).

Following (Khinchine 1933), we will prove the classical central limit theorem,

$$p_0\left(a\leqslant\frac{X_n}{\sqrt{n}}<b\right)\to\int_a^b\frac{e^{-x^2/2}}{\sqrt{2\pi}}\,dx \qquad \text{as}\quad n\to\infty.$$

Let $u(x, t)$ be the solution of the heat equation

$$(2.2.4)\qquad \frac{\partial u}{\partial t}=\frac{1}{2}\frac{\partial^2 u}{\partial x^2}\qquad t>0,\quad -\infty<x<\infty$$

$$u(x,0)=f(x);$$

that is,

$$u(x,t)=\frac{1}{\sqrt{2\pi t}}\int_{-\infty}^{\infty}e^{-(x-y)^2/2t}f(y)\,dy,$$

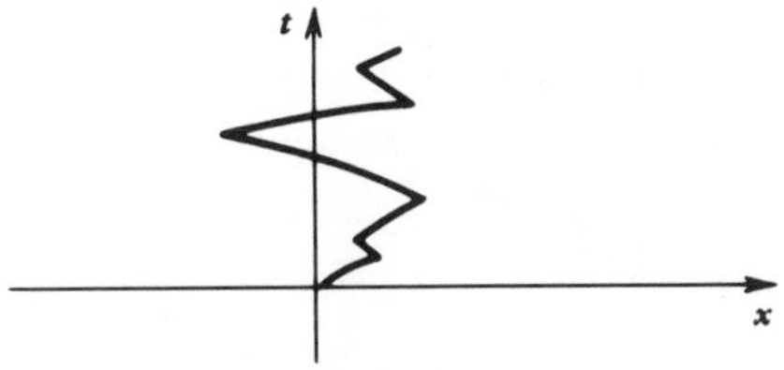

Figure 2.2.2

where f is a test function $u_n\left(\frac{k}{\sqrt{n}},\frac{m}{n}\right)=E_k f(\frac{X_m}{\sqrt{n}})=$ the expectation of $f(\frac{X_m}{\sqrt{n}})$ given that $X_0=k$; that is,

$$u_n\left(\frac{k}{\sqrt{n}},\frac{m}{n}\right)=\sum_{j=-\infty}^{\infty} p(m,k,j)f\left(\frac{j}{\sqrt{n}}\right).$$

The backward difference equation for u_n follows from (2.2.3),

$$u_n\left(\frac{k}{\sqrt{n}},\frac{m+1}{n}\right)-u_n\left(\frac{k}{\sqrt{n}},\frac{m}{n}\right)=\tfrac{1}{2}\left[u_n\left(\frac{k+1}{\sqrt{n}},\frac{m}{n}\right)-2u_n\left(\frac{k}{\sqrt{n}},\frac{m}{n}\right)\right.$$
$$\left.+u_n\left(\frac{k-1}{\sqrt{n}},\frac{m}{n}\right)\right]\equiv\tfrac{1}{2}\Delta_k u_n\left(\frac{k}{\sqrt{n}},\frac{m}{n}\right).$$

The analogous difference equation satisfied by the solution $u(x,t)$ of the heat equation (2.2.4) is

$$\frac{\left[u\left(\frac{k}{\sqrt{n}},\frac{m+1}{n}\right)-u\left(\frac{k}{\sqrt{n}},\frac{m}{n}\right)\right]}{(1/n)}+o(1)=\frac{\tfrac{1}{2}\Delta_k u\left(\frac{k}{\sqrt{n}},\frac{m}{n}\right)}{(1/\sqrt{n}\,)^2}+o(1)$$

as $n\to\infty$.

Since u is at least three times continuously differentiable with respect to t (verify this statement!), $o(1)$ is uniform in k and m as $n\to\infty$. So

$$\text{(2.2.5)}\quad u\left(\frac{k}{\sqrt{n}},\frac{m+1}{n}\right)=\tfrac{1}{2}\left[u\left(\frac{k+1}{\sqrt{n}},\frac{m}{n}\right)+u\left(\frac{k-1}{\sqrt{n}},\frac{m}{n}\right)\right]+o\left(\frac{1}{n}\right).$$

But

$$u_n\left(\frac{k}{\sqrt{n}},0\right)=u\left(\frac{k}{\sqrt{n}},0\right)=f\left(\frac{k}{\sqrt{n}}\right);$$

hence in m steps of the difference scheme,

$$u_n\left(\frac{k}{\sqrt{n}},\frac{m}{n}\right)=u\left(\frac{k}{\sqrt{n}},\frac{m}{n}\right)+o\left(\frac{m}{n}\right).$$

Letting $m/n \to t$, $k/\sqrt{n} \to x$, we see that $o(m/n) = o(1)$ as $n \to \infty$; that is, $u_n(k/\sqrt{n}, m/n) \to u(x,t)$. Hence, setting $k=0$, $m=n$, we obtain

$$E_0 f\left(\frac{X_n}{\sqrt{n}}\right) \to \frac{1}{\sqrt{2\pi}} \int_{-\infty}^{\infty} e^{-x^2/2} f(x)\,dx.$$

Replacing f by a sequence of test functions,

$$f_j \to f_0 = \begin{cases} 1 & a \leqslant x \leqslant b \\ 0 & \text{otherwise}, \end{cases}$$

we obtain

$$P_0\left(a \leqslant \frac{X_n}{\sqrt{n}} < b\right) = E_0 f\left(\frac{X_n}{\sqrt{n}}\right) \to \int_a^b e^{-x^2/2}\,dx.$$

We have shown that

$$E_k f\left(\frac{X_m}{\sqrt{n}}\right) \to (2\pi t)^{-1/2} \int_{-\infty}^{\infty} e^{-(x-y)^2/2t} f(y)\,dy$$

as $k/\sqrt{n} \to x$, and $m/n \to t$ as $n \to \infty$ for all test functions f. Once again, approximation of f_0 by test functions leads to

(2.2.6)
$$P_k\left(a \leqslant \frac{X_m}{\sqrt{n}} \leqslant b\right) = P_x(a \leqslant x_n(t) \leqslant b)$$
$$\to (2\pi t)^{-1/2} \int_a^b e^{-(x-y)^2/2t}\,dy,$$

where $k = [x\sqrt{n}\,]$, $m = [nt]$, $x_n(t) = X_m/\sqrt{n}$, and $n \to \infty$. Thus we see that the scaled random walk $x_n(t)$ converges in distribution to $w(t)$. Similarly, it can be shown that

$$(x_n(t_1), \ldots, x_n(t_j)) \to (w(t_1), \ldots, w(t_j))$$

as $n \to \infty$ in distribution for all finite sequences $t_1 < \cdots < t_j$. The properties of the scaled random walk $x_n(t)$ imply that $w(t)$ possesses the required properties of the Brownian motion as described in Section 2.1. Since

$(w(t_1),\dots,w(t_n))$ is Gaussian and $Ew(t_i)w(t_j)=t_i\wedge t_j$, we have

(2.2.7)

$$P\left[\bigcap_{i=1}^{n}(a_i\leqslant w(t_i)\leqslant b_i)\right]=\int_{a_1}^{b_1}dx_1\cdots\int_{a_n}^{b_n}dx_n$$

$$\times\prod_{i=1}^{n}\frac{\exp\{-(x_i-x_{i-1})^2/2(t_i-t_{i-1})\}}{\sqrt{2\pi(t_i-t_{i-1})}}$$

where $x_0=t_0=0$.

The joint probability density is given by

$$\prod_{i=1}^{n}\frac{\exp\{-(x_i-x_{i-1})^2/2(t_i-t_{i-1})\}}{\sqrt{2\pi(t_i-t_{i-1})}}. \tag{2.2.8}$$

In particular, we have, by (2.2.6) and (2.2.1),

$$P(w(t)\in B\,|\,w(s)=x)=\int_B\frac{e^{-(x-y)^2/2(t-s)}}{\sqrt{2\pi(t-s)}}\,dy$$

$$\equiv P(x,s,B,t)\qquad(t>s);$$

that is, the transition probability $P(x,s,B,t)$ has a density $p(x,s,y,t)$ such that

$$P(x,s,B,t)=\int_B p(x,s,y,t)\,dy$$

and

$$\frac{\partial p}{\partial t}=\frac{1}{2}\frac{\partial^2 p}{\partial x^2}=\frac{1}{2}\frac{\partial^2 p}{\partial y^2}$$

$$p(x,s,y,t)\to\delta(x-y)\qquad\text{as}\quad t\downarrow s.$$

From (2.2.1) we see that $w(t)$ has the *Markov property*; that is,

$$P(w(t)\in B\,|\,w(t_1)=x_1,w(t_1)=x_2,\dots,w(t_n)=x_n)$$

$$=P(w(t)=B\,|\,w(t_n)=x_n),$$

where $t_0<t_1<\cdots<t_n<t$, and B is any interval. The Markov property means that the motion begins afresh at any fixed time t. Indeed, the joint density (2.2.8) determines the conditional density

$$p(x|x_1,\ldots,x_n)=\frac{p(x,x_1,x_2,\ldots,x_n)}{p(x_1,x_2,\ldots,x_n)}$$

$$=\frac{e^{-(x-x_n)^2/2(t-t_n)}}{\sqrt{2\pi(t-t_n)}}=p(x|x_n).$$

The properties of the increments Δb_n of Section 2.1 and the properties of $\Delta w=w(t+\Delta t)-w(t)$ are thus the same.

2.3* CONSTRUCTION OF THE BROWNIAN MOTION

We shall construct the Brownian motion as an a.s. uniformly convergent series of continuous functions with random coefficients. We shall use properties (2.1.13)(i)–(iv) for the definition of the Brownian motion. The probability space Ω will be the unit interval with the uniform distribution. We construct first an infinite sequence of independent Gaussian variables $N(0,1)$ on Ω. Every number x in Ω can be represented by a binary expansion $x=0,x_1x_2\cdots$, where $\{x_i\}$ are independent random variables on Ω which take on the values 0 and 1 with probability $\frac{1}{2}$. The variables

$$y_1=0,x_1x_3x_6x_{10}\cdots$$

$$y_2=0,x_2x_5x_9\cdots$$

$$y_3=0,x_4x_8x_{13}\cdots$$

$$\vdots$$

are obviously independent and have the uniform distribution on Ω; that is,

$$P(y_i\leqslant\alpha)=\alpha,\qquad 0\leqslant\alpha\leqslant 1.$$

Next we define the variables g_i by

$$y_i=\int_{-\infty}^{g_i}\frac{e^{-s^2/2}}{\sqrt{2\pi}}ds;$$

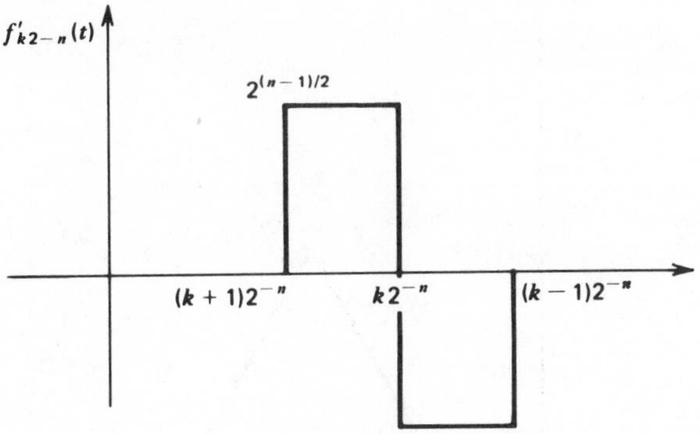

Figure 2.3.1

that is, $g_i = \text{erf}^{-1}(y_i)$. Then clearly the variables $\{g_i\}$ are independent and have a Gaussian distribution $N(0,1)$. Next, we construct a sequence of continuous functions on $[0,1]$ as follows. In Fig. 2.3.1, we construct their derivative by setting

$$f'_{k2^{-n}}(t) = \begin{cases} 0 & \text{if} \quad t<(k-1)2^{-n} \\ 2^{(n-1)/2} & \text{if} \quad (k-1)2^{-n} \leqslant t < k2^{-n} \\ -2^{(n-1)/2} & \text{if} \quad k2^{-n} \leqslant t < (k+1)2^{-n} \\ 0 & \text{if} \quad (k+1)2^{-n} \leqslant t \leqslant 1, \end{cases}$$

where $k = \text{odd} < 2^n$. The functions $\{f'_{k2^{-n}}\}$ are called the *Haar system* and they form an orthonormal basis in $L^2(0,1)$. More precisely,

(2.3.1) $$\langle f'_{i2^{-n}}, f'_{j2^{-n}} \rangle \equiv \int_0^1 f'_{i2^{-n}}(t) f'_{j2^{-n}}(t) = \delta_{ij},$$

where δ_{ij} is the Kronecker function, and each function $f(t)$ in $L^2(0,1)$ can be represented uniquely by

(2.3.2) $$f(t) = \sum_{n=1}^{\infty} \sum_{k=\text{odd}<2^n} a_{k,n} f'_{k2^{-n}}(t) + a_0 f'_0(t)$$

where $f'_0(t) \equiv 1$. The coefficients $a_{k,n}$ are given by

$$a_{k,n} = \int_0^1 f(t) f'_{k2^{-n}}(t)\, dt.$$

The series in (2.3.2) converges to $f(t)$ in the $L^2(0,1)$ sense (Natanson 1961).

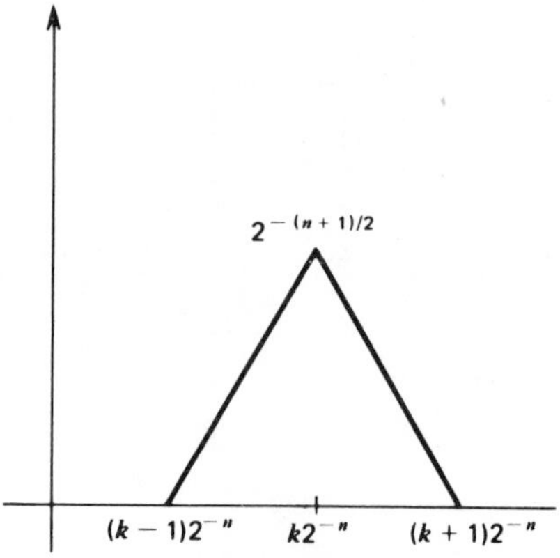

Figure 2.3.2

We define

$$f_{k2^{-n}}(t)=\int_0^t f'_{k2^{-n}}(s)\,ds.$$

We construct $w(t)$ by setting

$$w(t)=g_0f_0(t)+\sum_{n=1}^{\infty}\sum_{k=\text{odd}\leqslant 2^n} g_{k2^{-n}}f_{k2^{-n}}(t), \tag{2.3.3}$$

where $\{g_0, g_{k2^{-n}}\}$ is the sequence of Gaussian variables constructed above, and g_0 is another Gaussian variable, independent of $g_{k2^{-n}}$. Now we shall prove that the series in (2.3.3) converges uniformly a.s. Since the functions $f_{k2^{-n}}(t)$ are continuous and have disjoint supports for different k and fixed n, we have

$$\begin{aligned} a_n &\equiv \max_{0\leqslant t\leqslant 1}\left|\sum_{k=\text{odd}<2^n} g_{k2^{-n}}f_{k2^{-n}}(t)\right| \\ &\leqslant 2^{-(n+1)/2}\max_{k=\text{odd}<2^n}|g_{k2^{-n}}|. \end{aligned}$$

Hence

$$\begin{aligned} P\left(a_n>2^{-n/2}\sqrt{n}\right)&=P\left(2^{-(n+1)/2}\max_{k=\text{odd}<2^n}|g_{k2^{-n}}|>2^{-n/2}\sqrt{n}\right) \\ &\leqslant \sum_{k=\text{odd}<2^n} P\left(|g_{k2^{-n}}|>\sqrt{2n}\right)\leqslant 2^n\cdot 2\int_{\sqrt{2n}}^{\infty}\frac{e^{-x^2/2}}{\sqrt{2\pi}}dx \\ &\leqslant 2^{n+1}\int_{\sqrt{2n}}^{\infty}\frac{x}{\sqrt{2n}}\frac{e^{-x^2/2}}{\sqrt{2\pi}}dx=\left(\frac{2}{e}\right)^n\frac{1}{\sqrt{n\pi}}. \end{aligned}$$

Since

$$\sum_{n=1}^{\infty} P\left(a_n > 2^{-n/2}\sqrt{n}\right) \leqslant \sum_{n=1}^{\infty} \left(\frac{2}{e}\right)^n \frac{1}{\sqrt{\pi n}} < \infty,$$

we have by the Borel–Cantelli lemma,

$$P\left(a_n > 2^{-n/2}\sqrt{n} \text{ i.o.}\right) = 0.$$

It follows that

$$\sum_{n=1}^{\infty} a_n \leqslant \sum_{n=1}^{\infty} 2^{-n/2}\sqrt{n} < \infty \quad \text{a.s.}$$

Hence the series (2.3.3) converges uniformly a.s. We shall show now that $w(t)$ possesses the desired properties (2.1.13)(i)–(iv). First we note that all finite sums in (2.3.2) are linear combinations of independent Gaussian variables, so that $w(t)$ is the limit of Gaussian variables, and in fact the vector $(w(t_1), w(t_2), \dots, w(t_n)), 0 \leqslant t_1 < t_2 < \cdots < t_n \leqslant 1$ has a joint Gaussian distribution. Since obviously, $Ew(t_i) = 0$, the joint distribution of the vector is determined by the covariances $Ew(t_i)w(t_j)$. Now because of independence of the variables $g_{k2^{-n}}$, we have

$$\text{(2.3.4)} \quad Ew(t_i)w(t_j) = E\sum_n \sum_k g_{k2^{-n}} f_{k2^{-n}}(t_i) \sum_n \sum_k g_{k2^{-n}} f_{k2^{-n}}(t_j)$$

$$= f_0(t_i) f_0(t_j) + \sum_n \sum_k f_{k2^{-n}}(t_i) f_{k2^{-n}}(t_j).$$

To identify the last sum in (2.3.4) we consider the Haar expansion of a characteristic function of an interval

$$\chi_{(0,t)}(s) = \begin{cases} 1 & \text{if} \quad 0 < s < t \\ 0 & \text{if} \quad t \leqslant s \leqslant 1. \end{cases}$$

We have, by (2.3.2),

$$\chi_{(0,t)}(s) = \sum_n \sum_k a_{k,n} f'_{k2^{-n}}(s)$$

where

$$a_{k,n} = \int_0^1 \chi_{(0,t)}(s) f'_{k2^{-n}}(s)\, ds = f_{k2^{-n}}(t).$$

By (2.3.1),

$$\int_0^1 \chi_{(0,t_i)}(s)\chi_{(0,t_j)}(s)\,ds = \int_0^1 \sum_n \sum_k \int_0^{t_i} f'_{k2^{-n}}(u)\,du$$

$$f'_{k2^{-n}}(s)\sum_n \sum_k \int_0^{t_j} f'_{k2^{-n}}(v)\,dv\, f_{k2^{-n}}(s)$$

$$= \sum_n \sum_k f_{k2^{-n}}(t_i) f_{k2^{-n}}(t_j).$$

It follows that

$$Ew(t_i)w(t_j) = \int_0^1 \chi_{(0,t_i)}(s)\chi_{(0,t_j)}(s)\,ds = t_i \wedge t_j,$$

as required. We have proved properties (ii) and (iii). Property (i) is a consequence of (iii). Indeed, $\tilde{w}(s) = w(t+s) - w(t)$ is Gaussian, and in fact the vector $(\tilde{w}(s_1), \ldots, \tilde{w}(s_n))$ has a joint Gaussian distribution. Obviously, $E\tilde{w}(s) = 0$, and by (iii), $E\tilde{w}^2(s) = s$. The increment $\tilde{w}(s)$ is independent of $w(u)$ for all $u \leqslant t$, as by (iii)

$$E[w(t+s) - w(t)]w(u) = u - u = 0$$

and as we know, uncorrelated Gaussian variables are independent. Thus $\tilde{w}(s)$ is a Brownian motion and it is independent of t. The extension of $w(t)$ to $t > 1$ is achieved as follows. Using the Haar series we construct a sequence $\{w_j(t)\}$ of independent Brownian motions on $[0,1]$. Now we set

$$\begin{aligned} w(t) &= w_0(t) \quad \text{if} \quad 0 \leqslant t \leqslant 1 \\ &= w_0(1) + w_1(t-1) \quad \text{if} \quad 1 \leqslant t < 2 \\ &= w_0(1) + w_1(1) + w_2(t-2) \quad \text{if} \quad 2 \leqslant t < 3 \\ &\vdots \end{aligned}$$

It is easy to see that such a process has the properties (i)–(iv) for all $t \geqslant 0$. The paths of $w(t)$ are continuous, but not differentiable at any point. This fact is proved next. Assume that $w(t)$ is differentiable at some point $0 \leqslant t_0 \leqslant 1$; then in some interval we must have

$$|w(t) - w(t_0)| \leqslant K|t - t_0|$$

for some positive constant K. For points t_1 and t_2 sufficiently close to t_0,

we have

$$|w(t_1)-w(t_2)| \leqslant |w(t_1)-w(t_0)|+|w(t_0)-w(t_2)|$$
$$\leqslant K(|t_1-t_0|+|t_2-t_0|).$$

Setting $i=[nt_0]+1$, $i<j\leqslant i+3$, $t_1=j/n$ and $t_2=(j-1)/n$, we get for sufficiently large n

$$\left|w\left(\frac{j}{n}\right)-w\left(\frac{j-1}{n}\right)\right| \leqslant \frac{7K}{n}.$$

The event "$w(t)$ is differentiable at some t_0" is contained in the event

$$B=\bigcup_{k\geqslant 1}\bigcup_{m>1}\bigcap_{n\geqslant m}\bigcup_{0<i\leqslant n+1}\bigcap_{i<j\leqslant i+3}\left(\left|w\left(\frac{j}{n}\right)-w\left(\frac{j-1}{n}\right)\right|<\frac{7k}{n}\right).$$

We shall show that B is a countable union of events of probability zero; hence $P(B)=0$. In fact, because of (i) and (iii),

$$P\left[\bigcap_{n\geqslant m}\bigcup_{0<i\leqslant n+1}\bigcap_{i<j\leqslant i+3}\left(\left|w\left(\frac{j}{n}\right)-w\left(\frac{j-1}{n}\right)\right|<\frac{7k}{n}\right)\right]$$

$$\leqslant nP\left(\left|w\left(\frac{1}{n}\right)\right|<\frac{7k}{n}\right)^3 = \lim_{n\to\infty} n\left[2\int_0^{7k/n}\frac{e^{-x^2n/2}}{\sqrt{2\pi/n}}dx\right]^3$$

$$\leqslant \lim_{n\to\infty} Cn^{-1/2}=0,$$

where

$$C=\frac{8(7k)^3}{\sqrt{2\pi}}.$$

Two more subtle properties of the Brownian motion are the following properties (v) and (vi).

Let $\mathfrak{F}_t$ be the subfield of Ω generated by the events $\{w(s)<a\}$ for all $s\leqslant t$ and all real a. A random variable τ such that $0\leqslant\tau\leqslant\infty$ is called a *stopping time* (or *Markov time*) if the event $\{\tau<t\}\in\mathfrak{F}_t$ for each $t\geqslant 0$. For example, a *first passage time* τ_x, defined by

$$\tau_x=\min\{t\,|\,w(t)=x\},$$

is a stopping time, as

$$\{\tau_x \leqslant t\} = \bigcap_{m \geqslant 1} \bigcup_{r = \text{rational} \leqslant t} \left\{ w(r) > x - \frac{1}{m} \right\} \in \mathcal{F}_t;$$

hence

$$\{\tau_x < t\} = \bigcup_{n \geqslant 1} \left\{ \tau_x \leqslant t - \frac{1}{n} \right\} \in \mathcal{F}_t$$

(v) Let τ be a finite stopping time and set

$$\mathcal{F}_\tau = \bigcup_{t \geqslant 0} \mathcal{F}_t \cap \{B \in \Omega \,|\, B \cap \{\tau < t\} \in \mathcal{F}_t \quad \text{for all} \quad t \geqslant 0\}.$$

The *strong Markov property* of $w(t)$ is given by

$$P\{w(t+\tau) \in B \,|\, \mathcal{F}_\tau, w(s) = a\} = P\{w(t) \in B \,|\, w(0) = b\},$$

where $s < t$ and $b = w(\tau)$.

A consequence of (v) is that the Brownian motion begins afresh at stopping times. More precisely,

$$\tilde{w}(t) = w(t+\tau) - w(\tau)$$

is a Brownian motion independent of s for all $s \leqslant \tau$. The proofs of the strong Markov property and further description of the Brownian motion can be found in (Itô and McKean 1965; McKean 1969).

A stochastic process is called a *martingale* if

$$E|x(t)| < \infty \tag{2.3.5}$$

and

$$E\{x(t) | x(t_1) = a_1, \ldots, x(t_n) = a_n\} = a_n \tag{2.3.6}$$

for all $t_1 < t_2 < \cdots < t_n < t$ and $a_1, \ldots, a_n$.

(vi) *The Brownian motion is a martingale.* Inequality (2.3.5) is obviously satisfied. Now, by the Markov property of $w(t)$ and by property (iv),

$$E\{w(t) | w(s) = x\} = \int \frac{y e^{-(x-y)^2/2(t-s)}}{\sqrt{2\pi(t-s)}} dy$$

$$= -\int \frac{(x-y) e^{-(x-y)^2/2(t-s)}}{\sqrt{2\pi(t-s)}} dy + x \int \frac{e^{-(x-y)^2/2(t-s)}}{\sqrt{2\pi(t-s)}} dy = x.$$

The first integral vanishes since the integrand is an odd function of $x-y$, and the second integral equals 1.

Another important martingale is given by

$$x(t)=e^{\alpha w(t)-(\alpha^2/2)t}, \qquad \alpha>0.$$

Indeed,

$$(2.3.7)\quad E|x(t)|=\int\frac{e^{\alpha x-(\alpha^2/2)t}e^{-x^2/2t}}{\sqrt{2\pi t}}dx=\int\frac{e^{-(x-\alpha t)^2/2t}}{\sqrt{2\pi t}}dx=1;$$

hence inequality (2.3.5) follows. Now

$$\begin{aligned}x(t)&=e^{\alpha[w(t)-w(s)]-(\alpha^2/2)(t-s)}x(s)=e^{\alpha\tilde{w}(t-s)-(\alpha^2/2)(t-s)}x(s)\\&=\tilde{x}(t-s)x(s),\end{aligned}$$

where $\tilde{x}(t-s)$ has the same probability law as $x(t-x)$, but it is independent of $x(s)$, by property (i). Hence

$$\begin{aligned}E\{x(t)|x(s)=x\}&=E\{x(s)\tilde{x}(t-s)|x(s)=x\}\\&=xE\tilde{x}(t-s)=x,\end{aligned}$$

by (2.3.7).

EXERCISE 2.3.1

Show that the following are martingales:

(i) $w^2(t)-t$.

(ii) $e^{-\alpha t}\cosh[\sqrt{2\alpha}\, w(t)]$.

Martingales satisfy the following inequality of Kolmogorov: Let $x(t)$ be a continuous martingale; then

$$(2.3.8)\qquad P\Big(\max_{0\leqslant t\leqslant T}x(t)>\alpha\Big)\leqslant\frac{E(x(T)\vee 0)}{\alpha}$$

for any $\alpha>0$. Here $x(T)\vee 0=\max(0, x(T))$. Indeed, let $B=\{\max_{1\leqslant i\leqslant n}x(t_i)>\alpha\}$, where $0\leqslant t_1<t_2\cdots<t_n\leqslant T$ is any sequence in $[0,T]$. Setting $B_i=\{x(t_j)\leqslant\alpha$ for all $j<i$, $x(t_i)>\alpha\}$, we have $B=\cup_{i=1}^n B_i$, $B_i\cap B_j=0$ if $i\neq j$. Now let χ_B be the characteristic function of the event (set) B in Ω.

Then

$$E(x(T)\vee 0 \geqslant E(x(T)\vee 0)\chi_B$$

$$= \sum_{i=1}^{n} E\left[(x(T)\vee 0)\chi_{B_i}\right] \geqslant \sum_{i=1}^{n} E(x(T)\chi_{B_i})$$

$$= \sum_{i=1}^{n} E\{E(x(T)|x(t_i))\chi_{B_i}\}$$

$$= \sum_{i=1}^{n} E(x(t_i)\chi_{B_i}) > \alpha \sum_{i=1}^{n} P(B_i) = \alpha P(B);$$

hence (2.3.8) follows by continuity.

Finally, we define an *n-dimensional Brownian motion* by setting

$$\mathbf{w}(t) = (w_1(t), \ldots, w_n(t)), \qquad t \geqslant 0$$

where $w_1(t), \ldots, w_n(t)$ are independent one-dimensional Brownian motions.

The probability law of $\mathbf{w}(t)$ is given by

$$P\{\mathbf{w}(t) \in B | \mathbf{w}(s) = \mathbf{x}\} = \int_B \frac{e^{-|\mathbf{x}-\mathbf{y}|^2/2(t-s)}}{[2\pi(t-s)]^{n/2}} d\mathbf{y}. \tag{2.3.9}$$

CHAPTER 3

The Stochastic (Itô) Calculus

3.1 INTRODUCTION

In the simplest physical model of a Brownian motion the Langevin equation (2.1.1) was used to describe the motion of a particle driven by a *white noise* type of force (due to collisions with the smaller molecules of the fluid). The solution of Langevin's equation is given by (2.1.3)

$$\mathbf{y}(t)=\mathbf{y}_0 e^{-\beta t}+e^{-\beta t}\int_0^t e^{-\beta s}\mathbf{f}(s)\,ds \tag{3.1.1}$$

where $\mathbf{f}(s)$ is the white noise term. The Brownian motion $\mathbf{w}(t)$ is then, according to Section 2.1, given by

$$\mathbf{w}(t)=\frac{1}{q}\int_0^t \mathbf{f}(s)\,ds, \tag{3.1.2}$$

so that $\mathbf{f}(s)=q\,d\mathbf{w}(s)/ds$, but since $\mathbf{w}(t)$ is nowhere differentiable, $\mathbf{f}(s)$ is not a function; hence the solution (3.1.1) is not a well-defined function. In the simple case of (3.1.1), this difficulty can be overcome as follows. Using integration by parts in (3.1.1) and employing (3.1.2), we obtain

$$\mathbf{y}(t)=\mathbf{y}_0 e^{-\beta t}+q\mathbf{w}(t)-\beta q\int_0^t e^{-\beta(t-s)}\mathbf{w}(s)\,ds. \tag{3.1.3}$$

Since all functions in (3.1.3) are well defined and continuous we may interpret the solution (3.1.2) by giving it the meaning of (3.1.3). This procedure can be generalized as follows. Let the functions $f(t)$ and $g(t)$ be defined for $a \leqslant t \leqslant b$. For every partition $P: a \leqslant t_0 < t_1 < \cdots < t_n = b$, we set

$$S_P = \sum_{i=1}^{n} f(\xi_i)\left[g(t_i) - g(t_{i-1})\right],$$

where $t_i \leqslant \xi_i \leqslant t_i$. If there is a limit $\lim_{|P|\to 0} S_P = I$ where $|P| = \max_{1<i<n}(t_i - t_{i-1})$, we say that I is the *Strietjes integral* of $f(t)$ with respect to $g(t)$, and we denote

$$I = \int_a^b f(t)\, dg(t).$$

It is well known (Natanson 1961) that if $g(t)$ is a function of finite variation; that is, $g(t)$ is the difference of two finite monotone increasing functions, and if $f(t)$ is continuous, then I exists. Furthermore, it is well known that if $\int_a^b f(t)\, dg(t)$ exists, then $\int_a^b g(t)\, df(t)$ exists and

$$\int_a^b f(t)\, dg(t) = f(b)g(b) - f(a)g(a) - \int_a^b g(t)\, df(t).$$

If $g(t)$ is a differentiable function and $g'(t)$ and $f(t)$ are integrable, then

$$\int_a^b f(t)\, dg(t) = \int_a^b f(t) g'(t)\, dt.$$

Returning to the case of (3.1.1), we can define $\int_0^T g(s)\mathbf{f}(s)\, ds$ for every differentiable function $g(t)$ by setting

$$\int_0^t g(s)\mathbf{f}(s)\, ds \equiv \int_0^t g(s)\, d\mathbf{w}(s) \equiv g(t)\mathbf{w}(t) - \int_0^t \mathbf{w}(s) g'(s)\, ds.$$

Unfortunately, this procedure fails if $g(t)$ is a function of $\mathbf{w}(t)$. Even in the simplest one-dimensional case, $\int_0^t w(t)\, dw(t)$ cannot be defined by the foregoing procedure. We see that if the random force in Langevin's equation depends on the solution, for example,

$$\dot{\mathbf{y}} = -\beta \mathbf{y} + q(\mathbf{y})\mathbf{f}, \tag{3.1.4}$$

then (3.1.4), which is essentially equivalent (by definition) to the integral equation

$$\mathbf{y}(t) = \mathbf{y}_0 - \beta \int_0^t \mathbf{y}(s)\, ds + \int_0^t q(\mathbf{y}(s))\, d\mathbf{w}(s), \tag{3.1.5}$$

does not make any sense, because the last integral in (3.1.5) cannot be defined by the standard procedures. The first step toward the development of a theory of stochastic differential equation must therefore be a definition of a stochastic integral.

3.2 STOCHASTIC INTEGRATION IN THE SENSE OF ITÔ AND STRATONOVICH

The following definition of the stochastic integral is due to Itô. Let

$$\chi_{[a,b]}(t)=\begin{cases}1 & \text{if } a\leqslant t\leqslant b\\ 0 & \text{otherwise}\end{cases}$$

be *the characteristic function* of the interval $[a,b]$. We define, for $0\leqslant a<b\leqslant T$,

$$\int_0^T \chi_{[a,b]}(t)\,dw(t)\equiv w(b)-w(a).$$

If $f(t)$ is a *step function* on $[0,T]$, then

$$f(t)=\sum_{k=0}^{m-1} f(t_k)\chi_{(t_k,[t_{k+1})}(t),$$

say, where $0=t_0<t_1<\cdots<t_m=b$. We define

$$\int_0^T f(t)\,dw(t)\equiv\sum_{k=0}^{m-1} f(t_k)\left[w(t_{k+1})-w(t_k)\right].$$

Note that the value of $f(t)$ is taken at the left endpoint of the partition interval. The function $f(t)$ can be a random function; for example, $f(t)$ may be a function of $w(t)$. In the latter case, $f(t)$ is independent of the increment $w(t_{k+1})-w(t_k)\equiv\delta_k w$. A function $f(t)$ which is independent of the increments $w(t+s)-w(t)$ for all $s>0$ is called a *nonanticipating* function. It depends statistically on $w(u)$ for $u\leqslant t$, that is, on the "past" only. For a nonanticipating step function $f(t)$, the integral $\int_0^t f(s)\,dw(s)$ is therefore also a nonanticipating function. We shall assume that the jumps of the random step function $f(t)$ occur at nonrandom times t_k. We have the following simple properties of the stochastic integrals. Let f and g be two nonanticipating step functions; then

(i) $\int_0^T [f(t)+g(t)]\,dw(t)=\int_0^T f(t)\,dw(t)+\int_0^T g(t)\,dw(t)$

(ii) $\int_0^T cf(t)\,dw(t)=c\int_0^T f(t)\,dw(t)$

for any constant c.

(iii) If f and g satisfy

$$\int_0^T \left[Ef^2+Eg^2\right] dt<\infty,$$

then

$$E\int_0^T f(t)\,dw(t)=0 \tag{3.2.1}$$

and

$$E\left[\int_0^T f(t)\,dw(t)\int_0^T g(t)\,dw(t)\right]=\int_0^T Ef(t)g(t)\,dt. \tag{3.2.2}$$

Properties (i) and (ii) can be easily verified. Equation (3.2.1) follows from the assumption that $f(t)$ is not anticipating, for

$$E\int_0^T f(t)\,dw(t)=E\sum_{k=0}^{m-1} f(t_k)\,\delta_k w$$

$$=\sum_{k=0}^{m-1} E(f(t_k)\,\delta_k w)=\sum_{k=0}^{m-1} Ef(t_k)E\,\delta_k w=0.$$

To prove (3.2.2) we begin with the case $f=g$. Then

$$S=E\left[\int_0^T f(t)\,dw(t)\right]^2=E\left[\sum_{k=0}^{m-1} f(t_k)\,\delta_k w\right]^2$$

$$=\sum_{k<l<m-1} E[f(t_k)\,\delta_k w][f(t_l)\,\delta_l w].$$

If $k<l$, then

$$E[f(t_k)\,\delta_k w][f(t_l)\,\delta_l w]=E[f(t_k)\,\delta_k w f(t_l)]\,E\,\delta_l w=0,$$

as $\delta_l w$ is independent of $\delta_k w$ and of $f(t_k)$ and $f(t_l)$. Thus

$$S=\sum_{k=0}^{m-1} E\left[f^2(t_k)(\delta_k w)^2\right]=\sum_{k=0}^{m-1} Ef^2(t_k)E(\delta_k w)^2$$

$$=\sum_{k=0}^{m-1} Ef^2(t_k)\,\delta t_k=\int_0^T Ef^2(t)\,dt,$$

where $\delta t_k = t_{k+1}-t_k$. We have used the fact that $E[w(t_{k+1})-w(t_k)]^2 = t_{k+1}-t_k$.

To prove (3.2.2) for $g\neq f$, we set $f_1=g+f$; then

$$E\left[\int_0^T f_1\,dt\right]^2=\int_0^T Ef_1^2\,dt$$

$$=\int_0^T Ef^2\,dt+2\int_0^T\left[Ef(t)g(t)\right]dt+\int_0^T Eg^2\,dt,$$

but also

$$E\left[\int_0^T f_1\,dt\right]^2=E\left[\int_0^T f\,dw\right]^2+2E\left[\int_0^T f\,dw\int_0^T g\,dw\right]+E\left[\int_0^T g\,dw\right]^2$$

$$=\int_0^T Ef^2\,dt+2E\left[\int_0^T f\,dw\int_0^T g\,dw\right]+\int_0^T Eg^2\,dt.$$

Hence (3.2.2) follows.

We shall denote by $H_2[0,T]$ the class of all nonanticipating functions $f(t)$ such that

$$\int_0^T Ef^2(t)\,dt<\infty.$$

It will be shown (see Section 3.3) that for any function $f(t)$ in $H_2[0,T]$, there exists a sequence $\{g_n(t)\}$ of step functions such that

$$\int_0^T |f(t)-g_n(t)|^2\,dt\to 0 \quad \text{as } n\to\infty \quad \text{a.s.}$$

and

$$\int_0^t g_n(s)\,dw(s)\to\text{limit}=L(t) \quad \text{as } n\to\infty \tag{3.2.3}$$

uniformly for t is $[0, T]$ a.s. We define

$$\int_0^t f(s)\,dw(s) \equiv L(t) \qquad (0 \leqslant t \leqslant T). \tag{3.2.4}$$

Since $\int_0^t g_n(s)\,dw(s)$ is a continuous function for each n (why?) and the convergence in (3.2.3) is uniform, the integral $\int_0^t f(s)\,dw(s)$ is a continuous function of t a.s. It can be easily shown that the integral (3.2.4) is independent of the approximating sequence $\{g_n(t)\}$. The properties of the Itô integral (3.2.4) are very different from those of the familiar Riemann integral.

EXERCISE 3.2.1

(McKean 1969) Let $L^2[0,\infty]$ be the set of all square-integrable real functions on $[0,\infty]$. If f and g are elements of $L^2[0,\infty]$, then $\langle f,g\rangle = \int_0^\infty fg\,dt$ and $\|f\|^2 = \langle f,f\rangle$ are the inner product and norm, respectively. An *orthogonal transformation* ν on $L^2[0,\infty]$ is a transformation that preserves the scalar product; that is, $\langle \nu(f), \nu(g)\rangle = \langle f,g\rangle$ for all $f,g \in L^2[0,\infty]$. For example, $\nu f =$ the Fourier transform of f is such a transformation, by Parseval's identity. Show that if ν is an orthogonal transformation, then the mapping

$$\varphi(w(t)) = \int_0^\infty \nu(\chi_{[0,t]}(s))\,dw(s) \tag{3.2.5}$$

maps Brownian paths into Brownian paths. [*Hint.* Use (3.2.2), and assume the improper integral (3.2.5) exists.]

Obviously, if $f(t)$ is a deterministic smooth function, the integral $\int_0^T f(t)\,dw(t)$ is the Stieljes integral; hence

$$\int_a^b f(t)\,dw(t) = f(b)w(b) - f(a)w(a) - \int_a^b w(t)f'(t)\,dt \quad \text{a.s.} \tag{3.2.6}$$

EXERCISE 3.2.2

Find the probability density of the integral (3.2.6).

We shall compute next some useful integrals. We begin with the integral

$$\int_a^b w(t)\,dw(t). \tag{3.2.7}$$

Since $w(t)$ is continuous a.s. the step functions $w_n(t)$, defined by

$$w_n(t) = \sum_{i=0}^{n-1} w(t_i)\chi_{(t_{i+1}t_i]}(t),$$

where $t_i = a + i[(b-a)/n], i = 0, \ldots, n$, converge uniformly to $w(t)$ in $[a, b]$. The integrals of $w_n(t)$ are given by

$$I_n = \int_a^b w_n(t)\, dw(t) = \sum_{i=0}^{n-1} w(t_i)[w(t_{i+1}) - w(t_i)].$$

Hence

$$I_n = \tfrac{1}{2} \sum_{i=0}^{n-1} \left\{ w^2(t_{i+1}) - w^2(t_i) - [w(t_{i+1}) - w(t_i)]^2 \right\}$$

$$= \tfrac{1}{2}[w^2(b) - w^2(a)] - \tfrac{1}{2} \sum_{i=0}^{n-1} (\delta_i w)^2,$$

where $\delta_i w = w(t_{i+1}) - w(t_i)$. Setting

$$\eta_n = \sum_{i=1}^{n-1} (\delta_i w)^2,$$

we have

$$E\eta_n = \sum_{i=0}^{n-1} E(\delta_i w)^2 = \sum_{i=0}^{n-1} (t_{i+1} - t_i) = b - a.$$

We shall show that $\eta_n \to b - a$ in probability as $n \to \infty$. Indeed,

$$\text{Var } \eta_n = \sum_{i=0}^{n-1} \text{Var}(\delta_i w)^2,$$

since $(\delta_{iw})^2$ are independent random variables. Thus

$$\text{Var } \eta_n = \sum_{i=0}^{n-1} \left\{ E(\delta_i w)^4 - [E(\delta_i w)^2]^2 \right\}$$

$$\leqslant \sum_{i=0}^{n-1} E(\delta_i w)^4.$$

Since $\delta_i w$ is $N(0, \sqrt{t_{i+1}-t_i}\,)$, we have

$$E(\delta_i w)^4 = \int_{-\infty}^{\infty} \frac{x^4 e^{-x^2/2(t_{i+1}-t_i)}}{\sqrt{2\pi(t_{i+1}-t_i)}}\,dx$$

$$= 3(t_{i+1}-t_i)^2.$$

From Chebyshev's inequality (see Chapter 1),

$$P\{|\eta_n - E\eta_n| > \varepsilon\} \leqslant \frac{\operatorname{Var}\eta_n}{\varepsilon^2} \leqslant \frac{3}{\varepsilon^2}\sum_{i=0}^{n-1}(t_{i+1}-t_i)^2$$

$$= \frac{3}{\varepsilon^2}\frac{(b-a)^2}{n} \to 0 \qquad \text{as} \quad n\to\infty.$$

So $\eta_n \to E\eta_n = b-a$ in probability as $n\to\infty$. It follows that

$$\int_a^b w_n(t)\,dw(t) \to \tfrac{1}{2}\left[w^2(b) - w^2(a)\right] - \tfrac{1}{2}(b-a);$$

hence

$$\int_a^b w(t)\,dw(t) = \tfrac{1}{2}\left[w^2(t) - w^2(a)\right] - \tfrac{1}{2}[b-a], \tag{3.2.8}$$

unlike the classical integral.

Other types of integrals are also used in modeling physical phenomena. Consider, for example, the *backward integral*

$$(B)\int_0^T f(w(t))\,dw(t) = \lim_{\delta\to 0}\sum_{i=0}^{n-1} f(w(t_{i+1}))\left[w(t_{i+1}) - w(t_i)\right],$$

where

$$0 = t_0 < t_1 < \cdots < t_n = T,$$

$\delta = \max(t_{i+1}-t_i)$. If $f(x)$ is a smooth function, then

$$f(w(t_{i+1})) = f(w(t_i)) + \left[w(t_{i+1}) - w(t_i)\right] f'(w(t_i))$$

$$+ \frac{\left[w(t_{i+1}) - w(t_i)\right]^2 f''(\xi_i)}{2},$$

where ξ_i is a (random) number between $w(t_i)$ and $w(t_{i+1})$. The sum

$$S=\sum_{i=0}^{n-1}\left[w(t_{i+1})-w(t_i)\right]^2 f'(w(t_i))$$

converges to

$$\int_0^T f'(w(t))\,dt$$

as

$$\begin{aligned} ES &= \sum_{i=0}^{n-1} Ef'(w(t_i))E(w(t_{i+1})-w(t_i))^2 \\ &= \sum_{i=0}^{n-j}(t_{i+1}-t_i)Ef'(w(t_i)) \to \int_0^T Ef'(w(t))\,dt \\ &= E\int_0^T f'(w(t))\,dt \qquad \text{as} \quad \delta\to 0. \end{aligned}$$

Now Var $S\to 0$ as $\delta\to 0$, as was the case in the derivation of (3.2.8); hence $S-ES\to 0$ by Chebyshev's inequality. The sum

$$T=\sum_{i=0}^{n-1}\frac{\left[w(t_{i+1})-w(t_i)\right]^3 f''(\xi_i)}{2}$$

converges to zero, since

$$\begin{aligned} E|T| &\leqslant \sum_{i=0}^{n-1}\frac{E|w(t_{i+1})-w(t_i)|^3 |f''(\xi_i)|}{2} \\ &\leqslant \sum_{i=0}^{n-1}\left\{E|w(t_{i+1})-w(t_i)|^6 E|f''(\xi_i)|^2\right\}^{1/2} \\ &\leqslant c\sum_{i=0}^{n-1}\left\{(t_{i+1}-t_i)^3 E|f''(\xi_i)|^2\right\}^{1/2} \\ &\leqslant \delta^{1/2} C\sum_{i=0}^{n-1}(t_{i+1}-t_i)\left\{E|f''(\xi_i)|^2\right\}^{1/2}, \end{aligned}$$

where C is a constant independent of $f(x)$. Since the sum converges to

$$\int_0^T \{E|f''(w(t))|^2\}^{1/2}\,dt<\infty$$

and $\delta\to 0$, we see that $E|T|\to 0$ and similarly that $\operatorname{Var}|T|\to 0$. Hence

$$(B)\int_0^T f(w(t))\,dw(t)=\int_0^T f(w(t))\,dw(t)+\int_0^T f'(w(t))\,dt.$$

EXERCISE 3.2.3

Evaluate

$$(B)\int_0^T w(t)\,dw(t).$$

EXERCISE 3.2.4

Define the *second-order integral*

$$\int_0^T f(t)[\,dw(t)]^2=\lim_{\delta\to 0}\sum_{i=0}^{n-1} f(t_i)[\,w(t_{i+1})-w(t_i)]^2.$$

Show that

$$\int_0^T f(t)[\,dw(t)]^2=\int_0^T f(t)\,dt.$$

(See Jazwinski 1970 for a discussion.)

EXERCISE 3.2.5

Similarly, define $I_n=\int_0^T f(t)[dw(t)]^n$. Show that for $n>2$, $I_n=0$.

Another type of stochastic integral is that of Stratonovich (1965). It is defined for explicit functions of $w(t)$ by

$$(S)\int_0^T f(w(t),t)\,dw(t)=\lim_{\delta\to 0}\sum_{i=0}^{n-1} f\left(\frac{w(t_i)+w(t_{i+1})}{2},t_i\right)[\,w(t_{i+1})-w(t_i)].$$

EXERCISE 3.2.6

Use the discussion of the backward integral to show that

$$(S)\int_0^T f(w(t),t)\,dw(t)=\int_0^T f(w(t),t)\,dw(t)$$
$$+\tfrac{1}{2}\int_0^T f_x(w(t),t)\,dt,$$

where

$$f_x(x,t)=\frac{\partial f}{\partial x}(x,t).$$

Assume that $f(x,t)$ is a smooth function.

EXERCISE 3.2.7

Show that

$$(S)\int_0^T w(t)\,dw(t)=\frac{w^2(T)}{2}.$$

The Stratonovich integral satisfies all the formal rules of the classical calculus; for example, the fundamental theorem of the integral calculus,

$$(S)\int_a^b f'(w)\,dw=f(w(b))-f(w(a)), \tag{3.2.9}$$

integration by parts, and so on. We shall show that (3.2.9) holds. Assume that $f(x)$ has three continuous derivatives, f', f'', and f'''. Then, by the mean value theorem,

$$f'\left(\frac{w(t_i)+w(t_{i+1})}{2}\right)[w(t_{i+1})-w(t_i)] \tag{3.2.10}$$
$$=f(w(t_{i+1}))-f(w(t_i))-\frac{[w(t_{i+1})-w(t_i)]^2}{12}f'''(\xi_i),$$

where ξ_i is a point between $w(t_i)$ and $w(t_{i+1})$. Summing over i and taking the limit, we see that the first two terms on the right-hand side of (3.2.10) produce (3.2.9), while the sum of the last term converges to zero as in Exercise 3.2.5.

EXERCISE 3.2.8

Prove that the Stratonovich integral satisfies the rules of integration by parts and variable substitution.

EXERCISE 3.2.9

Show that the sums

$$\sum_{i=0}^{n-1} \frac{[f(w(t_i),t_i)+f(w(t_{i+1}),t_{i+1})][w(t_{i+1})-w(t_i)]}{2}$$

lead to the Stratonovich integral.

EXERCISE 3.2.10

Define an integral $(\lambda)\int_a^b f\,dw$, by using the sums

$$\sum_{i=0}^{n-1} f(w(t_i)\lambda+w(t_{i+1})(1-\lambda),t_i)[w(t_{i+1})-w(t_i)],$$

where $0\leqslant\lambda\leqslant 1$. For $\lambda=1$, obtain Itô's definition; for $\lambda=\frac{1}{2}$, obtain the Stratonovich definition; and for $\lambda=0$, recover the backward integral. Find the relationship between $(\lambda)\int_a^b f\,dw$ and $\int_a^b f\,dw$. Further properties of the Stratonovich integral and the question of modeling physical phenomena by stochastic integrals of Itô and Stratonovich will be discussed in the following sections.

3.3* THE CONSTRUCTION OF THE STOCHASTIC INTEGRAL

Let $\mathcal{F}_t$ be the σ-algebra of subsets of Ω generated by $w(s)$ for all $s\leqslant t$. Thus $\mathcal{F}_t$ is the smallest σ-algebra containing all the events $\{w(s)<a\}$, $0\leqslant s\leqslant t$, $a=$real number. We say that a random function $f(t)$ is nonanticipating if it is measurable with respect to $\mathcal{F}_t$. In particular, a nonanticipating function $f(t)$ is independent of $w(t+s)-w(t)$ for all $s>0$. If $f(t)\in H_2[0,T]$, then clearly $P[\int_0^T f^2(t)<\infty]=1$. For nonanticipating step functions in $H_2[0,T]$, the stochastic integral has been defined in Section 3.2. We shall show first that the stochastic integral of a simple function in $H_2(0,T]$ has the property that

$$h(t)=\exp\left[\int_0^t f(s)\,dw(s)-\tfrac{1}{2}\int_0^t f^2(s)\,ds\right]$$

is a martingale. Indeed, if $f(t)$ is constant, then $\int_0^t f(s)\,dw(s)=cw(t)$. It has been shown in Section 2.3* that $\exp[cw(t)-(c^2/2)t]$ is a martingale and that

$$E\exp\left[cw(t)-\frac{c^2}{2}t\right]=1.$$

If $f(t)$ is a step function, the same can be shown to be true by considering times in different intervals separately. Next we derive a useful inequality. Let $f(t)$ be a simple function in $H_2[0,T]$; then

$$P\left[\max_{t\leqslant 1}\left(\int_0^t f(s)\,dw(s)-\frac{\alpha}{2}\int_0^t f^2(s)\,ds\right)>\beta\right]\leqslant e^{-\alpha\beta}. \tag{3.3.1*}$$

For, replacing $f(t)$ by $\alpha f(t)$ we see that (3.2.1*) is equivalent to

$$P\left[\max_{t\leqslant 1}\left(\int_0^t \alpha f(s)\,dw(s)-\frac{1}{2}\int_0^t [\alpha f(s)]^2\,ds\right)>\alpha\beta\right]=P\left[\max_{t\leqslant 1}h(t)>e^{\alpha\beta}\right],$$

where $h(t)$ is a martingale. Using the martingale inequality, we obtain

$$P\left[\max_{t\leqslant 1}h(t)>e^{\alpha\beta}\right]\leqslant e^{-\alpha\beta}Eh(1)=e^{-\alpha\beta};$$

hence (3.3.1*) follows.

Lemma. If $\{f_n(t)\}$ is a sequence of simple functions in $H_2[0,T]$ and

$$P\left[\int_0^1 f_n^2(s)\,ds>2^{-n}\quad\text{i.o.}\right]=0,$$

then

$$P\left[\max_{t\leqslant 1}\left|\int_0^t f_n(s)\,dw(s)\right|\geqslant \nu(2^{-n+1}\log n)^{1/2}\quad\text{i.o.}\right]=0 \tag{3.3.2*}$$

for any $\nu>1$.

Proof. Setting $\alpha=(2^{n+1}\log n)^{1/2}$ and $\beta=\nu(2^{-n-1}\log n)^{1/2}$, we obtain

$$P\left[\max_{t<1}\int_0^t f_n(s)\,dw(s)\geqslant\nu(2^{-n+1}\log n)^{1/2}\right]$$

$$<P\left[\max_{t<1}\int_0^t f_n(s)\,dw(s)-\frac{\alpha}{2}\int_0^1 f_n^2(s)\,ds\geqslant 2\beta-\frac{\alpha}{2}\int_0^1 f_n^2(s)\,ds\right]$$

$$\leqslant P\left[\max_{t<1}\left(\int_0^t f_n(s)\,dw(s)-\frac{\alpha}{2}\int_0^t f_n^2(s)\,ds\right)\geqslant 2\beta-2^{-n-1}\alpha\right]$$

$$\leqslant P\left[\max_{t<1}\left(\int_0^t f_n(s)\,dw(s)-\frac{\alpha}{2}\int_0^t f_n^2(s)\,ds\right)\geqslant\nu(2^{-n+1}\log n)^{1/2}\right.$$

$$\left.-(2^{-n-1}\log n)^{1/2}=(2\nu-1)(2^{-n-1}\log n)^{1/2}\right]\leqslant e^{-(2\nu-1)\log n}$$

$$=n^{-(2\nu-1)}$$

for an infinite set of indices n. Now, since $2\nu-1>\nu>1$, we have $\Sigma_{n=1}^{\infty} n^{-(2\nu-1)}<\infty$; hence (3.3.2*) follows by the Borel–Cantelli lemma and from the same estimate for $f_n(t)$. Next we construct for every function $f(t)$ in $H_2[0,T]$ a sequence of step functions $f_n(t)$ such that $f_n\to f$ and such that $\int_0^t f_n(s)\,ds$ converge uniformly to a limit. Since $f(t)$ is integrable a.s. with respect to t, it is equal to the derivative of its integral almost everywhere. Hence, setting $\tilde{f}_k(t)=2^k\int_{t-2^{-k}}^t f(s)\,ds$, where $f(s)=0$ for $s<0$, we have $\tilde{f}_k(t)\to f(t)$ as $k\to\infty$ a.e. Since $\tilde{f}_k(t)$ is an absolutely continuous function, the approximating step functions $\tilde{f}_{m,k}(t)=\tilde{f}_k(2^{-m}[2^m t])$ converge to $\tilde{f}_k(t)$ uniformly as $m\to\infty$. Here $[x]$ is the greatest integer less than or equal to x. Since

$$\int_0^1\left[f(t)-\tilde{f}_{m,k}(t)\right]^2 dt\to 0 \quad \text{a.s.}$$

as $m\to\infty$ and $k\to\infty$, we can find for each n a pair of indices m,k so that

$$P\left[\int_0^1\left(f(t)-\tilde{f}_{m,k}(t)\right)^2 dt>2^{-n+1}\right]<2^{-n}.$$

We define $f_n(t)=\tilde{f}_{m,k}(t)$. It follows from the Borel–Cantelli lemma that

$$P\left[\int_0^1\left(f(t)-f_n(t)\right)^2 dt>2^{-n-1}\quad\text{i.o.}\right]=0;$$

hence

$$P\left[\int_0^1 (f_n(t)-f_{n-1}(t))^2\,dt>2^{-n}\quad \text{i.o.}\right]=0.$$

It follows from the lemma that

$$P\left[\max_{t\leqslant 1}\left|\int_0^t [f_n(s)-f_{n-1}(s)]\,dw(s)\right|\geqslant \nu(2^{-n+1}\log n)^{1/2}\quad \text{i.o.}\right]=0;$$

hence $\int_0^t f_n(t)\,dw(t)$ converges uniformly a.s. to a limit as $n\to\infty$, so we set

$$\int_0^t f(s)\,dw(s)\equiv \lim_{n\to\infty}\int_0^t f_n(s)\,dw(s). \qquad (3.3.3^*)$$

It can be easily shown that the limit is independent of the approximating sequence $f_n(t)$. Since each integral $\int_0^t f_n(s)\,dw(s)$ is a continuous function and the convergence in (3.3.3*) is uniform, the integral $\int_0^t f(s)\,dw(s)$ is a.s. a continuous function.

EXERCISE 3.3.1*

Show that the limit (3.3.3*) is independent of the approximating sequence.

EXERCISE 3.3.2*

Use step-function approximation to show that if

$$P\left[\int_0^\infty f^2(t)\,dt<\infty\right]=1,$$

then $\int_0^\infty f(t)\,dt$ can be defined in such a way that

$$P\left[\lim_{t\to\infty}\int_0^t f(s)\,ds=\int_0^\infty f(t)\,dt\right]=1.$$

EXERCISE 3.3.3*

Let $f\in H_2[0,T]$, show that

$$h(t)=\exp\left[\int_0^t f(s)\,dw(s)-\tfrac{1}{2}\int_0^t f^2(s)\,ds\right]$$

is a supermartingale and $Eh(t) \leqslant 1$. For an example where $Eh(t) < 1$, see McKean (1969). A supermartingale is defined by replacing the last equality in (2.3.6) by the inequality "less than or equal to."

EXERCISE 3.3.4*

Prove the lemma for all $f_n \in H_2[0, T]$.

EXERCISE 3.3.5*

Show that

$$E\exp\left[i\int_0^\infty f\,dw(t)+\tfrac{1}{2}\int_0^\infty f^2\,dt\right]=1$$

if

$$E\exp\left(\tfrac{1}{2}\int_0^\infty f^2\,dt\right)<\infty.$$

EXERCISE 3.3.6*

Show that

$$\zeta(t)=\int_0^t f(s)\,dw(s)$$

is a martingale.

3.4 STOCHASTIC DIFFERENTIALS AND ITÔ'S FORMULA

Let $a(t)$ and $b(t)$ be functions in $H_2[0, T]$. Let $x(t)$ be a stochastic (random) process satisfying

$$x(t_2)-x(t_1)=\int_{t_1}^{t_2} a(t)\,dt+\int_{t_1}^{t_2} b(t)\,dw(t)$$

for all $0 \leqslant t_1 < t_2 \leqslant T$. Then we say that $x(t)$ has a stochastic *differential*

(3.4.1) $$dx(t)=a(t)\,dt+b(t)\,dw(t).$$

If, for example, we set $x(t)=w^2(t)$, then, by (3.2.8),

(3.4.2) $$w^2(t_2)-w^2(t_1)=2\int_{t_1}^{t_2} w(t)\,dw(t)+\int_{t_1}^{t_2} 1\,dt;$$

hence

$$dw^2(t)=1\,dt+2w(t)\,dw(t).$$

Thus $a(t)\equiv 1$ and $b(t)=2w(t)$. If $f(t)$ is a deterministic smooth function then, by (3.2.6),

$$f(t_2)w(t_2)-f(t_1)w(t_1)=\int_{t_1}^{t_2} f(t)\,dw(t)+\int_{t_1}^{t_2} w(t)f'(t)\,dt;$$

hence

(3.4.3) $$d[\,f(t)w(t)]=f(t)\,dw(t)+w(t)\,df(t),$$

where $df(t)=f'(t)\,dt$.

EXERCISE 3.4.1

Show that if $x(t)$ has the differential

$$dx(t)=a(t)\,dt+b(t)\,dw(t)$$

and if $f(t)$ is a smooth deterministic function, then

$$d[\,x(t)f(t)]=x(t)\,df(t)+f(t)\,dx(t).$$

Next we derive the rule for computing the differential of a product. Let $dx_i(t)=a_i(t)\,dt+b_i(t)\,dw(t)(i=1,2)$, where $a_i(t)$ and $b_i(t)$ are functions in $H_2[0,T]$. We consider first the simplest case where a_i and b_i are constants. Then $x_i(t)=x_i(0)+a_it+b_iw(t)(i=1,2)$; hence

$$\begin{aligned} x_1(t)x_2(t)0=x_1(0)x_2(0)&\\ &+[\,x_1(0)a_2+x_2(0)a_1]t+[\,x_1(0)b_2+x_2(0)b_1]w(t)\\ &+a_1a_2t^2+[\,a_1b_2+a_2b_1]tw(t)+b_1b_2w^2(t). \end{aligned}$$

Taking differentials we see by (3.4.2) and (3.4.3) that

(3.4.4)

$$d[x_1(t)x_2(t)]=[x_1(0)a_2+x_2(0)a_1+2a_1a_2t+(a_1b_2+a_2b_1)w(t)+b_1b_2]\,dt+[(a_1b_2+a_2b_1)t+2b_1b_2w(t)]\,dw(t)=x_1(t)\,dx_2(t)+x_2(t)\,dx_1(t)+b_1b_2\,dt.$$

Formula (3.4.4) must hold for all step functions $a_i(t)$ and $b_i(t)$ in $H_2[0,T]$; therefore, using step-function approximation to any functions $a_i(t)$ and $b_i(t)$ in $H_2[0,T]$, we see that (3.3.4) holds for all functions $a_i(t)$ and $b_i(t)$ in $H_2[0,T]$. The product rule

(3.4.5) $$d[x_1(t)x_2(t)]=x_1(t)\,dx_2(t)+x_2(t)\,dx_1+b_1(t)b_2(t)\,dt$$

differs from the classical rule by the additional term $b_1(t)b_2(t)\,dt$.

Next, we compute the differential of any polynomial of $w(t)$. Using mathematical induction, we obtain from (3.4.2) and (3.4.5),

$$dw^m(t)=m[w(t)]^{m-1}\,dw(t)+\frac{m(m-1)}{2}[w(t)]^{m-2}\,dt$$

for any $m\geqslant 2$. It follows that for any polynomial $P(x)$, we have

(3.4.6) $$dP(w(t))=P'(w(t))\,dw(t)+\tfrac{1}{2}P''(w(t))\,dt.$$

Since any twice continuously differentiable function $f(x)$ can be uniformly approximated together with its derivatives $f'(x)$ and $f''(x)$ on any bounded interval, we see that (3.4.6) holds for all twice continuously differentiable functions $P(x)$.

EXERCISE 3.4.2

Supply the missing details in the previous argument.

Next we derive the chain rule for stochastic differentials. We begin with functions of the form $\Phi(x,t)=\phi(x)g(t)$, where g' and ϕ'' are continuous functions. Then, by (3.4.5),

$$d\Phi(w(t),t)=\phi(w(t))g'(t)\,dt+g(t)\,d\phi(w(t))=\left[\phi(w(t))g'(t)+\tfrac{1}{2}g(t)\phi''(w(t))\right]dt+g(t)\phi'(w(t))\,dw(t);$$

that is,

$$(3.4.7)\quad d\Phi(w(t),t)=\left[\frac{\partial\Phi}{\partial t}(w,t)+\frac{1}{2}\frac{\partial^2\Phi}{\partial x^2}(w,t)\right]dt+\frac{\partial\Phi(w,t)}{\partial w}dw(t).$$

Since any smooth function $\Phi(x,t)$ can be uniformly approximated on compact subsets of R^2 by "degenerate" functions

$$\Phi_n(x,t)=\sum_{k=1}^{n}g_k(t)\phi_k(x),$$

we conclude that (3.4.7) holds for all smooth functions $\Phi(x,t)$. Note that if $\Phi(x,t)$ is a nonanticipating random function, then (3.4.7) holds.

Finally, we replace $w(t)$ by any differentiable process. Let $x(t)$ be a stochastic process whose differential is given by

$$dx(t)=a(t)\,dt+b(t)\,dw(t),$$

$a,b\in H_2[0,T]$, and let $f(x,t)$ be a smooth function. Assume first that

$$x(t)=x_0+at+bw(t),$$

where x_0, a, and b are constants. Then

$$f(x(t),t)=f(x_0+at+bw(t),t)\equiv\Phi(w(t),t),$$

where $\Phi(x,t)=f(x_0+at+bx,t)$. We have

$$\frac{\partial\Phi}{\partial t}=\frac{\partial f}{\partial t}+a\frac{\partial f}{\partial x},$$

$$\frac{\partial\Phi}{\partial x}=b\frac{\partial f}{\partial x},\quad\text{and}\quad\frac{\partial^2\Phi}{\partial x^2}=b^2\frac{\partial^2 f}{\partial z^2}(z,t)\bigg|_{z=x_0+at+bx}.$$

From (3.4.7) we obtain

(3.4.8)

$$df(x(t),t)=d\Phi(w(t),t)=\left[\frac{\partial f}{\partial t}(x(t),t)\right.$$

$$\left.+a\frac{\partial f}{\partial x}(x(t),t)+\tfrac{1}{2}b^2\frac{\partial^2 f}{\partial x^2}(x(t),t)\right]dt+b\frac{\partial f}{\partial x}(x(t),t)\,dw(t).$$

Hence (3.4.8) holds if a and b are step functions in $H_2[0,T]$ and by approximation (3.4.8) holds for all functions $a(t)$ and $b(t)$ in $H_2[0,T]$. Equation (3.4.8) is called *Itô's formula*. It corrects the classical chain rule

$$df(x(t),t)=\frac{\partial f}{\partial x}dx+\frac{\partial f}{\partial t}dt$$
$$=\frac{\partial f}{\partial x}(a\,dt+b\,dw)+\frac{\partial f}{\partial t}dt$$

by the additional term

$$\tfrac{1}{2}b^2\frac{\partial^2 f}{\partial x^2}dt.$$

For functions of several variables, we have the following extension of Itô's formula. Let $\mathbf{a}(t)=[a_1(t),\ldots,a_n(t)]^T$ be a vector and let $\mathbf{B}(t)=\{b_{ij}(t)\}_{i,j<n}$ be a matrix whose elements are a function. Let $\mathbf{w}(t)=[w_1(t),\ldots,w_n(t)]^T$ be a vector of independent Brownian motion. Assume that $\mathbf{a}$ and $\mathbf{B}$ are nonanticipating with respect to $w(t)$. Let $x(t)=[x_1(t),\ldots,x_1(t)]^T$ be a vector of differentiable processes and assume that

$$d\mathbf{x}(t)=\mathbf{a}(t)\,dt+\mathbf{B}(t)\,d\mathbf{w}.$$

If $f(\mathbf{x},t)$ is a smooth function of the $n+1$ variables $\mathbf{x}=[x_1,\ldots,x_n]^T$ and t, then Itô's formula is given by

(3.4.9)

$$df(\mathbf{x}(t),t)=\left[\frac{\partial f}{\partial t}(\mathbf{x}(t),t)+\mathbf{a}(t)\cdot\nabla_{\mathbf{x}}f(\mathbf{x}(t),t)\right.$$
$$\left.+\sum_{i,j<n}\sigma_{ij}(t)\frac{\partial^2 f}{\partial x_i\partial x_j}(\mathbf{x}(t),t)\right]dt+\sum_{i,j<n}B_{ij}\frac{\partial f}{\partial x_i}(\mathbf{x}(t)t)\,dw_j$$

where

$$\sigma_{ij}=\tfrac{1}{2}(\mathbf{BB}^*)_{ij}.$$

EXERCISE 3.4.3

Supply the details in the derivation of (3.4.9).

EXERCISE 3.4.4

Show that

$$E\left[\int_0^t f(s)\,dw(s)\right]^{2m}=\frac{2m(2m-1)}{3}\int_0^t E\left[\int_0^s f(r)\,dw(r)\right]^{2m-2} f^2(s)\,ds$$

$$\leqslant\left[m(2m-1)\right]^{m-1}t^{m-1}\int_0^t Ef^{2m}(s)\,ds.$$

[*Hint.* Apply Itô's formula to $\phi(x)=x^{2m}$ and $x(t)=\int_0^t f(s)dw(s)$. Use Hölder's inequality.]

The chain rule for the Stratonovich integral is the classical rule

(3.4.10) $$d_S f(x(t),t)=f_t(x(t),t)\,dt+f_x(x(t),t)\,d_S x(t).$$

Indeed, let $x(t)$ be a differentiable process; that is, assume that

$$dx(t)=a(t)\,dt+b(t)\,dw(t)$$

in the sense of Itô. Then, by Itô's formula,

(3.4.11) $$f(x(t),t)=f(x(t_0),t_0)+\int_{t_0}^t\Big[f_t(x(s),s)+a(s)f_x(x(s),s)+\tfrac{1}{2}b^2(s)f_{xx}(x(s),s)\Big]\,ds+\int_{t_0}^t b(s)f_x(x(s),s)\,dw(s).$$

Using the result of Exercise 3.2.6, we have

(3.4.12)

$$\int_{t_0}^t b(s)f_x(x(s),s)\,dw(s)=(S)\int_{t_0}^t b(s)f_x(x(s),s)\,dw(s)-\frac{1}{2}\int_{t_0}^t\frac{\partial}{\partial w(s)}\left[b(s)f_x(x(s),s)\right]ds.$$

first we have $dt/dw(t)=0$, for

$$P\left(\left|\frac{\Delta t}{\Delta w}\right|>\varepsilon\right)=P\left(|\Delta w|<\frac{\Delta t}{\varepsilon}\right)$$

$$=\frac{1}{\sqrt{2\pi\Delta t}}\int_{-\Delta t/\varepsilon}^{\Delta t/\varepsilon}e^{-x^2/2\Delta t}\,dx=\frac{1}{\sqrt{2\pi}}\int_{-\sqrt{\Delta t}/\varepsilon}^{\sqrt{\Delta t}/\varepsilon}e^{-z^2/2}\,dz\to 0$$

as $\Delta t\to 0$. It follows that $\Delta t/\Delta w\to 0$ in probability as $\Delta t\to 0$. Consequently, for smooth $b(s)$

$$\frac{db(s)}{dw(s)}=\frac{db(s)}{ds}\,\frac{ds}{dw(s)}=0$$

and

$$\frac{df_x(x(s),s)}{\partial w(s)}=f_{xx}(x(s),s)\frac{dx(s)}{dw(s)}.$$

Now

$$\Delta x(t)=a(t)\,\Delta t+b(t)\,\Delta w(t)+o(\Delta t),$$

so that

$$\frac{\Delta x(t)}{\Delta w(t)}=a(t)\frac{\Delta t}{\Delta w(t)}+b(t)+\frac{o(\Delta t)}{\Delta w(t)}\to b(t)$$

as $\Delta t\to 0$. It follows that

$$\frac{dx(s)}{dw(s)}=b(s);$$

hence

$$\text{(3.4.13)}\qquad \frac{\partial b(s)f_x(x(s),s)}{\partial w(s)}=b^2(s)f_{xx}(x(s),s)+f_x(x(s),s)\frac{\partial b(s)}{\partial w(s)}.$$

Substituting (3.4.13) in (3.4.12) and then (3.4.12) in (3.4.11), we obtain

$$f(x(t),t)=f(x(t_0),t_0)+\int_{t_0}^{t}\left[f_t(x(s),s)\right.$$

$$\left.+\left(a(s)-\frac{\partial b(s)}{\partial w(s)}\right)f_x(x(s),s)\right]ds+(S)\int_{t_0}^{t}b(s)f_x(x(s),s)\,dw(s);$$

hence (3.4.10) follows.

EXERCISE 3.4.5

Assume that

$$(I)\ dx(t)=a(x(t),t)\,dt+b(x(t),t)\,dw(t).$$

Is (3.4.10) true in this case?

EXERCISE 3.4.6

Set $x(t)=e^{w(t)}$ and use Exercises 3.2.6 and (3.4.4) to show that $dx(t)=x(t)dw(t)$ in the sense of Stratonovich and $dx(t)=x(t)dw(t)+\frac{1}{2}x(t)dt$ in the sense of Itô. Obtain the same result from Itô's formula.

EXERCISE 3.4.7

Itô's formula provides a simple way to calculate $Ee^{w(t)}$ for

$$e^{w(t)}=1+\frac{1}{2}\int_0^t e^{w(s)}\,ds+\int_0^t e^{w(s)}\,dw(s);$$

hence

$$Ee^{w(t)}=1+\tfrac{1}{2}\int_0^t Ee^{w(s)}\,ds$$

so that

$$Ee^{w(t)}=e^{t/2}.$$

Use the same method to compute $\operatorname{Var} e^{w(t)}$, $Ee^{iw(t)}$, $\operatorname{Var}\sin w(t)$, and $\operatorname{Var}\cos w(t)$.

EXERCISE 3.4.8

Let $dx(t)=a(t)\,dt+b(t)\,dw(t)$, where $a(t)$ and $b(t)$ are deterministic functions. Find the distribution function of $x(t)$ and use Itô's formula to compute

$$Ex(t),\quad \operatorname{Var} x(t),\quad E[x(t)-Ex(t)]^n \qquad (n=3,4,\ldots).$$

EXERCISE 3.4.9

Let $w_1(t)$ and $w_2(t)$ be independent Brownian motions and let $x(t)$ be a nonanticipating process. Let $u_1(t)$ and $u_2(t)$ be differentiable processes such that

$$du_1(t) = -\sin x(t)\, dw_1 + \cos x(t)\, dw_2$$

$$du_2(t) = \cos x(t)\, dw_1 + \sin x(t)\, dw_2.$$

Show that $u_1(t)$ and $u_2(t)$ are independent Brownian motions (Viterbi 1966).

CHAPTER 4

Stochastic Differential Equations

4.1 ELEMENTARY THEORY AND LINEAR EQUATIONS

The stochastic differential equation

$$(4.1.1)\qquad dx(t)=a(x(t),t)\,dt+b(x(t),t)\,dw(t)\qquad x(0)=x_0$$

is defined by the Itô integral equation

$$(4.1.2)\qquad x(t)=x_0+\int_0^t a(x(s),s)\,ds+\int_0^t b(x(s),s)\,dw(s).$$

Thus a solution of (4.1.1) is a nonanticipating function $x(t)$ such that $|a(x(t),t)|^{1/2}$ and $b(x(t),t)$ are in $H_2[0,T]$ and such that (4.1.2) is satisfied. The same definition applies to the system

$$(4.1.3)\qquad d\mathbf{x}(t)=\mathbf{a}(\mathbf{x}(t),t)\,dt+\mathbf{B}(\mathbf{x}(t),t)\,d\mathbf{w}(t),$$

where

$$\mathbf{x}=[x_1,\ldots,x_n]^T,\qquad \mathbf{a}=[a_1,\ldots,a_n]^T$$

and $\mathbf{w}(t)=[w_1(t),\ldots,w_n(t)]^T$ is a vector of independent Brownian motions. The simplest example of a stochastic differential equation is the equation

$$(4.1.4)\qquad dx(t)=a(t)\,dt+b(t)\,dw(t),\qquad x(0)=x_0.$$

The solution is given by

$$x(t)=x_0+\int_0^t a(s)\,ds+\int_0^t b(s)\,dw(s).$$

To clarify the nature of the solution, we compute the transition *probability* density of $x(t)$, that is, a function $p(x,s,y,t)$ such that

$$P(x(t)\in A|x(s)=x)=\int_A p(x,s,y,t)\,dy \qquad (t>s)$$

where A is any set in R. We assume that $a(t)$ and $b(t)$ are deterministic functions. The stochastic integral $\zeta(t)=\int_0^t b(s)\,dw(s)$ is a limit of linear combinations of independent normal variables $\Sigma b(t_i)[w(t_{i+1})-w(t_i)]$; hence the integral is also a normal variable. Thus

$$\zeta(t)=x(t)-x_0-\int_0^t a(s)\,ds$$

is a normal variable; hence

$$p(x,s,y,t)=\frac{1}{\sqrt{2\pi v}}e^{-(y-\mu)^2/2v},$$

where

$$\mu=E(x(t)|x(s)=x).$$

We have $E(x(t)|x(s)=x)=x+\int_s^t a(u)\,du$ as the expectation of the stochastic integral vanishes. We also have

$$v=\mathrm{Var}\ x(t)=E\left[\int_s^t b(u)\,dw(u)\right]^2=\int_s^t b^2(u)\,du.$$

Hence

$$p(x,s,y,t)=\left[2\pi\int_s^t b^2(u)\,du\right]^{-1/2}\exp\left[\frac{-\left(y-x-\int_s^t a(u)\,du\right)^2}{2\int_s^t b^2(u)\,du}\right].$$

EXERCISE 4.1.1

Show that the transition probability density $p(x,s,y,t)$ of the solution of (4.1.4) is the solution of the parabolic *partial* differential equation

$$\frac{\partial p}{\partial t}=\frac{1}{2}b^2(t)\frac{\partial^2 p}{\partial y^2}-a(t)\frac{\partial p}{\partial y},\qquad t>s$$

$$p(x,s,y,t)\to\delta(x-y)\qquad \text{as}\quad t\downarrow s.$$

EXERCISE 4.1.2

Find the partial differential equation that $p(x,s,y,t)$ satisfies with respect to the "backward variables" (x,s).

Next we consider stochastic differential equations that can be reduced to the form (4.1.4) after a change of variables. Consider the change of variables $\zeta(t)=f(x(t),t)$, where $x(t)$ is the solution of (4.1.1); then, by Itô's formula,

(4.1.5)

$$d\zeta(t)=\left[\frac{\partial f}{\partial t}(x(t),t)+a(x(t),t)\frac{\partial f}{\partial x}(x(t),t)\right.$$
$$\left.+\tfrac{1}{2}b^2(x(t),t)\frac{\partial^2 f}{\partial x^2}(x(t),t)\right]dt+b(x(t),t)\frac{\partial f}{\partial x}(x(t),t)\,dw(t).$$

Assume that $f(x,t)$ has an inverse function (with respect to x) $g(x,t)$ such that

$$f(g(x,t),t)=x,\qquad g(f(x,t),t)=x.$$

Then $x(t)=g(\zeta(t),t)$; hence (4.1.5) can be written as

$$d\zeta(t)=\bar{a}(\zeta(t),t)\,dt+\bar{b}(\zeta(t),t)\,dw(t) \tag{4.1.6}$$

where

$$\bar{a}(x,t)=\frac{\partial f}{\partial t}(g(x,t),t)+a(g(x,t),t)\frac{\partial f}{\partial x}(g(x,t),t)$$

$$+\tfrac{1}{2}b(g(x,t),t)\frac{\partial^2 f}{\partial x^2}(g(x,t),t),$$

$$\bar{b}(x,t)=b(g(x,t),t)\frac{\partial f}{\partial x}(g(x,t),t).$$

If a function $f(x,t)$ can be found so that

$$\frac{\partial f}{\partial t}(x,t)+a(x,t)\frac{\partial f}{\partial x}(x,t)+\tfrac{1}{2}b^2(x,t)\frac{\partial^2 f}{\partial x^2}=\bar{a}(t) \tag{4.1.7}$$

(independent of x) and

$$b(x,t)\frac{\partial f}{\partial x}(x,t)=\bar{b}(t), \tag{4.1.8}$$

then, using $g(x,t)=x$ [as both sides of (4.1.7) and (4.1.8) are independent of x], equation (4.1.1) is reduced to the form (4.1.4) by (4.1.6)–(4.1.8).

To obtain the reducibility condition, we use (4.1.8) to get

$$\frac{\partial f}{\partial x}(x,t)=\frac{\bar{b}(t)}{b(x,t)}.$$

Next we differentiate (4.1.7) with respect to x to obtain

$$\frac{\partial^2 f}{\partial t\partial x}+\frac{\partial}{\partial x}\left[a(x,t)\frac{\partial f}{\partial x}(x,t)+\tfrac{1}{2}b^2(x,t)\frac{\partial^2 f}{\partial x^2}\right]=0.$$

Since

$$\frac{\partial^2 f}{\partial t\partial x}=\frac{\partial}{\partial t}\left[\frac{\bar{b}(t)}{b(x,t)}\right]$$

$$=\frac{[\partial\bar{b}(t)/\partial t]b(x,t)-\bar{b}(t)[\partial b(x,t)/\partial t]}{b^2(x,t)}$$

and

$$\frac{\partial^2 f}{\partial x^2}=\frac{\partial}{\partial x}\left[\frac{\bar{b}(t)}{b(x,t)}\right]$$

$$=-\bar{b}(t)\frac{[\partial b(x,t)/\partial x]}{b^2(x,t)}$$

we have

$$\frac{\bar{b}'(t)}{b(x,t)}-\bar{b}(t)\left[\frac{\partial b(x,t)/\partial x}{b^2(x,t)}-\frac{\partial}{\partial x}\left[\frac{a(x,t)}{b(x,t)}\right]+\frac{1}{2}\frac{\partial^2 b(x,t)}{\partial x^2}\right]=0;$$

hence

$$\frac{\bar{b}'(t)}{\bar{b}(t)}=b(x,t)\left\{\frac{\partial b(x,t)/\partial t}{b^2(x,t)}-\frac{\partial}{\partial x}\left[\frac{a(x,t)}{b(x,t)}\right]+\frac{1}{2}\frac{\partial^2 b(x,t)}{\partial x^2}\right\}. \tag{4.1.9}$$

It follows that we must have

(4.1.10)

$$\frac{\partial}{\partial x}\left(b(x,t)\left\{\frac{\partial b(x,t)/\partial t}{b^2(x,t)}-\frac{\partial}{\partial x}\left[\frac{a(x,t)}{b(x,t)}\right]+\frac{1}{2}\frac{\partial^2 b(x,t)}{\partial x^2}\right\}\right)=0.$$

Condition (4.1.10) is also sufficient for reducibility. For if (4.1.10) is satisfied, the right-hand side of (4.1.9) is independent of t, so $\bar{b}(t)$ can be obtained by integration. Now $f(x,t)$ can be determined from the relation $\bar{b}(t)/b(x,t)=\partial f(x,t)/\partial x$. Equation (4.1.9) is equivalent to

$$\frac{\partial}{\partial x}\left[\frac{\partial f(x,t)}{\partial t}+a(x,t)\frac{\partial f}{\partial x}+\tfrac{1}{2}b^2(x,t)\frac{\partial^2 f}{\partial x^2}\right]=0;$$

hence the expression in brackets is independent of x, so that it can be taken as $\bar{a}(t)$. Consider, for example, the linear equation with constant coefficients

$$dx=ax\,dt+bx\,dw.$$

Setting $z=\ln x$ and using Itô's formula, we obtain

$$dz=\left(a-\tfrac{1}{2}b^2\right)dt+b\,dw;$$

hence

$$z=z_0+\left(a-\tfrac{1}{2}b^2\right)t+bw$$

or

(4.1.11) $$x(t)=x_0\exp\left[\left(a-\tfrac{1}{2}b^2\right)t+bw(t)\right].$$

EXERCISE 4.1.3

Find the reducibility condition (4.1.10) in case $a(x,t)=a(x)$ and $b(x,t)=b(x)$. Show that in this case

$$\bar{b}(t)=e^{Ct}f(t,x)=e^{Ct}\int_0^x \frac{dy}{b(y)}$$

and

$$\bar{a}(t)=e^{Ct}\left[c\int_0^x \frac{dy}{b(y)}+\frac{a(x)}{b(x)}-\tfrac{1}{2}b'(x)\right].$$

EXERCISE 4.1.4

Let $a(x,t)=a(t)x$ and $b(x,t)=b(t)x$. Reduce (4.1.1) to (4.1.4) and find the transition probability density of the solution.

EXERCISE 4.1.5

Consider the linear equation

$$dx=\left[\alpha(t)+\beta(t)x\right]dt+\left[\gamma(t)+\delta(t)x\right]dw.$$

Let $x_0(t)$ be the solution of the homogeneous equation (with $\alpha(t)=\gamma(t)\equiv 0$; see Exercise 4.1.4). Solve the inhomogeneous equation by setting $x(t)=x_0(t)\zeta(t)$.

EXERCISE 4.1.6

Find the class of equations that are reducible to linear. Assume that the coefficients a and b are independent of t. Show that the reducibility condition is

$$\frac{d}{dx}\left[\frac{1}{c'(x)}\frac{d}{dx}b(x)c'(x)\right]=0,$$

where $c(x)=a(x)/b(x)-\frac{1}{2}b'(x)$. Show how reduction is accomplished if the reducibility condition is satisfied (Gihman and Skorohod 1972).

Before we proceed with the theory, we should note that the coefficients $a(x,t)$ and $b(x,t)$ must be defined for all $-\infty<x<\infty$, for if $x(t)$ is the solution, then there is in general a positive probability that $|x(t)|>M$ for any positive M and any positive t. If the coefficients $a(x,t)$ and $b(x,t)$ are defined only on a finite interval, then we can extend them to the whole line to make the equation meaningful. It is shown in Exercise 4.2*.3 that if a and b are defined on a finite interval $[\alpha,\beta]$, then the solution is independent of the extension of the coefficients as long as it does not leave the interval $[\alpha,\beta]$. After $x(t)$ leaves $[\alpha,\beta]$ for the first time, it becomes dependent on the extension.

The foregoing elementary theory has some important applications. The Langevin equation, discussed in Chapter 2, is the simplest linear equation

$$dv(t)=-\beta v(t)\,dt+q\,dw(t), \tag{4.1.12}$$

where $q=\sqrt{2kT\beta/m}$. The solution $v(t)$ is called the Ornstein–Uhlenbeck (1930) process, and it is a model for the velocity of a Brownian particle. The process $v(t)=v(t,\beta)$ can be used as a more physically acceptable approximation of a white noise process. The white noise process is the "derivative" of $w(t)$, which does not exist as a function (only as a distribution). The correlation between the white noise at times t and s should be given by

$$E\dot{w}(t)\dot{w}(s)=\frac{\partial^2}{\partial t\,\partial s}Ew(t)w(s)=\frac{\partial^2}{\partial t\,\partial s}(t\wedge s)=\delta(t-s).$$

Thus a δ-correlated white noise is only a mathematical idealization, since physical phenomena are assumed to vary continuously. Since

$$x(t)=x_0+\int_0^t v(s)\,ds$$

is the displacement of the particle, $x(t)$ is the physical Brownian motion; therefore, $v(t)$ is the physical noise.

The constant β, whose dimension is $[T^{-1}]$ (inverse of time i.e., frequency), is a parameter that describes the frequency of collisions. Between collisions the motion of the particle is perfectly correlated and can be assumed rectilinear. Therefore, β is a measure of physical correlation between velocities of a Brownian particle at various times. To compute the correlations of the velocities, we change the time scale to make it much coarser by setting $t=\beta s$ (i.e., small changes in s produce large changes in t). Taking into consideration the fact that $w(\beta s)/\sqrt{\beta}$ is a Brownian motion, we get in (4.1.12)

$$dv^{\beta}(s)=-\beta^{2}v^{\beta}(s)\,ds+\beta p\,dw_{1}(s),$$

where $v^{\beta}(s)=v(\beta s)$, $p=\sqrt{2kT/m}$, and $w_{1}(s)=w(\beta s)/\sqrt{\beta}$.

The solution is given by

$$v^{\beta}(s)=v_{0}^{\beta}e^{-\beta^{2}s}+p\beta\int_{0}^{s}e^{-\beta^{2}(s-u)}\,dw_{1}(u).$$

The physical velocity $\hat{v}^{\beta}(s)$ of the process $x^{\beta}(s)=x(\beta s)$ is βv^{β}, since $dx^{\beta}(s)/ds=\beta v^{\beta}(s)$. Hence

$$\hat{v}^{\beta}(s)=\frac{1}{\beta}\hat{v}_{0}^{\beta}e^{-\beta^{2}s}+p\int_{0}^{s}\beta^{2}e^{-\beta^{2}(s-u)}\,dw_{1}(u), \tag{4.1.13}$$

$$E\hat{v}^{\beta}(s)=\frac{1}{\beta}\hat{v}_{0}e^{-\beta^{2}s}\to 0 \qquad \text{as} \quad \beta\to\infty.$$

The correlation is given by

$$\begin{aligned}
E\{[\hat{v}^{\beta}(s_{1})-E\hat{v}^{\beta}(s_{1})][\hat{v}^{\beta}(s_{2})-E\hat{v}^{\beta}(s_{2})]\}\\
&=p^{2}\beta^{4}E\int_{0}^{s_{1}}e^{-\beta^{2}(s_{1}-u)}\,dw_{1}(u)\int_{0}^{s_{2}}e^{-\beta^{2}(s_{2}-u)}\,dw_{1}(u)\\
&=p^{2}\beta^{4}\int_{0}^{s_{1}\wedge s_{2}}e^{-\beta^{2}(s_{1}-u)-\beta^{2}(s_{2}-u)}\,du,
\end{aligned}$$

by (3.2.2). Assuming that $s_{2}<s_{1}$, we obtain

$$\operatorname{Cov}(\hat{v}^{\beta}(s_{1}),\hat{v}^{\beta}(s_{2}))=\frac{kT}{m}\beta^{2}e^{-\beta^{2}(s_{1}-s_{2})}.$$

Thus

$$\operatorname{Cov}\left(\hat{v}^{\beta}(s_1), \hat{v}^{\beta}(s_2)\right) \to \frac{kT}{m}\delta(s_1 - s_2) \qquad \text{as} \quad \beta\to\infty.$$

Indeed, if $\varphi(s)$ is a test function, then

$$\varphi(t) = \varphi(0) + t\varphi_1(t),$$

where $\varphi_1(t)$ is bounded. Hence

$$\beta^2 \int_0^\infty \varphi(t) e^{-\beta^2 t}\, dt = \beta^2 \varphi(0) \int_0^\infty e^{-\beta^2 t}\, dt + \beta^2 \int_0^\infty t\varphi_1(t) e^{-\beta^2 t}\, dt.$$

Now

$$\left| \beta^2 \int_0^\infty t\varphi_1(t) e^{-\beta^2 t}\, dt \right| \leqslant \beta^2 C \int_0^\infty t e^{-\beta^2 t}\, dt = \frac{C}{\beta^2} \to 0$$

as $\beta\to\infty$. Here C is a constant such that $|\varphi_1(t)| \leqslant C$. It follows that

$$\int_0^\infty \beta^2 e^{-\beta^2 t} \varphi(t)\, dt \to \varphi(0) \qquad \text{as} \quad \beta\to\infty;$$

hence $\beta^2 e^{-\beta^2 t} \to \delta(t)$ as $\beta\to\infty$. It follows from (4.1.13) that

$$\int_0^t v(s,\beta)\, ds \to \sqrt{\frac{2kT}{m}}\; w_1(t) \qquad \text{as} \quad \beta\to\infty.$$

The Ornstein–Uhlenbeck process $v(t,\beta)$ is therefore exponentially correlated, and thus it is a more physically acceptable model of the velocity process than is the white noise. The Ornstein–Uhlenbeck process is sometimes called *colored noise*.

Let us consider a simple example where the colored noise v is used instead of white noise, that is,

(4.1.14) $$\dot{x} = ax + bxv,$$

where

$$dv = -kv\, dt + \sqrt{2k}\, dw.$$

Equation (4.1.14) is an ordinary linear differential equation since v is a continuous function. The solution is given by

$$x(t)=x_0\exp\left[at+b\int_0^t v(s)\,ds\right].$$

Letting $k\to\infty$, we obtain

$$x(t)\to x_0 e^{at+bw(t)} \quad \text{as} \quad k\to\infty. \tag{4.1.15}$$

On the other hand, the solution of the "limit" equation

$$x=ax\,dt+bx\,dw \tag{4.1.16}$$

is given by (4.1.11):

$$x(t)=x_0\exp\left[\left(a-\tfrac{1}{2}b^2\right)t+bw(t)\right],$$

unlike (4.1.15). We see therefore that Itô's equation (4.1.16) cannot be used as a model of a physical phenomenon if the physical noise is colored. We shall show, following (Wong and Zakai 1965) that (4.1.16) should be interpreted in the Stratonovich sense in order to obtain the "right" physical result (4.1.15). Since the Stratonovich integral obeys all the rules of the classical integral calculus, the linear equation (4.1.16) can be solved by the classical procedure. Writing (4.1.16) in the form

$$d_S\left[xe^{-at+bw(t)}\right]=0,$$

we obtain the "correct" answer,

$$x(t)=x_0e^{at+bw(t)}.$$

We conclude that the Stratonovich equation should be used as an idealized model of a physical phenomenon rather than the Itô equation. Let

$$d_S x(t)=a(x(t),t)\,dt+b(x(t),t)\,d_S w(t) \tag{4.1.17}$$

be a Stratonovich equation. We derive next an equivalent Itô equation (Wong and Zakai 1965). If $x(t)$ is the solution of (4.1.17), we have, by (3.3.10),

$$(S)\int_{t_0}^t b(x(s),s)\,dw(s)=\int_{t_0}^t b(x(s),s)\,dw(s) + \frac{1}{2}\int_{t_0}^t b(x(s),s)b_x(x(s),s)\,ds. \tag{4.1.18}$$

Hence, by (4.1.17),

$$x(t)=x_0+\int_{t_0}^{t}a(x(s),s)\,ds+(S)\int_{t_0}^{t}b(x(s),s)\,dw(s) \tag{4.1.19}$$

$$=x(t_0)+\int_{t_0}^{t}\left[a(x(s),s)+\tfrac{1}{2}b(x(s),s)b_x(x(s),s)\right]ds$$

$$+\int_{t_0}^{t}b(x(s),s)\,dw(s).$$

Equation (4.1.19) can be written in the differential form

$$dx(t)=\left[a(x(t),t)+\tfrac{1}{2}b(x(t),t)b_x(x(t),t)\right]dt \tag{4.1.20}$$

$$+b(x(t),t)\,dw(t).$$

Equation (4.1.20) is the Itô equivalent of (4.1.17). Note that if $b(x,t)$ is independent of x, then (4.1.17) and (4.1.20) are identical and no correction is necessary. The term $\frac{1}{2}b(x,t)b_x(x,t)$ is called the *Zakai–Wong correction*.

EXERCISE 4.1.7

Let $\mathbf{x}(t)$ be the solution of the Itô system

$$d\mathbf{x}(t)=\mathbf{a}(\mathbf{x}(t),t)\,dt+\mathbf{b}(\mathbf{x}(t),t)\,d\mathbf{w}(t).$$

Show that for any continuous function $f(t)$,

$$\sum_{n=0}^{m} f(t_n)\left[x_i(t_{n+1})-x_i(t_n)\right]\left[x_j(t_{n+1})-x_j(t_n)\right]$$

$$\to\int_{\alpha}^{\beta}f(t)\sum_{k}b_{ik}(x(t),t)b_{jk}(x(t),t)\,dt$$

as

$$\max(t_{n+1}-t_n)\to 0$$

where

$$\alpha=t_0<t_1<\cdots<t_{m+1}=\beta.$$

EXERCISE 4.1.8

Derive the multidimensional Zakai–Wong correction to the Stratonovich system

$$d_S\mathbf{x}(t)=\mathbf{a}(\mathbf{x}(t),t)\,dt+\mathbf{b}(\mathbf{x}(t),t)\,d_S\mathbf{w}(t).$$

(*Answer*

$$(4.1.21)\quad dx_i(t)=\left[a_i(\mathbf{x}(t),t)+\tfrac{1}{2}\sum_{k,j}b_{kj}(\mathbf{x}(t),t)\frac{\partial b_{ij}(\mathbf{x}(t),t)}{\partial x_k}\right]dt$$

$$+\sum_j b_{ij}(\mathbf{x}(t),t)\,dw_j(t)$$

($i=1,2,\ldots,n$) (Jazwinski 1970; Stratonovich 1965).)

EXERCISE 4.1.9

Write down the equivalent Itô form of the Stratonovich linear system $d_S\mathbf{x}=\mathbf{A}\mathbf{x}\,dt+\mathbf{x}^T\mathbf{B}\,d_S\mathbf{w}$, where $\mathbf{B}$ is a vector.

4.2* EXISTENCE AND UNIQUENESS OF SOLUTIONS

We shall prove existence and uniqueness of solutions under the following simplified conditions:

(i) a and b are functions of x and $a,b\in C^1(R)$.
(ii) $\max(|da/dx|+|db/dx|)=K<\infty$.

We proceed to construct a solution by the method of successive approximations.

Writing the Itô equation in the integral form

$$(4.2.1^*)\qquad x(t)=x_0+\int_0^t a(x(s))\,ds+\int_0^t b(x(s))\,dw(s)$$

we set

$$x_0(t)=x_0$$

and

$$x_n(t)=x_0+\int_0^t a(x_{n-1}(s))\,ds+\int_0^t b(x_{n-1}(s))\,dw(s).$$

Clearly, $x_n(t)$ is a nonanticipating continuous process. Using the inequality

$$(A+B)^2 \leqslant 2(A^2+B^2),$$

we obtain

$$E[x_{n+1}(t)-x_n(t)]^2 \leqslant 2E\left\{\left[\int_0^t (a(x_n(s))-a(x_{n-1}(s)))\,ds\right]^2\right.$$

$$\left.+\left[\int_0^t (b(x_n(s))-b(x_{n-1}(s)))^2\,dw(s)\right]^2\right\}$$

$$\leqslant 2tE\left\{\int_0^t [a(x_n(s))-a(x_{n-1}(s))]^2\,ds\right.$$

$$\left.+\int_0^t [b(x_n(s))-b(x_{n-1}(s))]^2\,ds\right\}.$$

We have used the Cauchy–Schwarz inequality and (3.2.2). Using the mean value theorem, we obtain

$$E[x_{n+1}(t)-x_n(t)]^2 \leqslant 4K^2t\int_0^t E[x_n(s)-x_{n-1}(s)]^2\,ds.$$

Hence, for $t \leqslant T$,

$$E[x_{n+1}(t)-x_n(t)]^2 \leqslant \frac{M(4K^2Tt)^{n+1}}{(n+1)!},$$

where $M=TE[a^2(x_0)+b^2(x_0)]$. Now, since

$$y_n(t)=\int_0^t [b(x_n(s))-b(x_{n-1}(s))]\,dw(s)$$

is a martingale, the function $y_n^2(t)$ is a submartingale, so that

Kolmogorov's inequality gives

$$P\Big[\max_{t\leqslant T}|y_n(t)|\geqslant v\Big]\leqslant E\left[\frac{y_n(T)}{v}\right]^2$$

$$=\frac{E\left[\int_0^T b(x_n(t))-b(x_{n-1}(t))\right]^2 dt}{v^2}$$

$$\leqslant\left(\frac{K}{v}\right)^2\int_0^T E[x_n(t)-x_{n-1}(t)]^2\,dt$$

$$\leqslant\frac{M(4K^2T^2)^nK^2}{(n)!v^2}\equiv\frac{C_1C_2^n}{v^2n!}.$$

Similarly,

$$P\left[\max_{t\leqslant T}\left|\int_0^t[a(x_n(s))-a(x_{n-1}(s))]\,ds\right|\geqslant v\right]\leqslant\frac{C_1C_2^n}{v^2n!}.$$

If T is sufficiently small, $C_2\leqslant 1$, so choosing $v=1/\sqrt{(n-2)!}$, we obtain

$$P\left[\max_{t\leqslant T}|x_n(t)-x_{n-1}(t)|\geqslant\frac{1}{\sqrt{(n-2)!}}\right]\leqslant\frac{C_1}{(n-1)n}.$$

Since $\Sigma_{n=2}^{\infty}1/[n(n-1)]<\infty$ we have, by the Borel–Cantelli lemma,

$$P\left[\max_{t\leqslant T}|x_n(\mathrm{t})-x_{n-1}(t)|\geqslant\frac{1}{\sqrt{(n-2)!}}\text{ i.o.}\right]=0.$$

It follows that $x_n(t)$ converges uniformly a.s. to a solution of (4.2.1*).

To prove uniqueness of solutions, let $x_1(t)$ and $x_2(t)$ be two solutions of (4.2.1*) and set

$$\tau=\min(t\,|\,|x_1(t)|\text{ or }|x_2(t)|\text{ equals }n).$$

Then

$$[x_2(t)-x_1(t)]\chi_{[0,\tau]}(t)$$

$$=\int_0^t [a(x_2(s)\chi_{[0,\tau]}(s))-a(x_1(s)\chi_{[0,\tau]}(s))]\,ds$$

$$+\int_0^t [b(x_2(s)\chi_{[0,\tau]}(s))-b(x_1(s)\chi_{[0,\tau]}(s))]\,dw(s)$$

for all $t\leqslant\tau$. Since

$$E[x_2(t)\chi_{[0,\tau]}(t)-x_1(t)\chi_{[0,\tau]}(t)]^2\leqslant 4n^2<\infty,$$

we have

$$E|[x_2(t)-x_1(t)]\chi_{[0,\tau]}(t)|$$

$$\leqslant K\int_0^t E|[x_2(s)-x_1(s)]\chi_{[0,\tau]}(s)|\,ds.$$

It follows by Gronwall's inequality that

$$\chi_{0,\tau}(t)[x_2(t)-x_1(t)]=0 \qquad \text{for all} \quad t\leqslant\tau \text{ a.s.}$$

Letting $n\to\infty$, we obtain $x_1(t)=x_2(t)$ a.s.

Further results concerning existence and uniqueness are given in McKean (1969) and Gihman and Skorohod (1972).

EXERCISE 4.2.1*

Use Itô's formula to show that if $Ex_0^{2m}<\infty$, then the solution of (4.2.1*) satisfies

$$E[x(t)]^{2m}\leqslant E(1+x_0^{2m})e^{Ct},$$

where C is a constant.

EXERCISE 4.2.2*

Show that

$$E[x(t)-x_0]^{2m} \leqslant DE(1+x_0^{2m})e^{Ct}t^m,$$

where D is a constant.

EXERCISE 4.2.3*

Prove the following *Localization principle*: let $a_1(x,t)=a_2(x,t)$ and $b_1(x,t)=b_2(x,t)$ for $c \leqslant x \leqslant d$ and $t \geqslant 0$. Assume that $a_i(x,t)$ and $b_i(x,t)$ $(i=1,2)$ satisfy the conditions of the existence and uniqueness theorem in R for the stochastic differential equations

$$dx_i(t)=a_i(x,t)\,dt+b_i(x,t)\,dw(t) \qquad (i=1,2)$$

$$x(0)=x_0 \in [c,d].$$

Let $\tau_i=\inf\{t \mid x_i(t) \notin (c,d)\}$ $(i=1,2)$. Then $\tau_1=\tau_2$ a.s. and $x_1(t)=x_2(t)$ for all $t \leqslant \tau_1$ a.s. (Gihman and Skorohod 1973).

4.3 STOCHASTIC DIFFERENTIAL EQUATIONS AND DIFFUSION PROCESSES

(a) Markov Processes

A stochastic process $\zeta(t)$ on $[0,T]$ is called a *Markov process* if for $n=1,2,3,\ldots$ and any sequences $0 \leqslant t_0 < \cdots < t_n \leqslant T$ and $x_0, x_1, \ldots, x_n$, the equality

$$(4.3.1) \quad P(\zeta(t_n)<x_n \mid \zeta(t_{n-1})=x_{n-1}, \zeta(t_{n-2})=x_{n-2},\ldots,\zeta(t_0)=x_0)$$
$$=P(\zeta(t_n)<x_n \mid \zeta(t_{n-1})=x_{n-1})$$

is satisfied. Equation (4.3.1) means that the process "forgets" the past, provided that t_{n-1} is regarded as the present.

EXERCISE 4.3.1*

Show that (4.3.1) is equivalent to the following statement. Let $\mathcal{F}_t$ be the minimal σ-algebra with respect to which $\zeta(s)$ are measurable for $s \leqslant t$. Let

$$F(t,x,s,A) = P(\zeta(s) \in A \mid \zeta(t) = x) \qquad \text{for} \quad 0 \leqslant t \leqslant s \leqslant T$$

(A = Borel set). The process $\zeta(t)$ is a Markov process if

$$P(\zeta(s) \in A \mid \mathcal{F}_t) = F(t, \zeta(t), s, A) = P(\zeta(s) \in A \mid \zeta(t))$$

for any $0 \leqslant t < s < T$ and A.

Assume that the transition probability function of a Markov process

$$\zeta(t) = F(t,x,s,y) = P(\zeta(s) < y \mid \zeta(t) = x) \qquad (s > t)$$

has a density $p(t,x,s,y)$, with respect to y; that is,

$$F(t,x,s,y) = \int_{-\infty}^{y} p(t,x,s,z)\,dz.$$

Then the Markov property (4.3.1) implies that

$$p(t,x,s,y) = \int_{-\infty}^{\infty} p(t,x,\tau,z)\,p(\tau,z,s,y)\,dz \qquad (t < \tau < s); \tag{4.3.2}$$

that is, the probability that $\zeta(\cdot)$ goes from x to y in the time interval $[t,s]$ is the probability that $\zeta(\cdot)$ goes to any point z at any time τ and then, *independently* of the way it reached z, it goes to y. Equation (4.3.2) is called the *Chapman–Kolmogorov equation* for Markov processes. It should be noted that there are non-Markov processes satisfying (4.3.2) (Feller 1971).

If changes in $\zeta(t)$ occur at times $t_1, t_2, \ldots$ only (e.g., at times $1, 2, \ldots$), then $\{\zeta_n\}$ is a *Markov chain*. The values ζ_n takes are called *states*. Let $E_1^{(n)}, E_2^{(n)}, \ldots$ be the possible states at time (or generation) n; then the matrix whose elements are given by

$$P_{ij}^{(n)} = P\left(\zeta^n = E_j^{(n)} \mid \zeta^{n-1} = E_i^{(n-1)}\right)$$

is called the *transition probability matrix*.

(b) Diffusion Processes

A Markov process $\zeta(t)$ is called a *diffusion process* if its transition probability satisfies the following two conditions:

(i) For every $\epsilon>0$, t, and x,

$$\lim_{h\to 0}\frac{1}{h}\int_{|x-y|>\delta} F(t,x,t+h,y)\,dy=0.$$

(ii) There exist functions $a(x,t)$ and $b(x,t)$ such that for all $\epsilon>0$, t, and x,

(a) $$\lim_{h\to 0}\frac{1}{h}\int_{|x-y|<\epsilon}(y-x)p(t,x,t+h,y)\,dy=a(x,t).$$

(b) $$\lim_{h\to 0}\frac{1}{h}\int_{|x-y|<\epsilon}(y-x)^2 p(t,x,t+h,y)\,dy=b(x,t).$$

The function $a(x,t)$ is called the (*infinitesimal*) *drift coefficient* of $\zeta(t)$ and $b(x,t)$ is called the (*infinitesimal*) *diffusion coefficient*. The intuitive meaning of conditions (i) and (ii) and of $a(x,t)$ and $b(x,t)$ is the following. In a short time interval, h the displacement of $\zeta(\cdot)$ from a point x at time t is given by $a(x,t)h+\delta x+0(h)$, where $a(x,t)$ is the velocity of the medium in which the particle [whose motion is described by $\zeta(\cdot)$] drifts, δx is the random fluctuation of the particle due to random collisions or thermal fluctuations, and so on. $E\delta x=0$, $\operatorname{Var}\delta x=b(x,t)h$; that is, $b(x,t)$ is proportional to the average energy of the fluid molecules in the neighborhood of the particle. The following conditions imply (i) and (ii):

(i*) $\frac{1}{h}E_{x,t}|\zeta(t+h)-\zeta(t)|^{2+\epsilon}\to 0$

(ii*) (a) $\frac{1}{h}E_{x,t}[\zeta(t+h)-\zeta(t)]\to a(x,t)$

(b) $\frac{1}{h}E_{x,t}[\zeta(t+h)-\zeta(t)]^2\to b(x,t)$

as $h\to 0$, where ϵ is some postive number.

EXERCISE 4.3.2*

Prove the last assertions.

EXERCISE 4.3.3

Show that if $g(x,t)$ is a smooth function, monotone in x, the process $\eta(t)=g(\zeta(t),t)$, where $\zeta(t)$ is a diffusion process, is also a diffusion process. Show that

$$\bar{a}(x,t)=\frac{\partial g}{\partial t}(g^{-1}(x,t),t)+a(g^{-1}(x,t),t)\frac{\partial g}{\partial x}(g^{-1}(x,t),t)$$
$$+\tfrac{1}{2}b(g^{-1}(x,t),t)\frac{\partial^2}{\partial x^2}g(g^{-1}(x,t),t)$$

(differentiate first!) and

$$\bar{b}(x,t)=g(g^{-1}(x,t),t)\left[\frac{\partial}{\partial x}g(g^{-1}(x,t),t)\right]$$

are the drift and diffusion coefficients of $\eta(t)$, respectively. Here a and b are the coefficients of $\zeta(t)$. Take

$$g(x,t)=\int_0^x \frac{du}{\sqrt{b(u,t)}}.$$

(c) Diffusion Processes and Stochastic Differential Equations

Let $\eta(t)$ be the solution of the stochastic differential equations $d\eta=a\,dt+\sigma\,dw$ and assume that a and σ satisfy the conditions of the existence theorem. Then $\eta(t)$ is a diffusion process, and

$$F(t,x,s,y)=P(\eta(s)\leqslant y\,|\,\eta(t)=x)$$

Indeed, conditions (i*) and (ii*) are consequences of Exercise 4.2.2*:

$$E_{x,t}[\eta(t+h)-\eta(t)]^4=E_{x,t}[\eta(t+h)-x]^4\leqslant K_1h^2(1+x^4)$$

Hence (i*) follows.

$$\frac{1}{h}E_{x,t}(\eta(t+h)-x)=\frac{1}{h}\int_t^{t+h}E_{x,t}a(\eta(u),u)\,du$$

$$=\int_0^1 E_{x,t}a(\eta(t+sh),t+sh)\,ds$$

$$\to\int_0^1 E_{x,t}a(\eta(t),t)\,ds=a(x,t)\qquad \text{as}\quad h\to 0.$$

Hence (ii*)(a) is satisfied.

Next, using Itô's formula with $f(y)=y^2$, we get

$$E_{x,t}(\eta(t+h)-x)^2=E_{x,t}[\eta(t+h)]^2-x^2-2xE_{x,t}[\eta(t+h)-x]$$

$$=E_{x,t}\left\{\int_t^{t+h}\left[2\eta(u)a(\eta(u),u)+\sigma^2(\eta(u),u)\right]du\right.$$

$$\left.+\int_t^{t+h}2\eta(u)\sigma(\eta(u),u)\,dw(u)\right\}$$

$$-2xa(x,t)h+o(h).$$

Hence

$$\lim_{h\to 0}\frac{1}{h}E_{x,t}[\eta(t+h)-x]^2=\lim_{h\to 0}\int_0^1 2E_{x,t}[\eta(t+sh)a(\eta(t+sh),t+sh)$$

$$+\sigma^2\eta(t+sh,t+sh)\big]\,ds-2xa(x,t)=\sigma^2(x,t),$$

so (ii*)(b) holds.

The converse is also true; namely, if $\zeta(t)$ is an a.s. continuous diffusion process with smooth drift and diffusion coefficients a and b, $b\geqslant\delta>0$, and whose transition probability function satisfies certain continuity conditions, then $\zeta(t)$ is a solution of the stochastic differential equation

$$d\zeta=a\,dt+\sqrt{b}\,dw$$

for some Brownian motion $w(t)$ (Gihman and Skorohod 1972).

CHAPTER 5

Stochastic Differential Equations and Partial Differential Equations

5.1 THE FORMULAS OF KOLMOGOROV, FEYNMAN, AND KAC

Let $\eta(t)$ be the solution of the stochastic differential equation $d\eta = a\,dt + \sigma\,dw$ and set

$$u(x,t) = E_{x,t} f(\eta(s)) \qquad (0 \leqslant t \leqslant s).$$

If a and σ are smooth and satisfy the conditions of the existence theorem, then

$$\frac{\partial u(x,t)}{\partial t} + a(x,t)\frac{\partial u(x,t)}{\partial x} + \tfrac{1}{2}\sigma^2(x,t)\frac{\partial^2 u(x,t)}{\partial x^2} = 0$$

and

$$\lim_{t \uparrow s} u(x,t) = f(x)$$

for any test function $f(x)$.

Indeed, using Itô's formula,

(5.1.1)

$$\lim_{h\to 0}\frac{1}{h}E_{x,t-h}[f(\eta(t))-f(x)]=\lim_{h\to 0}\frac{1}{h}E_{x,t-h}\left\{\int_{t-h}^{t}\left[a(\eta(s),s)f'(\eta(s))\right.\right.$$

$$\left.+\tfrac{1}{2}\sigma^2(\eta(s),s)f''(\eta(s))\right]ds$$

$$\left.+\int_{t-h}^{t}\sigma(\eta(s),s)f'(\eta(s))\,dw(s)\right\}$$

$$=a(x,t)f'(x)+\tfrac{1}{2}\sigma^2(x,t)f''(x)$$

as the expectation of the stochastic integral vanishes. Now

(5.1.2) $$u(x,t-h)=E_{x,t-h}f(\eta(s))$$

$$=E_{x,t-h}E_{\eta(t),t}f(\eta(s))=E_{x,t-h}u(\eta(t),t).$$

Using (5.1.1), we find that

$$\lim_{h\to 0}\frac{1}{h}E_{x,t-h}[u(\eta(t),t)-u(x,t)]=a(x,t)\frac{\partial u}{\partial x}+\tfrac{1}{2}\sigma^2(x,t)\frac{\partial^2 u}{\partial x^2}$$

and from (5.1.2),

$$0=\frac{u(x,t)-u(x,t-h)}{h}+E_{x,t-h}\frac{u(\eta(t),t)-u(x,t)}{h}.$$

Hence

(5.1.3) $$\frac{\partial u}{\partial t}+a\frac{\partial u}{\partial x}+\tfrac{1}{2}\sigma^2\frac{\partial^2 u}{\partial x^2}=0.$$

Equation (5.1.3) is called *Kolmogorov's equation* and the representation of the solution $u(x,t)$ of (5.1.3) such that

(5.1.4) $$u(x,t)\to f(x)\quad\text{as}\quad t\uparrow s,$$

given by

$$u(x,t)=E_{x,t}f(\eta(s)),$$

is called *Kolmogorov's formula*.

EXERCISE 5.1.1

Prove (5.1.4).

Denote by L the operator

$$a\frac{\partial}{\partial x}+\frac{1}{2}\sigma^2\frac{\partial^2}{\partial x^2}.$$

If a and σ are time independent, then Kolmogorov's formula shows that L generates a semigroup G on $C(R)$:

$$G(t)f(x)=E_x f(\eta(t)).$$

EXERCISE 5.1.2

Set

$$v(x,t)=E_{x,t}f(\eta(T))\exp\int_0^T g(\eta(s),s)\,ds \tag{5.1.5}$$

Using similar arguments, show that

$$\frac{\partial v}{\partial t}+Lv+gv=0,\qquad 0\leqslant t\leqslant T \tag{5.1.6}$$

and

(5.1.7) $v(x,t)\to f(x)$ as $t\uparrow T$ (Gihman and Skorohod 1972).

Formula (5.1.5) is the *Feynman–Kac formula* for the solution of (5.1.6), (5.1.7).

EXERCISE 5.1.3

Derive the representation

$$v(x,t)=E_{x,t}\int_t^T f(\eta(s),s)\,ds$$

for the solution of the equation

$$\frac{\partial v}{\partial t}+Lv+f=0 \qquad t<T, \quad -\infty<x<\infty$$

$$v(x,T)=0.$$

We can interpret the Feynman–Kac formula as describing a process with "killing" as follows. Let $x(t)$ be a Brownian motion with killing; that is, the Brownian particle may disappear at a random time T=killing time (e.g., a neutron may be absorbed by the medium in which it moves). The probability of getting killed in the time interval $(t,t+dt)$ is equal to $k(x(t))\,dt+o(dt)$. The probability of surviving until time T, given that the particle moves along a path $x(t)$, is equal to

$$(1-k(x(t_1))\,dt)(1-k(x(t_2))\,dt\cdots(1-kx(t_n)\,dt)+o(1)$$

where $0=t_0<t_2<\cdots<t_n=T$, $dt\approx t_{j+1}-t_j$. As $dt\to 0$, the probability of surviving becomes equal to $\exp[-\int_0^T k(x(t))\,dt]$. Hence

$$u(x,t)=E_x(f(x(t)),T>t)$$

$$\equiv E_x[f((t))P_x(T>t\,|\,x(\cdot))]$$

$$=E_x f((t))\exp\left[-\int_0^t k(x(s))\,ds\right].$$

EXERCISE 5.1.4

Derive the representation

$$\psi(x,t)=E_{x,t}\left\{\exp\left[\int_t^T g(\eta(s),s)\,ds+\int_t^T h(\eta(s),s)\,dw(s)\right]f(\eta(t))\right\}$$

for the solution of the equation

$$\frac{\partial\psi}{\partial t}+L\psi+h\frac{\partial\psi}{\partial x}+\left(g+\tfrac{1}{2}h^2\right)\psi=0 \qquad (t<T)$$

$$\psi(x,t)\to f(x) \qquad \text{as} \quad t\uparrow T$$

(Gihman and Skorohod 1972).

5.2 THE FOKKER–PLANCK AND KOLMOGOROV'S FORWARD AND BACKWARD EQUATIONS

In classical mechanics the evolution of the density D of a system of particles in phase space is governed by the Liouville equation (10.1.3). If random collisions are introduced to the system, the density, or the transition probability density p in phase space, is governed by a generalized Liouville's equation, that was derived first by Fokker and Planck.

Let $\zeta(t)$ be a diffusion process whose transition density is $p(t, x, s, y)$ and whose drift and diffusion coefficients are $a(x, t)$ and $b(x, t)$, respectively. We show that p satisfies the partial differential equation

$$\text{(5.2.1)}\quad \frac{\partial p}{\partial s}(t, x, s, y)+\frac{\partial}{\partial y}[a(y, s)p(t, x, s, y)] - \frac{1}{2}\frac{\partial^2}{\partial y^2}[b(y, s)p(t, x, s, y)]=0$$

$$p(t, x, s, y)\to\delta(x-y) \qquad \text{as} \quad s\downarrow t.$$

Equation (5.2.1) is called the *Fokker–Planck* or the *forward Kolmogorov equation* (in the "forward" variables s and y). For any bounded smooth function $g(y)$, we have, by properties (i*) and (ii*) of Section 4.3,

(5.2.2)

$$\frac{1}{h}\int [g(y)-g(x)]\,p(t, x, t+h, y)\,dy$$

$$=\frac{1}{h}\int \left[g'(x)(y-x)+\tfrac{1}{2}g''(x)(y-x)^2+(y-x)^3R\right]$$

$$p(t, x, t+h, y)\,dy\to a(x, t)g'(x)+\tfrac{1}{2}b(x, t)g''(x) \qquad \text{as} \quad h\downarrow 0,$$

where R is the remainder in Taylor's expansion of $g(y)$ about x. Hence, using the Chapman–Kolmogorov equation,

$$\text{(5.2.3)}\quad \int_{-\infty}^{\infty} p(t, x, s+h, z)g(z)\,dz$$

$$=\int_{-\infty}^{\infty} p(t, x, s, y)\left[\int_{-\infty}^{\infty} p(s, y, s+h, z)g(z)\,dz\right]dy.$$

It follows that

$$\int_{-\infty}^{\infty}[p(t,x,s+h,z)-p(t,x,s,z)]g(z)\,dz$$

$$=\int_{-\infty}^{\infty}p(t,x,s,y)\left[\int_{-\infty}^{\infty}p(x,y,s+h,z)(g(z)-g(y))\,dz\right]dy.$$

Dividing by h and taking the limit as $h\to 0$, we get from (5.2.2),

$$\lim_{h\to 0}\frac{1}{h}\int[p(t,x,s+h,z)-p(t,x,s,z)]g(z)\,dz$$

$$=\int g(z)\frac{\partial p}{\partial s}(t,x,s,z)\,dz$$

$$=\int p(t,x,s,y)\left[a(y,s)g'(y)+\tfrac{1}{2}b(y,s)g''(y)\right]dy.$$

Hence the function $p(\cdot,\cdot,s,y)$ is the solution (in the weak sense) of the parabolic equation (5.2.1).

EXERCISE 5.2.1

Derive a formula analogous to (5.2.2) for $g=g(y,t)$; that is, find the limit

$$\lim_{h\to 0}\frac{1}{h}E_{x,t}[g(\zeta(t+h),t+h)-g(\zeta(t),t)].$$

EXERCISE 5.2.2

Derive the *backward Kolmogorov equation*

(5.2.4)

$$\frac{\partial p(t,x,\cdot,\cdot)}{\partial t}+a(x,t)\frac{\partial p}{\partial x}(t,x,\cdot,\cdot)+\frac{1}{2}b(x,t)\frac{\partial^2 p(t,x,\cdot,\cdot)}{\partial x^2}=0.$$

If a and b are independent of t, we say that the equation

$$d\zeta(t)=a(\zeta(t))\,dt+\sigma(\zeta(t))\,dw(t)\qquad(\sigma^2=b)$$

is *time homogeneous*. The equation is invariant under translations in t; hence

(5.2.5) $$p(t,x,s,y)=p(0,x,s-t,y).$$

Since now $\frac{\partial p}{\partial t}=-\frac{\partial p}{\partial s}$, we can write the backward Kolmogorov equation (5.2.4) in the form

$$\frac{\partial p}{\partial t}=Lp$$

and the Fokker–Planck equation (5.2.1) in the form

$$\frac{\partial p}{\partial t}=L^*p,$$

where L^* is the adjoint operator to L.

We can use the backward Kolmogorov equation to derive partial differential equations for the moments of the solution of the stochastic differential equation

$$dx(t)=a(x(t))\,dt+b(x(t))\,dw(t)$$

$$x(0)=x.$$

Since

(5.2.6) $$u(x,t)=E_x\varphi(x(t))=\int_{-\infty}^{\infty}\varphi(y)p(x,t,y)\,dy$$

we have, by (5.2.4),

(5.2.7) $$\frac{\partial u}{\partial t}=a(x)\frac{\partial u}{\partial x}+\tfrac{1}{2}b^2(x)\frac{\partial^2 u}{\partial x^2}$$

$$u(x,0)=\varphi(x).$$

Setting $\varphi(x)=x^n$, we see that $u(x,t)$ is the conditional moment (5.2.6).

5.3 SYSTEMS OF STOCHASTIC DIFFERENTIAL EQUATIONS AND BOUNDARY CONDITIONS

(a) Itô's Formula for Systems

Let

$$\mathbf{w}(t)=\begin{bmatrix} w_1(t) \\ \vdots \\ w_n(t) \end{bmatrix}$$

be a vector of independent Brownian motions, let $\boldsymbol{\sigma}(\mathbf{x}, t)$ be an $n\times n$ matrix, and let $\mathbf{b}(\mathbf{x}, t)$ be a vector,

$$b(\mathbf{x}, t)=\begin{bmatrix} b_1(x, t) \\ \vdots \\ b_n(x, t) \end{bmatrix}, \qquad \mathbf{x}=\begin{bmatrix} x_1 \\ \vdots \\ x_n \end{bmatrix}.$$

The system of stochastic differential equations

(5.3.1) $$d\mathbf{x}=\mathbf{b}\,dt+\boldsymbol{\sigma}\,d\mathbf{w}$$

leads to the following Itô's formula:

(5.3.2) $$df(\mathbf{x}(t), t)=Lf\,dt+\nabla_x f^T\boldsymbol{\sigma}\,d\mathbf{w}$$

where

(5.3.3) $$Lf=\frac{\partial f}{\partial t}+\mathbf{b}\cdot\nabla_{\mathbf{x}} f+\tfrac{1}{2}\sum_{i,j=1} a_{ij}\frac{\partial^2 f}{\partial x_i\,\partial x_j}\equiv\frac{\partial f}{\partial t}+Mf$$

and

(5.3.4) $$a_{ij}=(\boldsymbol{\sigma}\boldsymbol{\sigma}^T)_{ij}.$$

The Kolmogorov and Feynman–Kac formulas of Section 5.1 remain unchanged. Kolmogorov's backward equation takes the form

$$\frac{\partial p}{\partial t}+Mp=0$$

and the Fokker–Planck (forward) equation is given by

$$\frac{\partial p}{\partial s}+\nabla_{\mathbf{y}}\cdot(\mathbf{b}p)-\tfrac{1}{2}\sum_{i,j=1}^{n}\frac{\partial^2(a_{ij}p)}{\partial y_i\,\partial y_j}=0.$$

(b) Absorbing Boundary

Let $\mathbf{x}(t)$ be the solution of the system (5.3.1) in a certain domain Ω in R^n whose boundary $\partial\Omega$ is smooth. Assume that a perfectly absorbing barrier is placed on the boundary $\partial\Omega$ of Ω so that $P[\mathbf{x}(t)=\mathbf{x}(\tau)|\mathbf{x}(0)=\mathbf{x}_0\in\Omega]=1$ for all $t\geqslant\tau$, where $\tau=\inf\{s\geqslant 0|\mathbf{x}(s)\in\partial\Omega\}$ [i.e., τ is the first exit time of $\mathbf{x}(t)$ from Ω].

Then $p(t,\mathbf{x},s,\mathbf{y})=0$ for all $\mathbf{x}\in\partial\Omega$, $\mathbf{y}\in\Omega$, as the transition probability from a point $\mathbf{x}$ on $\partial\Omega$ to any point $\mathbf{y}$ in Ω is zero [$\mathbf{x}(t)$ is absorbed!]. Thus $p(t,\mathbf{x},s,\mathbf{y})$ is Green's function for the backward Kolmogorov equation and satisfies the boundary conditions

$$p(t,\mathbf{x},s,\mathbf{y})=\delta(\mathbf{x}-\mathbf{y})\qquad t\leqslant s,\ \mathbf{x}\in\partial\Omega$$

$$p(t,\mathbf{x},s,\mathbf{y})\to\delta(\mathbf{x}-\mathbf{y})\ \text{as}\ t\uparrow s,\ \mathbf{x},\mathbf{y}\in\Omega$$

(see Section 5.2).

We shall consider as an example the Brownian motion with an absorbing barrier. The forward Kolmogorov equation or the Fokker–Planck equation for a Brownian motion on $x>0$ with an absorbing boundary at $x=0$ is given by

$$\frac{\partial p}{\partial t}=\frac{1}{2}\frac{\partial^2 p}{\partial y^2}\qquad\text{in}\ \ y>0$$

$$p(0,t,y)=0,\qquad t>0,\ \ y>0$$

$$p(x,t,y)\to\delta(x-y)\qquad\text{as}\ \ t\downarrow 0,\ \ x>0,\ \ y>0.$$

The solution of this initial boundary value problem is given by

$$p(x,t,y)=\frac{1}{\sqrt{2\pi t}}\left[e^{-(x-y)^2/2t}-e^{-(x+y)^2/2t}\right].$$

We see that by symmetry $p(x,t,0)=0$. To see that, $p(x,t,y)\to\delta(x-y)$ as

$t\downarrow 0$, let $\varphi(x)$ be any test function such that $\varphi(x)=0$ for all $x<0$. Then

$$\int_{-\infty}^{\infty} p(x,t,y)\varphi(x)\,dx=\frac{1}{\sqrt{2\pi t}}\int_{-\infty}^{\infty} e^{-(x-y)^2/2t}\varphi(x)\,dx$$
$$-\frac{1}{\sqrt{2\pi t}}\int_{-\infty}^{\infty} e^{-(x+y)^2/2t}\varphi(x)\,dx.$$

We have shown (see Section 1.5) that

$$\int_{-\infty}^{\infty} e^{-(x-y)^2/2t}\varphi(x)\,dx\to\varphi(y) \qquad \text{as} \quad t\downarrow 0.$$

It follows that

$$\frac{1}{\sqrt{2\pi t}}\int_{-\infty}^{\infty} e^{-(x+y)^2/2t}\varphi(x)\,dx\to\varphi(-y)=0 \qquad \text{as} \quad t\downarrow 0 \quad \text{if} \quad y>0.$$

Thus $p(x,t,y)\to\delta(x-y)$ as $t\downarrow 0$ for all $x>0, y>0$.

EXERCISE 5.3.1

Consider a random walk (Chapter 1) with a perfectly absorbing barrier at $m=m_1$; that is,

$$P(X_{n+1}=m\,|\,X_n=m_1)=\delta_{m,m_1}.$$

Show that the probability $W(m,N,m_1)$ of arriving at $m\leqslant m_1$ after taking N steps in given by

$$W(m,N,m_1)=W(m,N)-W(2m_1-m,N),$$

where $W(m,N)=P(X_N=m\,|\,X_0=0)$ (use the symmetry of the random walk with respect to m_1). Rescale m, m_1, and n by letting

$$\frac{m}{\sqrt{n}}\to x, \qquad \frac{m_1}{\sqrt{n}}\to x_1, \qquad \frac{2N}{n}\to t$$

as $n\to\infty$ and show that

$$W(m,N,m_1)\to\frac{1}{\sqrt{\pi t}}\left(e^{-x^2/4t}-e^{-(2x_1-x)^2/2t}\right)\equiv w(x,t,x_1).$$

Show that w satisfies the Fokker–Planck equation

$$\frac{\partial w}{\partial t}=\frac{1}{2}\frac{\partial^2 w}{\partial x^2} \qquad \text{for} \quad x<x_1 \quad \text{or} \quad x>x_1$$

with the boundary condition $w(x_1,t,x_1)=0$ and the initial condition

$$w(x,0,x_1)=\delta(x).$$

EXERCISE 5.3.2

Use the backward Kolmogorov equation with appropriate boundary conditions to find the expected value of a Brownian motion with absorption at the boundaries of the interval $[0,\pi]$. Use (5.2.7) and separation of variables to obtain

$$E(w(t)|w(s)=x)=x,\ t>s,\ x\in(0,\pi)$$

$$E\big(w(t)^2|w(s)=x\big)=\sum_{n=0}^{\infty} a_n T_n(t-s)X_n(x)+x^2.$$

Obtain $T_n(t)=e^{-n^2t/2}$ and $X_n(x)=\sin nx$. Compute a_n.

EXERCISE 5.3.3

Obtain a similar result for

$$E(\varphi(w(t))|w(s)=x),$$

where $\varphi(x)$ is a smooth function in $[0,\pi]$.

(c) Reflecting Boundary

Let $\mathbf{x}(t)$ be the solution of the system (5.3.1) in Ω. After $\mathbf{x}(t)$ reaches $\partial\Omega$, we modify $\mathbf{x}(t)$ as follows. We extend $\mathbf{b}(\mathbf{x},t)$ and $\boldsymbol{\sigma}(\mathbf{x},t)$ into $R^n-\Omega$ (in a neighborhood of $\partial\Omega$) by reflection in $\partial\Omega$. Then we let $\tilde{\mathbf{x}}(t)$, the reflected process, be the reflection *into* Ω of the solution $\mathbf{x}(t)$ of the extended system (5.3.1). It should be noted that reflection is instantaneous at $\partial\Omega$, and after $\tilde{\mathbf{x}}(t)$ returns into Ω, the original equation (5.3.1) determines the motion. Thus the process of reflection is limited to the boundary, so that we have to extend $\mathbf{a}$ and $\boldsymbol{\sigma}$ near $\partial\Omega$ only (this extension determines the speed and direction of the reflection). The boundary conditions for the backward Kolmogorov equation are determined as follows. Let $\mathbf{x}_0 \in \partial\Omega$ and let N be a neighborhood of $\mathbf{x}_0$. We can change variables $\boldsymbol{\zeta} = \mathbf{f}(\mathbf{x})$ in the neighborhood of $\mathbf{x}_0$ so that $N \cap \partial\Omega$ is mapped into the plane $\zeta_n = 0$, $\mathbf{x}_0$ is mapped into the origin, and $N \cap \Omega$ is mapped into $\zeta_n > 0$. Thus the extension of $\mathbf{b}$ and $\boldsymbol{\sigma}$ by reflection leads to

$$\mathbf{b}(\zeta_n, \ldots) = -\mathbf{b}(-\zeta_n, \ldots) \qquad \text{and} \qquad \boldsymbol{\sigma}(\zeta_n, \ldots) = \boldsymbol{\sigma}(-\zeta_n, \ldots)$$

if $\zeta_n < 0$. It follows that the solution of the backward Kolmogorov equation (in the variables $\boldsymbol{\zeta}$) near $\zeta_n = 0$ must satisfy $p(-\zeta_n, \ldots) = p(\zeta_n, \ldots)$. Hence

$$\frac{\partial p}{\partial \zeta_n} = 0 \quad \text{at} \quad \zeta_n = 0. \tag{5.3.5}$$

Mapping back into the variables $\mathbf{x}$, we see that (5.3.5) takes the form

$$\frac{\partial p}{\partial \nu_{\mathbf{x}}}(t, x, s, y) = 0 \quad \text{for} \quad \mathbf{x} \in \partial\Omega, \tag{5.3.6}$$

where $\nu_{\mathbf{x}}$ is the outer normal to $\partial\Omega$ at the point $\mathbf{x}$.

Note. The argument presented in (c) precludes neither the possibility of the particle staying at the boundary for a while and then being reflected, nor the possibility of oblique reflection. For a full discussion, see Mandl (1968) and Anderson and Orey (1976).

Once again, let us consider the Brownian motion on $x > 0$ with reflection at the origin. The Fokker–Planck equation is given by

$$\frac{\partial p}{\partial t} = \frac{1}{2}\frac{\partial^2 p}{\partial y^2}, \qquad y > 0$$

$$\frac{\partial p}{\partial y}(x, t, y)\Big|_{y=0} = 0$$

$$p(x, t, y) \to \delta(x - y) \quad \text{as} \quad t \downarrow 0, \quad x > 0, \quad y > 0.$$

The solution is given by

$$p(x, t, y) = \frac{1}{\sqrt{2\pi t}}\left[e^{-(x-y)^2/2t} + e^{-(x+y)^2/2t}\right].$$

We also have

$$\frac{\partial p}{\partial x}(x,t,y)|_{x=0}=0.$$

EXERCISE 5.3.4

Consider again a random walk with a reflecting barrier at $m_1>0$; that is, $P(X_{n+1}=m_1-1|X_n=m_1)=1$. Using symmetry, show that

$$W(m,N,m_1)=W(m,N)+W(2m_1-m,N)$$

(see Exercise 5.3.1 for notation). In the limit, obtain

$$w(x,t,x_1)=\frac{1}{\sqrt{\pi t}}\left(e^{-x^2/2t}+e^{-(2x_1-x)^2/2t}\right);$$

hence

$$\frac{\partial w}{\partial x}(x,t,x_1)|_{x=x_1}=0.$$

In summary, the transition probability density $p(t,\mathbf{x},s,\mathbf{y})$ is Green's function for the backward (or forward) Kolmogorov equation with the boundary condition

$$\frac{\partial p}{\partial \nu_{\mathbf{x}}}=0,\qquad \mathbf{x}\in\partial\Omega,\mathbf{y}\in\Omega\quad\text{(backward)}$$

or

$$\frac{\partial p}{\partial \nu_{\mathbf{y}}}=0,\qquad x\in\Omega,\ y\in\partial\Omega\quad\text{(forward)}$$

and the initial condition $p(t,\mathbf{x},t,\mathbf{y})=\delta(\mathbf{x}-\mathbf{y})$.

EXERCISE 5.3.5

Obtain a representation for $E(w(t)|w(s)=x)$ with reflection at the boundary of the interval $[0,\pi]$, $x\in(0,\pi)$.

5.4 APPLICATIONS OF ITÔ'S FORMULA

(a) Exit (First Passage) Times and Dynkin's Equation

Let $\mathbf{x}(t)$ be the solution of (5.3.1) in a domain $\Omega \subset R^n$ and let $\tau_{\mathbf{x},s} = \inf\{t \geqslant s \mid \mathbf{x}(t) \in \partial\Omega, \mathbf{x}(s) = \mathbf{x}\}$ be the first exit time of $\mathbf{x}(t)$ from Ω. We can find the *expected exit time* $E\tau_{\mathbf{x},s}$ as follows. Let $u(\mathbf{x}, t)$ be the solution of the problem

$$\frac{\partial u}{\partial t} + Mu = -1, \qquad t \geqslant s, \mathbf{x} \in \Omega \tag{5.4.1}$$

$$u(\mathbf{x}, t) = 0, \qquad \mathbf{x} \in \partial\Omega,$$

where

$$M = \mathbf{b}(\mathbf{x}, t) \cdot \nabla_{\mathbf{x}} + \frac{1}{2} \sum_{i,j=1}^{n} (\boldsymbol{\sigma}\boldsymbol{\sigma}^*)_{ij} \frac{\partial^2}{\partial x_i \partial x_j}$$

is the operator defined in Itô's formula [see (5.3.2) and (5.3.3)].

If $u(\mathbf{x}, t)$ is a bounded smooth function, then by Itô's formula

$$u(\mathbf{x}(t), t) = u(\mathbf{x}, s) + \int_s^t \left(\frac{\partial u}{\partial t} + Mu \right) dt' + \int_s^t (\nabla u)^T \cdot \boldsymbol{\sigma}\, d\mathbf{w}(t') \tag{5.4.2}$$

for all $s \leqslant t \leqslant \tau_{\mathbf{x},s}$. Setting $t = \tau_{\mathbf{x},s}$ and taking expectation in (5.4.2), we obtain

$$Eu(\mathbf{x}(\tau_{\mathbf{x},s}), \tau_{\mathbf{x},s}) = u(\mathbf{x}, s) + E\int_s^{\tau_{\mathbf{x},s}} (-1)\, dt'.$$

Since $\mathbf{x}(\tau_{\mathbf{x},s}) \in \partial\Omega$ and $u = 0$ on $\partial\Omega$, we have

$$0 = u(\mathbf{x}, s) - E\tau_{\mathbf{x},s} + s.$$

Thus

$$E\tau_{\mathbf{x},s} = s + u(\mathbf{x}, s). \tag{5.4.3}$$

If $\mathbf{b} = \mathbf{b}(\mathbf{x})$ and $\boldsymbol{\sigma} = \sigma(\mathbf{x})$, then (5.3.1) is time homogeneous and $u = u(\mathbf{x})$. Thus, in this case

$$E\tau_{\mathbf{x}} = u(\mathbf{x}), \tag{5.4.4}$$

where $u(\mathbf{x})$ is the solution of the *elliptic* boundary value problem

$$Mu = -1 \quad \text{in} \quad \Omega \tag{5.4.5}$$
$$u = 0 \quad \text{on} \quad \partial\Omega.$$

We see that if (5.4.1) [or (5.4.5)] has a smooth bounded solution, then

$$\tau_{\mathbf{x},s} < \infty \text{ a.s.}$$

[See Friedman and Schuss (1971) and Schuss (1972, 1977) for the discussion of backward parabolic equations of type (5.4.1).] (5.4.5) is called *Dynkin's equation*. (Dynkin 1965).

Another derivation of (5.4.5) is based on the backward Kolmogorov equation. Let $p(\mathbf{x},\mathbf{y},t)$ be the transition probability density of the solution $\mathbf{x}(t)$ of (5.3.1), with absorption at $\partial\Omega$. Then $p(\mathbf{x},\mathbf{y},t)=0$ if $\mathbf{x}\in\partial\Omega$, and $p(\mathbf{x},\mathbf{y},t)\to\delta(\mathbf{x}-\mathbf{y})$ as $t\to 0$. Integrating the backward Kolmogorov equation with respect to t from 0 to ∞ and then with respect to $\mathbf{y}$ over Ω, we obtain for the function

$$v(\mathbf{x}) = \int_\Omega \int_0^\infty p(\mathbf{x},\mathbf{y},t)\,dt\,d\mathbf{y}$$

the equation (5.4.5). On the other hand, we have

$$E_{\mathbf{x}}\tau = \int_0^\infty t\,d_t P_{\mathbf{x}}(\tau \leqslant t).$$

Integrating by parts and noting that $P_{\mathbf{x}}(\tau > t)$ is the probability that by time t the trajectory $\mathbf{x}(t)$ is still in Ω, that is,

$$P_{\mathbf{x}}(\tau > t) = \int_\Omega p(\mathbf{x},\mathbf{y},t)\,d\mathbf{y},$$

we have

$$E_{\mathbf{x}}\tau = \int_0^\infty \int_\Omega p(\mathbf{x},\mathbf{y},t)\,d\mathbf{y}\,dt.$$

Changing the order of integration, we see that $E_{\mathbf{x}}\tau = v(\mathbf{x})$. Let $q(\mathbf{x},t)$ be the solution of the problem

$$\begin{aligned} &\frac{\partial q}{\partial t}+Mq=0, && T>t\geqslant s,\ \mathbf{x}\in\Omega \\ (5.4.6)\quad &q=1, && \mathbf{x}\in\partial\Omega\times[s,T] \\ &q(\mathbf{x},T)=0, && \mathbf{x}\in\Omega \end{aligned}$$

Using Itô's formula, we find that

$$Eq(\mathbf{x}(\tau_{\mathbf{x},s}\wedge T),\tau_{\mathbf{x},s}\wedge T)=q(\mathbf{x},s)\equiv q(\mathbf{x},s,T).$$

But

$$Eq(\mathbf{x}(\tau_{\mathbf{x},s}\wedge T),\tau_{\mathbf{x},s}\wedge T)=q(\mathbf{x},T)P(\tau_{\mathbf{x},s}\wedge T=T)$$

$$+\int_{\partial\Omega}\left[\int_s^T q(\mathbf{y},z)P(\mathbf{x}(\tau_{\mathbf{x},s})=\mathbf{y},\tau_{\mathbf{x},s}=z)\,dz\right]dS_{\mathbf{y}}=P(\tau_{\mathbf{x},s}<T).$$

Thus

$$(5.4.7)\qquad P(\tau_{\mathbf{x},s}<T)=q(\mathbf{x},s,T).$$

If (5.3.1) is time homogeneous, then

$$P(\tau_{\mathbf{x}}<T)=q(\mathbf{x},T).$$

(Without loss of generality, we may assume that $s=0$ in this case.) Thus $q(\mathbf{x},T)$ can be found by separation of variables.

(b) The Distribution of Exit Points

Let $f(\mathbf{x})$ be any smooth function on $\partial\Omega$ and let $u(\mathbf{x},t)$ be the solution of the problem.

$$\begin{aligned} (5.4.8)\quad &\frac{\partial u}{\partial t}+Mu=0, && t\geqslant s,\quad \mathbf{x}\in\Omega \\ &u(\mathbf{x},t)=f(\mathbf{x}), && \mathbf{x}\in\partial\Omega. \end{aligned}$$

Then, using Itô's formula as above, we obtain *Kolmogorov's formula*

$$(5.4.9)\qquad Eu(\mathbf{x}(\tau_{\mathbf{x},s}),\tau_{\mathbf{x},s})=u(\mathbf{x},s)$$

or

$$(5.4.9')\qquad \int_{\partial\Omega}f(\mathbf{y})P(\mathbf{x}(\tau_{\mathbf{x},s})=\mathbf{y})\,dS_{\mathbf{y}}=u(\mathbf{x},s).$$

Thus the probability *density* $p=P(\mathbf{x}(\tau_{\mathbf{x},s})=\mathbf{y})$ is Green's function for the problem (5.4.8). If (5.3.1) is time homogeneous, then p is Green's function for the Dirichlet problem for $Mu=0$.

Remark. Sometimes the solution to the Dirichlet problem $Mu=0$ is written in the form

$$u(\mathbf{x})=\int_{\partial\Omega}\frac{\partial}{\partial\nu}G(\mathbf{x},\mathbf{y})f(\mathbf{y})\,dS_{\mathbf{y}}$$

and $G(\mathbf{x},\mathbf{y})$ is called Green's function. In this case $p=\frac{\partial}{\partial\nu}G$.

The probability of exit at a given part Γ of the boundary is therefore given by

$$u(\mathbf{x})=\int_{\Gamma}p(\mathbf{x},\mathbf{y})\,dS_{\mathbf{y}}.$$

Thus $u(\mathbf{x})$ is the solution of the problem

$$\begin{aligned} Mu&=0 && \text{in } \Omega\\ u&=1 && \text{on } \Gamma\\ u&=0 && \text{on } \partial\Omega-\Gamma \end{aligned}$$

[$\mathbf{b},\boldsymbol{\sigma}=\mathbf{b}(\mathbf{x}),\boldsymbol{\sigma}(\mathbf{x})$ is assumed].

We shall consider, as an example, the expected exit time of a Brownian motion $w(t)$ from an interval $[a,b]$. Let

$$v(x)=E_x\tau_{[a,b]},$$

where $\tau_{[a,b]}=\inf\{t\,|\,w(t)=a \text{ or } w(t)=b,\ w(0)=x\}$. Then, by (5.4.1),

$$\tfrac{1}{2}v''(x)=-1$$

$$v(a)=v(b)=0.$$

Hence

$$v(x)=(x-a)(b-x).$$

Thus $w(t)$ hits the boundary of the interval in a finite time.

If $a=-\infty$, then $v(x)=\infty$, then the expected time until the Brownian motion hits a point b is infinite even though the time is finite. For, let

$q(x,t)$ be the solution of

$$\frac{\partial q}{\partial t}=\frac{1}{2}\frac{\partial^2 q}{\partial x^2}, \qquad x<b$$

$$q(b,t)=1$$

$$q(x,0)=0.$$

Then $P(\tau_b<t\,|\,w(0)=x)=q(x,t)$, by (5.4.7). Setting $r(x,t)=1-q(x,t)$, we obtain

(5.4.10)
$$\frac{\partial r}{\partial t}=\frac{1}{2}\frac{\partial^2 r}{\partial x^2}, \qquad -\infty<x<b$$

$$r(b,t)=0$$

$$r(x,0)=1, \qquad -\infty<x<b.$$

To show that $q(x,t)\to 1$ as $t\to\infty$, we have to show that $r\to 0$. This will show that

$$P(\tau_b<\infty\,|\,w(0)=x)=1,$$

so that $w(t)$ hits b in finite time.

To this end we solve the initial value problem (5.4.10) with the initial condition

$$r(x,0)=1, \qquad -\infty<x<b$$

$$r(x,0)=-1, \qquad b<x<\infty.$$

Setting $b-x=\xi$, we have

$$\frac{\partial r}{\partial t}=\frac{1}{2}\frac{\partial^2 r}{\partial \xi^2}$$

$$r(0,t)=0$$

$$r(\xi,0)=\pm 1 \qquad \text{for} \quad \pm\xi>0.$$

The solution is given by (5.2.6),

$$r(\xi,t)=\frac{1}{\sqrt{2\pi t}}\int_{-\infty}^{\infty}\varphi(\eta)e^{-(\xi-\eta)^2/2t}\,d\eta$$

where $\varphi(\eta)=r(\eta,0)$. Hence

$$r(\xi,t)=\frac{1}{\sqrt{2\pi t}}\int_0^{\infty}\left[e^{-(\xi-\eta)^2/2t}-e^{-(\xi+\eta)^2/2t}\right]d\eta$$

$$=\frac{1}{\sqrt{2\pi}}\int_{-\xi/\sqrt{t}}^{\xi/\sqrt{t}}e^{-\zeta^2/2}d\zeta\to 0 \qquad \text{as} \quad t\to\infty.$$

Obviously $r(0,t)=0$ for all $t>0$. Let us consider now the problem of exit from a finite interval. In this case we have to consider the problem (5.4.10) in (a,b) with the boundary conditions $r(a,t)=r(b,t)=0$ and the initial condition $r(x,0)=1$ in (a,b). We shall show that $r(x,t)$ decays exponentially fast in time. We multiply (5.4.10) by $r(x,t)$ and integrate, to obtain

$$\frac{1}{2}\frac{\partial}{\partial t}\int_0^1 r^2(x,t)\,dx=\frac{1}{2}\int_0^1 r(x,t)\frac{\partial^2 r}{\partial x^2}(x,t)\,dx=-\frac{1}{2}\int_0^1\left(\frac{\partial r}{\partial x}\right)^2 dx.$$

For any differentiable function $r(x\,t)$ such that $r(a,t)=r(b,t)=0$, we have, by the Cauchy–Schwarz inequality,

$$|r(x,t)|^2=\left|\int_a^x \frac{\partial r}{\partial x}\,dx\right|^2\leqslant(b-a)\int_a^b\left(\frac{\partial r}{\partial x}\right)^2 dx.$$

Integrating between a and b, we obtain

$$\int_0^1 r^2(x,t)\,dx\leqslant(b-a)^2\int_a^b\left(\frac{\partial r}{\partial x}\right)^2 dx.$$

Setting $\int_0^1 r^2(x,t)\,dx=\|r(x,t)\|^2$, we have

$$\frac{d}{dt}\|r(x,t)\|^2\leqslant-\frac{1}{(b-a)^2}\|r(x,t)\|^2;$$

hence

$$\|r(x,t)\|^2\leqslant\|r(x,0)\|^2e^{-t/(b-a)^2}\to 0 \qquad \text{as} \quad t\to\infty.$$

It follows that the probability of a Brownian particle staying in a finite interval for a time interval of length t decays exponentially in t.

To find the probability that a Brownian motion will reach a point a before it reaches the point b given $w(0)=x$ and $a<x<b$, we have to solve

(5.4.8):

$$\tfrac{1}{2}u''(x)=0$$

$$u(a)=1$$

$$u(b)=0;$$

that is, $u(x)=(b-x)/(b-a)$. Then

$$P(w(\tau_{[a,b]})=a\,|\,w(0)=x)=\frac{b-x}{b-a}.$$

We shall see that the probabilities of exit and exit times have important physical interpretations and are not easily computed in higher dimensions.

EXERCISE 5.4.1

Let $\zeta(t)$ be the solution of the equation $d\zeta(t)=a(\zeta)\,dt+b(\zeta)\,dw(t)$ in $[\alpha,\beta]$ and let α and β be reflecting boundaries. Show that

$$E\tau_{x,y}=V_0(y)-V_0(x)\qquad(\alpha<x<y<\beta),$$

where

$$\frac{\sigma^2(x)}{2}V_0''(x)+a(x)V_0'(x)=1,\qquad V_0'(\alpha)=0.$$

If $a<y<x<b$, then

$$E\tau_{x,y}=V_1(y)-V_1(x),$$

where V_1 satisfies the same equation as $V_0(x)$ with $V_1'(\beta)=0$.

EXERCISE 5.4.2

Let $\zeta(t)$ be as in Exercise 5.4.1.

$$f(x,y,t)=p\{\zeta(t)<y\,|\,\zeta(0)=x\}.$$

Show that f satisfies the Fokker–Planck equation in (α,β) and the initial and boundary conditions $f(x,\alpha,t)=0$, $f(x,\beta,t)=1$, $f(x,y,0)=\theta_x(y)$, where $\theta_x(y)=0$ if $y<x$ and $\theta_x(y)=1$ if $y \geqslant x$. Show that $f(x, y, t) \to f(y)$ as $t \to \infty$. Find the differential equation and boundary conditions that $f(y)$ satisfies. Solve explicitly for $f(y)$.

EXERCISE 5.4.3

Find the expected time until a Brownian particle, starting at $x<0$, hits the line $x=at+b$ for the first time.

(c) Stability of Solutions of Stochastic Differential Equations: Lyapunov's Criterion

The concept of stability of solutions of stochastic differential equations differs in character from that of ordinary (deterministic) differential equations. For a deterministic system of differential equations, we say that the solution $\xi(t)$ of the system

(5.4.11) $$\dot{\mathbf{x}}=\mathbf{b}(\mathbf{x})$$

is stable if for any positive number ε there exists two numbers, $\delta>0$ and T, such that for any solution $x(t)$ of (5.4.11) we have

$$|\mathbf{x}(t)-\xi(t)|<\varepsilon \qquad t \geqslant T$$

whenever

(5.4.12) $$|\mathbf{x}(t_0)-\xi(t_0)|<\delta$$

for some $t_0 \leqslant T$. The solution $\xi(t)$ is said to be asymptotically stable if it is stable and, in addition,

(5.4.13) $$\lim_{t \to \infty} |\mathbf{x}(t)-\xi(t)|=0$$

for any solution $\mathbf{x}(t)$ satisfying (5.4.10). If (5.4.13) holds for all solutions of (5.4.11), then $\xi(t)$ is said to be *globally stable*.

By setting $\mathbf{x}(t)=\xi(t)+\mathbf{y}(t)$, and assuming that $\mathbf{b}(x)$ is smooth, we can reduce the problem of stability of $\xi(t)$ to that of the problem of stability of the solution $\mathbf{y}(t)\equiv\mathbf{0}$ for the system

(5.4.14) $$\dot{\mathbf{y}}=\mathbf{B}(t)\mathbf{y}+\mathbf{C}(t,\mathbf{y})$$

where

$$[\mathbf{B}(t)]_{ij} = \left.\frac{\partial b_i(\mathbf{x})}{\partial x_j}\right|_{\mathbf{x}=\boldsymbol{\xi}(t)}$$

and where $\mathbf{C}(t,\mathbf{y}) = o(|\mathbf{y}|)$ as $|\mathbf{y}| \to 0$. If the matrix $\mathbf{B}(t)$ is constant, independent of t, and the real parts of the eigenvalues of $\mathbf{B}$ are negative, then the solution $\mathbf{y}(t) = \mathbf{0}$ is asymptotically stable; hence $\boldsymbol{\xi}(t)$ is asymptotically stable for (5.4.11).

A point $\mathbf{x}_0$ is called a *critical point* of (5.4.11) if $\mathbf{b}(\mathbf{x}_0) = \mathbf{0}$. In this case $\boldsymbol{\xi}(t) = \mathbf{x}_0$ is a solution of (5.4.11). If $\mathbf{x}_0$ is a critical point of (5.4.11), then $\mathbf{B}(t)$ in (5.4.14) is constant. Another method for examining the stability of the critical point $\mathbf{x}_0 = \mathbf{0}$, say, is due to Lyapunov. A function $V(\mathbf{x})$ is called a Lyapunov function for (5.4.11) at $\mathbf{x}_0 = \mathbf{0}$ if (i) $V(\mathbf{x})$ is defined, continuous, and differentiable in a neighborhood of $\mathbf{0}$; (ii) $V(\mathbf{x}) > 0$ if $\mathbf{x} \neq \mathbf{0}$ and $V(\mathbf{0}) = 0$; and (iii) $b(\mathbf{x}) \cdot \nabla V(\mathbf{x}) \leqslant 0$ in a neighborhood of $\mathbf{0}$. If the system (5.4.11) possesses a Lyapunov function, then $\mathbf{x} \equiv \mathbf{0}$ is a stable solution, given $\mathbf{b}(\mathbf{0}) = \mathbf{0}$. If

$$b(\mathbf{x}) \cdot \nabla V(\mathbf{x}) < 0 \qquad \text{for} \quad \mathbf{x} \neq \mathbf{0},$$

then $\mathbf{x} \equiv \mathbf{0}$ is asymptotically stable.

Consider, for example, the linear system

$$\dot{\mathbf{x}} = \mathbf{A}\mathbf{x}. \tag{5.4.15}$$

If the equation $\mathbf{A}^T\mathbf{X} + \mathbf{X}\mathbf{A} = -\mathbf{Q}$ has a positive definite solution $\mathbf{X}$ for some positive definite matrix $\mathbf{Q}$, then the function

$$V(\mathbf{x}) = \mathbf{x}^T \mathbf{X} \mathbf{x}$$

is a Lyapunov function for (5.4.15). Indeed,

$$\nabla V(\mathbf{x}) \cdot \mathbf{A}\mathbf{x} = \mathbf{x}^T \mathbf{X}\mathbf{A}\mathbf{x} + \mathbf{x}^T \mathbf{A}^T \mathbf{X}\mathbf{x} = -\mathbf{x}^T \mathbf{Q}\mathbf{x} < 0$$

for $\mathbf{x} \neq \mathbf{0}$ (Coddington and Levinson 1955; Hale 1969).

EXERCISE 5.4.4

Let $\mathbf{A}$ be a 2×2 matrix with real entries. Draw the trajectories of the solutions of the system

$$\frac{d}{dt}\begin{pmatrix}x\\y\end{pmatrix}=\mathbf{A}\begin{pmatrix}x\\y\end{pmatrix}$$

near $x=y=0$ in the (x, y) plane. Consider the possibilities for the eigenvalues λ_1 and λ_2 of $\mathbf{A}$.

(i)	$\lambda_1<\lambda_2<0$	(node)	
(ii)	$0<\lambda_1<\lambda_2$	(node)	
(iii)	$\lambda_1<0<\lambda_2$	(a saddle point)	
(iv)	$0\neq\lambda_1=\lambda_2$	(regular node)	
(v)	$\operatorname{Im}\lambda\neq 0$,	$\operatorname{Re}\lambda\neq 0$	(spiral or focal point)
(iv)	$\operatorname{Im}\lambda\neq 0$,	$\operatorname{Re}\lambda=0$	(center).

Examine the stability of the origin in each case.

EXERCISE 5.4.5

Discuss the stability of the critical points of the system

$$\dot{\mathbf{x}}=-\nabla\phi(\mathbf{x})$$

according to the geometry of the surface $\mathbf{y}=\phi(\mathbf{x})$.

We shall assume for simplicity that (5.3.1) is time homogeneous. If $\boldsymbol{\sigma}\neq\mathbf{0}$ for all $\mathbf{x}\in\bar{\Omega}\equiv\Omega\cup\partial\Omega$, then the operator M is uniformly elliptic. It is well known that in this case equation (5.4.5) has a bounded solution. It follows, by (5.4.4), that $\tau_{\mathbf{x},s}<\infty$ a.s. Thus $\mathbf{x}(t)$ will leave any bounded domain in finite time a.s.; that is, the stochastic system (5.3.1) is unstable regardless of the size of $\boldsymbol{\sigma}$ even if the *deterministic* system $d\mathbf{x}/dt=\mathbf{b}(\mathbf{x})$ is globally stable.

We see that even the smallest white noise type of perturbation of a deterministic system will drive the particle out of every bounded set. We will therefore consider the case $\mathbf{b}(\mathbf{0})=\mathbf{0}$ and $\boldsymbol{\sigma}(\mathbf{0})=\mathbf{0}$, so that $\mathbf{x}=\mathbf{0}$ is an equilibrium point (a trap). Let $V(\mathbf{x})>0$ for $\mathbf{x}\neq 0$, $V(\mathbf{0})=0$, and let $MV\leqslant 0$. Then V is called a *Lyapunov function*. From Itô's formula, we obtain for

(5.3.1),

$$0 \leqslant V(\mathbf{x}(t)) = V(\mathbf{x}) + \int_0^t MV(\mathbf{x}(s))\,ds + \int_0^t \nabla V \cdot \boldsymbol{\sigma}\,d\mathbf{w}(s)$$

$$\leqslant V(\mathbf{x}) + \int_0^t \nabla V \cdot \boldsymbol{\sigma}\,d\mathbf{w}(s) \equiv V(\mathbf{x}) + f(t).$$

From Kolmogorov's inequality (2.3.8), we obtain

$$p\Big\{ \sup_{t>0} \left[V(\mathbf{x}) + f(t) \right] > V(\mathbf{x}) + \varepsilon \Big\} \leqslant \frac{V(\mathbf{x})}{\varepsilon + V(\mathbf{x})} \leqslant \frac{V(\mathbf{x})}{\varepsilon};$$

hence

$$P\Big\{ \sup_{t\geqslant 0} V(\mathbf{x}(t)) < V(\mathbf{x}) + \varepsilon \Big\} \geqslant 1 - \frac{V(\mathbf{x})}{\varepsilon}.$$

For any positive ε_1 and ε_2, a positive number $\delta > 0$ can be found so that

(5.4.16) $$P_{\mathbf{x}}\Big\{ \sup_{t>0} |\mathbf{x}(t)| < |\mathbf{x}| + \varepsilon_1 \Big\} \geqslant 1 - \varepsilon_2 \quad \text{if} \quad |\mathbf{x}| < \delta.$$

The meaning of (5.4.16) is that all trajectories that begin sufficiently close to the origin remain for all times in the neighborhood of the origin with the exception of a set of trajectories of arbitrarily small positive probability.

We say that the origin is *stochastically stable* if (5.4.16) holds. It follows in this case that

$$\lim_{|\mathbf{x}|\to 0} P_{\mathbf{x}}\Big\{ \sup_{t>0} |\mathbf{x}(t)| \geqslant \varepsilon \Big\} = 0.$$

If in addition to stability we also have

$$\lim_{|\mathbf{x}|\to 0} P_{\mathbf{x}}\Big\{ \lim_{t\to\infty} \mathbf{x}(t) = \mathbf{0} \Big\} = 1,$$

we say that the origin is stochastically asymptotically stable. If

(5.4.17) $$P_{\mathbf{x}}\Big\{ \lim_{t\to\infty} \mathbf{x}(t) = \mathbf{0} \Big\} = 1$$

for all $\mathbf{x} \in R^n$, we say that the origin is globally asymptotically stable. It can be easily shown that if the Lyapunov function satisfies the sharp inequality $MV < 0$ for $\mathbf{x} \neq 0$, then the origin is stochastically asymptotically stable. If, in addition, $V(\mathbf{x})\to\infty$ as $|\mathbf{x}|\to\infty$, then the origin is globally asymptotically stable in the sense of (5.4.17). If, in particular,

$$MV \leqslant -kV \qquad \text{for some} \quad k>0,$$

then we have global stability.

Consider, for example, the one-dimensional linear equation

$$dx = ax\,dt + bx\,dw.$$

The solution is given by

$$x(t) = x(0)e^{(a-b^2/2)t+bw(t)}.$$

Since $w(t)/t\to 0$ as $t\to\infty$ [$sw(1/s)$ being a Brownian motion $s=1/t$], we have

$$x(t)\to 0 \qquad \text{if } a-\frac{b^2}{2}<0$$

$$x(t)\to\infty \qquad \text{if } a-\frac{b^2}{2}>0.$$

If $a=b^2/2$, then $x(t)=x(0)e^{bw(t)}$; therefore, $P(\limsup_{t\to\infty} x(t)=+\infty)=1$.

The Lyapunov function approach yields the same result as follows. Let $V=|x|^\alpha$; then

$$MV=\left[a+\tfrac{1}{2}b^2(\alpha-1)\right]\alpha|x|^\alpha.$$

If $a-b^2/2<0$, we can choose $0<\alpha<1-2a/b^2$ and obtain a Lyapunov function V such that $MV\leqslant -kV, k>0$.

EXERCISE 5.4.6

Construct the Lyapunov function in the remaining two cases and show stochastic instability.

Note that the stability of the system is not determined by a but by $a-b^2/2$, that is, by the drift coefficient in the equivalent Stratonovich equation

$$(S)\,dx=\left(\frac{a-b^2}{2}\right)x\,dt+bx\,d_s w.$$

This again shows that the Stratonovich equation is often a more adequate model for a physical phenomenon than the Itô equation. Consider a system of homogeneous linear equations

$$dx = A(t)x\,dt + \sum_{i=1}^{n} B_i(t)x\,dw_i, \tag{5.4.18}$$

where $w_i(t)$ are independent Brownian motions. Taking expectations in (5.4.18), we obtain for $y(t) = E_x x(t)$ the linear equation

$$\dot{y} = A(t)y; \tag{5.4.19}$$

hence $y \equiv 0$ must be a stable solution if $x \equiv 0$ is to be stable for (5.4.18). Also the covariance matrix of $x(t)$ should not increase as $t \to \infty$ if the origin is to be stable. This happens if the origin is stable for the system

$$\dot{Z}(t) = A(t)Z(t) + Z(t)A^T(t) + \sum_{i=1}^{n} B_i(t)Z(t)B_i^T(t) \tag{5.4.20}$$

where

$$(Z(t))_{ij} = E_x x_i(t) x_j(t)$$

If A and B_i are independent of time, then there is stability in (5.4.20) if the algebraic system of equations

$$AZ + ZA^T + \sum B_i Z B_i^T = -C$$

has a symmetric positive definite solution Z for some symmetric positive definite matrix C.

EXERCISE 5.4.7

Consider the question of stability for the one-dimensional homogeneous linear equation with time-independent coefficients from the point of view of partial differential equations. More specifically, consider the existence of bounded solutions for the partial differential equations for the mean exit time and the exit time distribution from an interval $[\varepsilon, 1]$ and compute the limit as $\varepsilon \to 0$. Find in this case the probability of hitting 1 before ε.

EXERCISE 5.4.8

Consider the second-order linear equation with noisy coefficients

$$\ddot{x}+(a+b\dot{w}_1)\dot{x}+(c+d\dot{w}_2)x=0,$$

where w_1 and w_2 are independent Brownian motions. Transform the equation into Itô form and use the Routh–Hurwitz criterion to show that the stability condition is

$$b^2<2a, \qquad d^2<(2a-b^2)c$$

[see (Khasminski 1969) for a further discussion of stability].

EXERCISE 5.4.9

Let $a(0)=\sigma(0)=0$ and let $x(t)$ be the solution of the equation

$$dx(t)=a(x)\,dt+\sigma(x)\,dw$$
$$w(0)=x_0>0.$$

Show that for every $\varepsilon>0$, there exists $\delta>0$ such that

$$P_{x_0}\Big(\lim_{t\to\infty} x(t)=0\Big)\geqslant 1-\varepsilon$$

whenever $0<x_0<\delta$ if and only if

$$\int_0^\delta \exp\left\{\int_0^s \frac{2a(t)}{\sigma^2(t)}\,dt\right\} ds<\infty.$$

Conclude that if $\sigma(x)=\sigma_0 x+o(x)$ and $a(x)=a_0 x+o(x)$, then the stability condition is $a_0/\sigma_0^2<\frac{1}{2}$.

EXERCISE 5.4.10

Investigate the stability of the origin for the equation (5.4.21) $dx=ax\,dt+\sqrt{1+x^2}\;dw$ for various values of the constant a. In the stable case find the stationary distribution. Interpret (5.4.21) in the Itô and in the Stratonovich sense. (This example is due to Zakai.)

CHAPTER 6

Asymptotic Analysis of Stochastic Differential Equations

6.1 THE SMOLUCHOWSKI–KRAMERS APPROXIMATION

Consider the Langevin equation for a particle in a force field

$$\text{(6.1.1)} \qquad \frac{d\mathbf{x}}{dt} = \mathbf{y}$$

$$\frac{d\mathbf{y}}{dt} = -\beta\mathbf{y} + \mathbf{K}(\mathbf{x}) + \sqrt{\frac{2\beta kT}{m}}\,\frac{d\mathbf{w}}{dt}.$$

In a typical case $\beta \approx 10^9$ to 10^{10}, while $kT/m \approx 1$. We may therefore consider β to be a large parameter and try to expand the solution as $\beta \to \infty$.

Rescaling time by setting $t = \beta s$, noting that $(1/\sqrt{\beta}\,)\mathbf{w}(\beta s) = \tilde{\mathbf{w}}(s)$ is a Brownian motion, and setting $\mathbf{y}^{\beta}(s) = \mathbf{y}(\beta s)$ and $\mathbf{x}^{\beta}(s) = \mathbf{x}(\beta s)$ in (6.1.1), we get

$$\text{(6.1.2)} \qquad d\mathbf{x}^{\beta}(s) = \beta\mathbf{y}^{\beta}(s)\,ds$$

$$d\mathbf{y}^{\beta}(s) = \beta p\,d\tilde{\mathbf{w}}(s) - \beta^2\mathbf{y}^{\beta}(s)\,ds + \beta\mathbf{K}(\mathbf{x}^{\beta}(s))\,ds$$

where $p = \sqrt{2kT/m}$. Now let $\mathbf{z}^{\beta}(s)$ be the solution of

$$\text{(6.1.3)} \qquad d\mathbf{z}^{\beta}(s) = \beta p\,d\tilde{\mathbf{w}}(s) - \beta^2\mathbf{z}^{\beta}(s)\,ds$$

$$\mathbf{z}(0) = \mathbf{y}_0$$

and set $\mathbf{y}^\beta(s)=\mathbf{z}^\beta(s)+\mathbf{v}^\beta(s)$. We will determine $\mathbf{v}^\beta(s)$:

$$\text{(6.1.4)}\quad d\mathbf{y}^\beta(s)=d\mathbf{z}^\beta(s)+d\mathbf{v}^\beta(s)=\beta p\,d\tilde{\mathbf{w}}(s)-\beta^2\mathbf{z}^\beta(s)\,ds+d\mathbf{v}^\beta(s)$$
$$=\beta d\tilde{\mathbf{w}}(s)-\beta^2\mathbf{y}^\beta(s)\,ds+\beta^2\mathbf{y}^\beta(s)\,ds+d\mathbf{v}^\beta(s).$$

Substituting (6.1.3) in (6.1.4), we obtain an ordinary differential equation for $\mathbf{v}^\beta(s)$; that is,

$$\frac{d\mathbf{v}^\beta(s)}{ds}=-\beta^2\mathbf{v}^\beta(d)+\beta\mathbf{K}(\mathbf{x}^\beta(s)).$$

The solution is given by

$$\mathbf{v}^\beta(s)=\beta\int_0^s e^{-\beta^2(s-r)}\mathbf{K}(\mathbf{x}^\beta(r))\,dr.$$

The solution equation of (6.1.3) is given by

$$\mathbf{z}^\beta(s)=\mathbf{y}_0e^{-\beta^2 s}+p\int_0^s \beta e^{-\beta^2(s-r)}\,d\tilde{\mathbf{w}}(r).$$

Setting $\hat{\mathbf{y}}(s)=\beta\mathbf{y}^\beta(s)$, we obtain from (6.1.1),

$$\hat{\mathbf{y}}(s)=\hat{\mathbf{y}}_0e^{-\beta^2 s}+p\int_0^s \beta^2 e^{-\beta^2(s-r)}\,d\tilde{\mathbf{w}}(r)$$
$$+\int_0^s \beta^2 e^{-\beta^2(s-r)}\mathbf{K}(\mathbf{x}^\beta(r))\,dr.$$

Integrating $\hat{\mathbf{y}}(s)$, we obtain after some manipulations

$$\mathbf{x}^\beta(s)=\mathbf{x}_0+\int_0^s\hat{\mathbf{y}}(s)\,ds=\frac{1}{\beta^2}\hat{\mathbf{y}}_0(1-e^{-\beta^2 s})+p\beta^2\int_0^s\tilde{\mathbf{w}}(r)e^{-\beta^2(s-r)}\,dr$$
$$+\int_0^s\int_0^t\beta^2e^{-\beta^2(t-r)}\mathbf{K}(\mathbf{x}^\beta(r))\,dr\,dt.$$

Since $\beta^2e^{-\beta^2(s-r)}\to\delta(s-r)$ as $\beta\to\infty, s\geqslant r$, we have $\mathbf{x}^\beta(s)\to\tilde{\mathbf{x}}(s)$, where

$$\tilde{\mathbf{x}}(s)=\tilde{\mathbf{x}}_0+p\tilde{\mathbf{w}}(s)+\int_0^s K(\tilde{\mathbf{x}}(r))\,dr,$$

which is equivalent to the stochastic differential equation

$$d\tilde{\mathbf{x}}(s)=\mathbf{K}(\tilde{\mathbf{x}}(s))\,ds+\sqrt{\frac{2kT}{m}}\;d\tilde{\mathbf{w}}(s), \tag{6.1.5}$$

where $t=\beta s$. Equation (6.1.5) is called the Smoluchowski–Kramers approximation. The Fokker–Planck equation corresponding to (6.1.1) is given by

$$\frac{kT}{m}\Delta u-\nabla\cdot(\mathbf{K}u)=\frac{\partial u}{\partial s}.$$

Several derivations of (6.1.5) exist in the mathematical literature: Chandrasekhar (1954), Papanicolaou (1975), and Larsen and Schuss (1978).

We shall present two additional derivations. The next derivation was given by Papanicolaou (1975) in a more general setting of stochastic differential equations with more general noise. It is based on the asymptotic expansion of the solution of the backward Kolmogorov equation in terms of β^{-1}. This can be done in time intervals of order β. Thus we set $t=\beta s$, and noting that $(w/\sqrt{\beta}\,)(\beta s)=\tilde{w}(s)$ is a Brownian motion, we obtain from (6.1.1) the equation

$$dy^{\beta}(s)=\{-\beta^2 y^{\beta}(s)+\beta K(x^{\beta}(s))\}\,ds+\beta\sqrt{\frac{2kT}{m}}\;d\tilde{w}(s),$$

where $y^{\beta}(s)=y(\beta s)$. We shall consider for simplicity the one-dimensional case.

Setting $\beta=1/\varepsilon$, we obtain the backward Kolmogorov equation

$$\frac{\partial u^{\varepsilon}}{\partial s}=\frac{1}{\varepsilon^2}\left\{\frac{kT}{m}\frac{\partial^2 u}{\partial y^2}-y\frac{\partial u^{\varepsilon}}{\partial y}\right\}+\frac{1}{\varepsilon}y\frac{\partial u^{\varepsilon}}{\partial x}+\frac{1}{\varepsilon}K(x)\frac{\partial u^{\varepsilon}}{\partial y}. \tag{6.1.6}$$

Equation (6.1.6) is a typical singular perturbation problem. Setting $u^{\varepsilon}=u_0+\varepsilon u_1+\cdots$ and comparing coefficients of like powers of ε in (6.1.6), we see that

$$\mathcal{L}_1 u_0\equiv\frac{kT}{m}\frac{\partial^2 u_0}{\partial y^2}-y\frac{\partial u_0}{\partial y}=0, \tag{6.1.7}$$

where u_0 is assumed bounded. The solutions of (6.1.7) is given by

$$u_0(x,y,s)=C_1(x,s)+C_2(x,s)e^{my^2/kT}.$$

We must therefore conclude that $C_2(x,s)\equiv 0$; hence $u_0(x,y,s)=u_0(x,s)$. Next, we get

$$\mathfrak{L}_1 u_1 = \left\{ -y\frac{\partial u_0}{\partial x} + K(x)\frac{\partial u_0}{\partial y} \right\} \equiv -\mathfrak{L}_2 u_0 \tag{6.1.8}$$

and

$$\mathfrak{L}_1 u_2 = -\mathfrak{L}_2 u_1 + \frac{\partial u_0}{\partial s}. \tag{6.1.9}$$

It is well known (Courant and Hilbert 1962) that the solvability condition for (6.1.8) is $\mathfrak{L}_2 u_0 \perp u^*$; that is,

$$\int_{-\infty}^{\infty} u^* \mathfrak{L}_2 u_0 \, dy = 0$$

for any solution u^* of the adjoint equation $\mathfrak{L}_1^* u^* = 0$. This condition is satisfied since the transformation $y \to -y$ leaves the equation

$$\mathfrak{L}_1^* u^* = \frac{kT}{m}\frac{\partial u^*}{\partial y^2} + \frac{\partial y u^*}{\partial y} = 0$$

invariant. Hence $u^*(-y)=u^*(y)$ and

$$\int_{-\infty}^{\infty} u^* \mathfrak{L}_2 u_0 \, dy = \int_{-\infty}^{\infty} u^*(y)\left(-y\frac{\partial u_0}{\partial x} + K(x)\frac{\partial u_0}{\partial y} \right) dy$$

$$= -\frac{\partial u_0}{\partial x}\int_{-\infty}^{\infty} y u^*(y)\, dy = 0.$$

Thus $u_1 = \mathfrak{L}_1^{-1}\mathfrak{L}_2 u_0$ is well defined (up to normalization). The solvability condition for (6.1.9) implies that

$$\left\{ -\mathfrak{L}_2 \mathfrak{L}_1^{-1} \mathfrak{L}_2 u_0 + \frac{\partial u_0}{\partial s} \right\} \perp u^*;$$

$$\int \frac{e^{-(m/2kT)y^2}}{(2\pi kT/m)^{1/2}} \left\{ -y\frac{\partial}{\partial x} + K(x)\frac{\partial}{\partial y} \right\} \mathfrak{L}_1^{-1}\left(-y\frac{\partial}{\partial x} \right.$$

$$\left. + K(x)\frac{\partial}{\partial y} \right) u_0(x,s)\, dy = \frac{\partial u_0}{\partial s}. \tag{6.1.7}$$

First, we find that

$$v=\mathcal{L}_1^{-1}\left(-y\frac{\partial}{\partial x}+K(x)\frac{\partial}{\partial y}\right)u_0=\mathcal{L}_1^{-1}\left(-y\frac{\partial u_0}{\partial x}\right).$$

The function v is the solution of the equation

$$\frac{kT}{m}\frac{\partial^2 v}{\partial y^2}-y\frac{\partial v}{\partial y}=-Cy,$$

where the constant C is equal to $\partial u_0(x,s)/\partial x$. The solution of this equation is $v=Cy$; hence (6.1.7) takes the form

$$\int\frac{e^{-(m/2kT)y^2}}{\sqrt{2\pi kT/m}}\left[\left(-y\frac{\partial}{\partial x}+K(x)\frac{\partial}{\partial y}\right)\left(-y\frac{\partial u_0}{\partial x}\right)\right]dy=\frac{\partial u_0}{\partial s}$$

or

$$\int\frac{e^{-(m/2kT)y^2}}{\sqrt{2\pi kT/m}}\left[y^2\frac{\partial^2 u_0}{\partial x^2}+K(x)\frac{\partial u_0}{\partial x}\right]dy=\frac{\partial u_0}{\partial s}.$$

Hence

$$\frac{kT}{m}\frac{\partial^2 u_0}{\partial x^2}+K(x)\frac{\partial u_0}{\partial x}=\frac{\partial u_0}{\partial s}. \tag{6.1.10}$$

Equation (6.1.10) is the Smoluchowski–Kramers approximation to the backward Kolmogorov equation. This asymptotic analysis shows that in time intervals of order β^{-1}, the displacement distribution is governed by the stochastic differential equation (6.1.5); hence the displacement process $x(t)$ becomes Markovian and independent of the velocity process $y(t)$. It follows that the joint density function P of $(x(t), y(t))$ must have the form $P(x,y,t)\approx p(y)u(x,t)$. Substituting this expression in the Fokker–Planck equation, we see that

$$\frac{\partial(yp)}{\partial y}+\frac{kT}{m}\frac{\partial^2 p}{\partial y^2}=0;$$

hence

$$p(y)=\sqrt{\frac{m}{2\pi kT}}\;e^{-my^2/2kT},$$

so that the velocity distribution is Maxwellian. For short times $(x(t), y(t))$ is Markovian, but $x(t)$ and $y(t)$ are not; moreover, $x(t)$ and $y(t)$ are dependent variables.

The third derivation of the Smoluchowski–Kramers equation was given by Larsen and Schuss (1978). The procedure adopted in this derivation is based on the asymptotic expansion of the solution of the Fokker–Planck equation, viewed as a transport equation. The same method, called homogenization, can be used to derive diffusion approximations to transport equations (see the Exercises in this section). We shall assume that the force $\mathbf{K}(\mathbf{x})$ is a gradient of a potential; that is,

$$\mathbf{K}(\mathbf{x}) = -\nabla\Phi(\mathbf{x}).$$

The solution of the Fokker–Planck equation may have quite large gradients if

$$\max_{\mathbf{x}} \Phi(\mathbf{x}) \gg \frac{kT}{m}.$$

This is the case, for example, if $\Phi(\mathbf{x})$ is the potential for a chemical reaction. Also, if $\Phi(\mathbf{x})$ is periodic with periodic cell C, as is the case in atomic migration in crystals, the previous procedure did not take into consideration the possibility of large gradients while using the first term $u_0(\mathbf{x}, s)$ in the expansion of the solution. We shall assume that $\Phi(\mathbf{x})$ is periodic with a cell C, whose characteristic dimension is l.

We begin by introducing the following dimensionless variables:

$$\mathbf{r} = \left(\frac{1}{l}\right)\mathbf{x}$$

$$\omega = \left(\frac{1}{v_0}\right)\mathbf{v}$$

$$\tau = \left(\frac{v_0}{l}\right)t.$$

Here l is a typical diameter of a cell C and v_0 is a typical particle velocity. Thus $\mathbf{r}$ is a dimensionless position variable, in terms of which a typical cell diameter is $O(1)$, ω is a dimensionless velocity variable in terms of which a typical velocity is $O(1)$, and τ is a dimensionless time variable in terms of which the time required for a typical particle unaffected by Φ to traverse a cell is $O(1)$. We also define

(6.1.11)
$$\varepsilon = \sqrt{\frac{v_0}{\beta l}}\,,$$
$$\lambda = \frac{kT}{mv_0^2}\,,$$

and

(6.1.12)
$$\phi(\mathbf{r},\varepsilon) = \frac{1}{\varepsilon}\phi_0(\mathbf{r}) + \phi_1(\mathbf{r}) = \left(\frac{1}{v_0^2}\right)\Phi(\mathbf{x}),$$

(6.1.13)
$$\psi(\mathbf{r},\omega,\tau) = \left(\frac{1}{l^3 v_0^3}\right)P(\mathbf{x},\mathbf{v},t),$$

where the transition probability density

$$P(\mathbf{x},\mathbf{v},\boldsymbol{\xi},\boldsymbol{\eta},t) = \Pr\{\mathbf{x}(t)=\mathbf{x},\mathbf{v}(t)=\mathbf{v}|\mathbf{x}(0)=\boldsymbol{\xi},\mathbf{v}(0)=\boldsymbol{\eta}\}$$

of the point $(\mathbf{x},\mathbf{v})$ in phase space can be shown, by a derivation from the Langevin equation, to satisfy the Fokker–Planck equation

$$0 = \frac{\partial P}{\partial t} + \mathbf{v}\cdot\nabla_{\mathbf{x}}P - \nabla_{\mathbf{x}}\Phi\cdot\nabla_{\mathbf{v}}P - \beta\nabla_{\mathbf{v}}(\mathbf{v}P) - \frac{\beta kT}{m}\Delta_{\mathbf{v}}P,$$
$$P(\mathbf{x},\mathbf{v},\boldsymbol{\xi},\boldsymbol{\eta},0) = \delta(\mathbf{x}-\boldsymbol{\xi})\,\delta(\mathbf{v}-\boldsymbol{\eta}).$$

In (6.1.12) we require

$$\min_{\mathbf{r}\in\partial C}\phi_i(\mathbf{r}) = O(1), \qquad i=0,1$$

where ∂C in the boundary of the cell C. Also, we allow $\phi_0=0$ or $\phi_1=0$ (or both). Thus, for $\varepsilon \ll 1$, this analysis accounts for the possibilities that the activation energy, related to

$$\min_{\mathbf{x}\in\partial C}\Phi(\mathbf{x}),$$

is large [i.e., $O(\varepsilon^{-1}) = O(\sqrt{\beta}\,)$], is $O(1)$, or is zero.

The starting point of our analysis is the Fokker–Planck equation. Thus we introduce the foregoing equations into the Fokker–Planck equation and get

$$(6.1.14)\qquad 0=\varepsilon^2\left[\frac{\partial\psi}{\partial\tau}+\omega\cdot\nabla_{\mathbf{r}}\psi-\nabla_{\mathbf{x}}\phi_1\cdot\nabla_\omega\psi\right]-\varepsilon\left[\nabla_{\mathbf{x}}\phi_0\cdot\nabla_\omega\psi\right]$$
$$-\left[\nabla_\omega\cdot(\omega\psi)+\lambda\Delta\omega\psi\right].$$

We shall study this equation for $\varepsilon\ll 1$. By (6.1.11), this means that the viscosity β must be large compared to the frequency v_0/l by which particles unaffected by ϕ pass through a cell. Equation (6.1.14) has the equilibrium solution

$$\psi=\exp\left[-\frac{1}{\lambda}\left(\frac{1}{\varepsilon}\phi_0+\phi_1+\tfrac{1}{2}\omega^2\right)\right],\qquad \omega^2=\omega\cdot\omega.$$

After a finite relaxation time, ψ should have the form shown above modulo certain spatial and time variations. Thus our assumption for (6.1.14) is

$$(6.1.15)\qquad \psi\sim\exp\left[-\frac{1}{\lambda}\left(\frac{1}{\varepsilon}\phi_0+\phi_1+\tfrac{1}{2}\omega^2\right)\right]\sum_{n=0}^{\infty}\varepsilon^n\psi_n(\mathbf{y};\mathbf{r},\omega,\tau;s,\sigma),$$

where

$$(6.1.16)\qquad \mathbf{y}=\frac{1}{\varepsilon}\mathbf{r},$$
$$s=\varepsilon\tau,$$
$$\sigma=\varepsilon^2\tau.$$

The spatial variable $\mathbf{y}$ describes boundary layers, if any exist, and s,σ are slow time variables.

Introducing (6.1.15) and (6.6.16) into (6.1.14), and equating the coefficients of different powers of ε, we obtain the following system of equations:

(6.1.17)

$$\omega\cdot\nabla_\omega\psi_n-\lambda\Delta_\omega\psi_r=\left[\nabla_r\phi_0\cdot\nabla_\omega\psi_{n-1}-\omega\cdot\nabla_{\mathbf{y}}\psi_{n-1}\right]$$
$$-\left[\omega\cdot\nabla_r\psi_{n-2}+\nabla_{\mathbf{r}}\phi_1\cdot\nabla_\omega\psi_{n-2}+\frac{\partial}{\partial\tau}\psi_{n-2}\right]-\frac{\partial}{\partial s}\psi_{n-3}-\frac{\partial}{\partial\sigma}\psi_{n-4}.$$

We reject exponential growth in ω and require each ψ_n to have at most polynomial growth in ω as $\omega\to\infty$. Then, by (6.1.15), ψ will essentially decay exponentially as $\omega\to\infty$.

Setting $n=0$ in (6.1.17) gives

$$\omega\cdot\nabla_\omega\psi_0-\lambda\Delta_\omega\psi_0=0.$$

This equation has the bounded solution

$$\psi_0=A(\mathbf{y};\mathbf{r},\theta;s;\sigma), \tag{6.1.18}$$

where A is (at this point) undetermined.

Next we set $n=1$ in (6.1.17) and get

$$\omega\cdot\nabla_\omega\psi_1-\lambda\Delta_\omega\psi_1=-\omega\cdot\nabla_\mathbf{y}A.$$

This equation has the polynomial solution

$$\psi_1=-\omega\cdot\nabla_\mathbf{y}A(\mathbf{y};\mathbf{r},\tau;s;\sigma).$$

We could include a solution of the homogeneous equation in ψ_1, but do not, since this can be incorporated into A.

Setting $n=2$ in (6.1.17) and rearranging gives

$$\begin{aligned}\omega\cdot\nabla_\omega\psi_2-\lambda\Delta_\omega\psi_2&=\{-\omega\cdot\nabla_\mathbf{r}A\}\\&+\{\nabla_\mathbf{y}\cdot[\omega\omega-\lambda I]\cdot\nabla_\mathbf{y}A\}\\&+\left\{\lambda\Delta_\mathbf{y}A-\nabla_\mathbf{r}\phi_0\cdot\nabla_\mathbf{y}A-\frac{\partial A}{\partial\tau}\right\}.\end{aligned} \tag{6.1.19}$$

Here $\omega\omega^T$ is the matrix whose entries are $\{\omega_i\omega_j\}$. Since the inhomogeneous equation

$$[\omega\cdot\nabla_\omega-\lambda\Delta_\omega]f(\omega)=1$$

has the particular solution

$$f(\omega)=-\frac{1}{\lambda}\int_{r=0}^{\omega}\frac{e^{r^2/2\lambda}}{r^2}\int_{s=0}^{t}s^2e^{-s^2/2\lambda}\,ds\,dr,$$

which is an exponentially growing function, the solution ψ_2 is a polynomial in ω only if the last term in (6.1.19) is zero; otherwise, ψ_2 will grow exponentially in ω. Hence, we set this last term equal to zero. This is the "solvability condition" for (6.1.19):

$$\frac{\partial A}{\partial\tau}=\lambda\Delta_\mathbf{y}A-\nabla_\mathbf{r}\phi_0\cdot\nabla_\mathbf{y}A. \tag{6.1.20}$$

Then u_2 is given by

$$u_2 = -\omega \cdot \nabla_{\mathbf{r}} A(\mathbf{y}; \mathbf{r}, \tau; s; \sigma) + \tfrac{1}{2}(\omega \cdot \nabla_{\mathbf{y}})^2 A(\mathbf{y}; \mathbf{r}, \tau; s; \sigma).$$

As above, we choose not to include a solution of the homogeneous equation in u_2.

We continue the foregoing procedure by solving (6.1.17) for $n=3$ and 4. Since only the solvability conditions are of interest to us and the calculations follow exactly as above, we just state these conditions. For $n=3$, we get

$$\frac{\partial A}{\partial s} = 2\lambda \nabla_{\mathbf{y}} \cdot \nabla_{\mathbf{r}} A - \nabla_{\mathbf{r}} \phi_0 \cdot \nabla_{\mathbf{r}} A - \nabla_{\mathbf{r}} \phi_1 \cdot \nabla_{\mathbf{y}} A, \tag{6.1.21}$$

and for $n=4$,

$$\frac{\partial A}{\partial \sigma} = \lambda \nabla_{\mathbf{r}} A - \nabla_{\mathbf{r}} \phi_1 \cdot \nabla_{\mathbf{r}} A. \tag{6.1.22}$$

Equations (6.1.20)–(6.1.22) describe the evolution of A according to the three time variables τ, s, and σ, and the two position variables $\mathbf{r}$ and $\mathbf{y}$. We can combine these equations into a single equation for

$$a(\mathbf{r}, \sigma) = A(\mathbf{y}; \mathbf{r}, \tau; s; \sigma), \tag{6.1.23}$$

where the dependent variables are related by (6.1.16). By (6.1.12), (6.1.16), and (6.1.20)–(6.1.22),

$$\begin{aligned}
\varepsilon^2 \frac{\partial a}{\partial \sigma} &= \frac{\partial A}{\partial \tau} + \varepsilon \frac{\partial A}{\partial s} + \varepsilon^2 \frac{\xi A}{\partial \sigma} \\
&= \varepsilon^2 \lambda \left(\nabla_{\mathbf{r}} + \frac{1}{\varepsilon} \nabla_{\mathbf{y}} \right) \cdot \left(\nabla_{\mathbf{r}} + \frac{1}{\varepsilon} \nabla_{\mathbf{y}} \right) A \\
&\quad - \varepsilon^2 \nabla_{\mathbf{r}} \left(\frac{1}{\varepsilon} \phi_0 + \phi_1 \right) \cdot \left(\nabla_{\mathbf{r}} + \frac{1}{\varepsilon} \nabla_{\mathbf{y}} \right) A \\
&= \varepsilon^2 \lambda \Delta_{\mathbf{r}} a - \varepsilon^2 \nabla_{\mathbf{r}} \phi \cdot \nabla_{\mathbf{r}} a.
\end{aligned}$$

Therefore, by (6.1.12), (6.1.15), (6.1.18), and (6.1.23),

$$\psi \sim \exp\left[-\frac{1}{\lambda} \left(\phi(\mathbf{r}) + \tfrac{1}{2}\omega^2 \right) \right] a(\mathbf{r}, \sigma), \tag{6.1.24}$$

where

$$\frac{\partial a}{\partial \sigma} = \lambda \Delta_{\mathbf{r}} a - \nabla_{\mathbf{r}} \phi \cdot \nabla_{\mathbf{r}} a. \tag{6.1.25}$$

Equations (6.1.24) and (6.1.25) describe the asymptotic solution of the Fokker–Planck equation (6.1.14) for large times. It is essential to keep in mind that, by (6.1.12), ϕ can be $O(1/\varepsilon)$, and by (6.1.16) and (6.1.23), $\partial a/\partial\sigma$ can be $O(1/\varepsilon^2)$, and $\nabla_{\mathbf{r}} a$ can be $O(1/\varepsilon)$. In other words, ϕ can be (suitably) large, and a can have (suitable) boundary layers.

Using (6.1.11), (6.1.12), and (6.1.16), we may write (6.1.24) as

$$\psi \sim e^{-mv^2/2kT} p(\mathbf{x}, t)\left(\frac{1}{l^3 v_0^3}\right), \tag{6.1.26}$$

where

$$p(\mathbf{x}, t) = e^{[-(m/kT)\Phi](\mathbf{x})} a\left(\frac{1}{l}\mathbf{x}, \frac{v_0^2}{\beta l^2} t\right) l^3 v_0^3. \tag{6.1.27}$$

Thus using (6.1.25) and (6.1.27), one can show that p satisfies the Smoluchowski–Kramers equation

$$\beta \frac{\partial p}{\partial t} = \frac{kT}{m} \Delta_{\mathbf{x}} p + \nabla_{\mathbf{x}}(p \cdot \nabla_{\mathbf{x}} \Phi) \tag{6.1.28}$$

$$p(\mathbf{x}, \xi, 0) = \delta(\mathbf{x} - \xi). \tag{6.1.29}$$

Also, (6.1.13) and (6.1.26) imply that

$$P(\mathbf{x}, \mathbf{v}, \xi, \eta, t) \sim \left(\frac{m}{2\pi KT}\right)^{3/2} e^{-mv^2/2kT} p(\mathbf{x}, \xi, t). \tag{6.1.30}$$

To summarize, in this section we have derived the Smoluchowski–Kramers equation (6.1.28) from the Fokker–Planck equation (6.1.14). In so doing, we have shown that the Smoluchowski–Kramers equation remains valid if the potential and spatial derivatives of the solution are suitable large. We have not derived the initial condition (6.1.29) from that of the Fokker–Planck equation, but we remark that it can be derived by means of an asymptotic "initial layer" analysis of (6.1.14) and the initial condition

$$P(\mathbf{x}, \mathbf{v}, \xi, \eta, t) \to \delta(\mathbf{x} - \xi, \mathbf{y} - \eta) \quad \text{as} \quad t \downarrow 0.$$

(Larsen and Schuss 1978; Larsen and Keller 1974).

Equation (6.1.28) is the Fokker–Planck equation for the transition probability density of the solution of the stochastic differential equation (6.1.5).

EXERCISE 6.1.1

Write down the Smoluchowski equation for a Brownian harmonic oscillator. Solve the equation and compute the mean and the variance of the displacement and of the velocity. Compare the results with those of Exercise 2.1.6.

EXERCISE 6.1.2

Consider the Langevin equation for a charged Brownian particle in an electromagnetic field

$$\frac{d\mathbf{x}}{dt}=\mathbf{y}$$

$$\frac{d\mathbf{y}}{dt}=-\beta\mathbf{y}+\mathbf{K}(\mathbf{x})-\mathbf{H}(\mathbf{x})\times\mathbf{y}+\sqrt{\frac{2\beta kT}{m}}\,\frac{d\mathbf{w}}{dt},$$

where $\mathbf{x}=(x_1,x_2,x_3)^T$, $\mathbf{K}(\mathbf{x})=(K_1(\mathbf{x}),K_2(\mathbf{x}),K_3(\mathbf{x}))^T$. $\mathbf{H}(\mathbf{x})$ is the magnetic field and $\times$ denotes vector product. Derive the Smoluchowski–Kramers equation.

EXERCISE 6.1.3

Generalize the asymptotic analysis of the Fokker–Planck equation to the following transport equation:

$$\frac{\partial u}{\partial t}=y\frac{\partial u}{\partial x}+(K(x)-\beta y)\frac{\partial u}{\partial y}+qLu$$

where L is an operator acting on the variable y and $q=q(\beta)$ is a parameter. Find conditions under which $u(x,y,t,\beta)\to u_0(x,s)$ as $\beta\to\infty$, $t\to\infty$, $t=\tau(\beta)$ (use the right scaling!)$\to s$ as $\beta\to\infty$. The function $u_0(x,s)$ is the solution of an appropriate Smoluchowski equation.

EXERCISE 6.1.4

Consider the Goldstein model for a one-dimensional diffusion. Particles are distributed along the x-axis and move to right or to the left with a fixed speed α, colliding elastically. The velocity $\{y(t), t \geq 0\}$ is a process that takes the values $\pm\alpha$ with mean time between collisions$=1/\beta, \beta>0$. The time between collisions is exponentially distributed so that the transition probability of the velocity process $y(t)$ is given by

$$(6.1.31)\quad P(y(t+s)=\pm\alpha \mid y(s)=\pm\alpha)$$
$$=\frac{1}{2}\begin{pmatrix} 1+e^{-2\beta t} & 1-e^{-2\beta t} \\ 1-e^{-2\beta t} & 1+e^{-2\beta t} \end{pmatrix} \rightarrow \frac{1}{2}\begin{pmatrix} 1 & 1 \\ 1 & 1 \end{pmatrix}$$

as $t\rightarrow\infty$. The displacement process $x(t)$ satisfies $dx/dt=y(t)$, $x(0)=x$, and $(x(t), y(t))$ is a Markov process. Let $\mathbf{u}(t,x,y)+E_{x,\mathbf{y}}\mathbf{f}(x(t),\mathbf{y}(t))$, where $\mathbf{y}=\binom{\alpha}{-\alpha}$, $x=x(0)$. Using (6.1.31), get

$$\frac{\partial \mathbf{u}}{\partial t}=y\frac{\partial \mathbf{u}}{\partial x}+\mathbf{A}\mathbf{u}$$

where

$$\mathbf{u}=\begin{pmatrix} u(t,x,\alpha) \\ u(t,x,-\alpha) \end{pmatrix} \quad \text{and} \quad \mathbf{A}=\frac{1}{2}\begin{pmatrix} 1 & 1 \\ 1 & 1 \end{pmatrix}.$$

The limit, as the collisions are sped up, can be found by rescaling $s=\varepsilon^{-2}t$, $y^{\varepsilon}(t)=y(t/\varepsilon^2)$. Get

$$\frac{dx^{\varepsilon}(t)}{dt}=\frac{1}{\varepsilon}y^{\varepsilon}(t)$$

$$\frac{\partial \mathbf{u}^{\varepsilon}}{\partial s}=\frac{1}{\varepsilon}\frac{\partial \mathbf{u}}{\partial x}+\frac{1}{\varepsilon^2}\mathbf{A}\mathbf{u}^{\varepsilon}$$

$$\mathbf{u}^{\varepsilon}(0,x,y)=\mathbf{f}(x,y),$$

where

$$\mathbf{u}^{\varepsilon}(t,x,y)=E_{x,y}\mathbf{f}(x^{\varepsilon}(t),y^{\varepsilon}(t)).$$

The vector $\mathbf{f}(x, y)$ has components $f_+(x)=f(x,\alpha)$ and $f_-(x)=f(x,-\alpha)$. Assume that $f_+=f_-=f$, to avoid initial layers. Write

$$\frac{\partial}{\partial t}\begin{pmatrix}u_+\\ u_-\end{pmatrix}=\frac{1}{\varepsilon}\begin{pmatrix}\alpha & 0\\ 0 & -\alpha\end{pmatrix}\frac{\partial}{\partial x}\begin{pmatrix}u_+\\ u_-\end{pmatrix}+\frac{1}{\varepsilon^2}\begin{pmatrix}-\beta & \beta\\ \beta & -\beta\end{pmatrix}\begin{pmatrix}u_+\\ u_-\end{pmatrix},$$

$\mathbf{u}(0,x)=f_{\pm}(x)$. Show that $w=u_+ + u_-$ satisfies the equation

$$\varepsilon^2\frac{\partial^2 w}{\partial t^2}=\alpha^2\frac{\partial^2 w}{\partial x^2}-2\beta\frac{\partial w}{\partial t}$$

$$w(0,x)=2f(x), \qquad \frac{\partial w}{\partial t}(0,x)=0$$

(this is the telegrapher's equation). Apply the previous expansion as $\varepsilon\to 0$, taking

$$\mathcal{L}_1=A, \qquad \mathcal{L}_2=\begin{pmatrix}\alpha & 0\\ 0 & -\alpha\end{pmatrix}\frac{\partial}{\partial x}$$

$$\mathcal{L}_1^{-1}=\frac{1}{4\beta}\begin{pmatrix}-1 & 1\\ 1 & -1\end{pmatrix}-P\mathcal{L}_2\mathcal{L}_1^{-1}\mathcal{L}_2P=\frac{\alpha^2}{4\beta}\begin{pmatrix}1 & 1\\ 1 & 1\end{pmatrix}\frac{\partial^2}{\partial x^2},$$

where P is the projection into the null space of $\mathcal{L}_1$; that is,

$$P=\lim_{t\to\infty}e^{\mathcal{L}_1 t}=\frac{1}{2}\begin{pmatrix}1 & 1\\ 1 & 1\end{pmatrix}.$$

Hence show that

$$\frac{\partial u_0(t,x)}{\partial t}=\frac{\alpha^2}{4\beta}\frac{\partial^2 u_0(t,x)}{\partial x^2} \tag{6.1.32}$$

$$u_0(0,x)=f(x).$$

Equation (6.1.32) is the one-dimensional diffusion equation with diffusion coefficient$=\alpha^2/4\beta$, where $\alpha=$velocity of particles, and $\beta=$mean time between collisions (Papanicolaou 1975).

EXERCISE 6.1.5*

Consider the linear transport equation

$$\frac{\partial u}{\partial t}+\nabla_{\mathbf{x}}\cdot\mathbf{y}u+\nabla\mathbf{y}\cdot\frac{1}{m}Fu=Lu+S.$$

Here $\mathbf{x}=(x_1, x_2, x_3)^T$ is the space variable, $\mathbf{y}=(y_1, y_2, y_3)^T$ is the velocity vector, and $u(\mathbf{x},\mathbf{y}, t)$ is the concentration of particles in phase space.

$$\mathbf{F}=\mathbf{G}(\mathbf{x}, t)+e\mathbf{E}(\mathbf{x}, t)+\frac{e}{c}\mathbf{y}\times\mathbf{B}(\mathbf{x}, t)$$

represents the external forces acting on the system; **G**, **E**, and **B** represent gravitational, electrical, and magnetic forces, respectively. The constant m is the mass of the rarefied particles, and $S(\mathbf{x},\mathbf{y}, t)$ is a source term representing the concentration of rarefied particles spontaneously emitted into a volume element in phase space. L is an operation governing collisions between the gas and the medium. After nondimensionalizing the transport equation, assume that the mean free path and the mean free time between collisions of the gas (e.g., neutrons) and the medium are small [i.e., $x\to\varepsilon x$ and $t\to\varepsilon^2 t(\varepsilon\ll 1)$]. Let $\varepsilon w(\mathbf{x}, t)$ be the relatively small velocity of the medium, so that $\mathbf{y}\to\mathbf{y}-\varepsilon\mathbf{w}(\mathbf{x}, t)$ (the new **x**!). Replace $\mathbf{E}+\mathbf{G}\to\varepsilon\mathbf{E}$, $\mathbf{B}\to\mathbf{B}$, $S\to\varepsilon^2 S$, $Lu\to Lu^\varepsilon$, where u^ε is the density with respect to the new variables. Obtain the transport equation

$$\varepsilon^2\frac{\partial u^\varepsilon}{\partial t}+\nabla_{\mathbf{x}}(\varepsilon\mathbf{y}+\varepsilon^2\mathbf{w})u+\nabla_{\mathbf{y}}\Big[\mathbf{y}\times\mathbf{B}+\varepsilon(\mathbf{E}+\mathbf{w}\times\mathbf{B})$$
$$-\varepsilon^2(\mathbf{y}\cdot\nabla_{\mathbf{x}}\mathbf{w})-\varepsilon^3\left(\frac{\partial}{\partial t}\mathbf{w}+\mathbf{w}\cdot\nabla_{\mathbf{x}}\mathbf{w}\right)\Big]u=Lu+\varepsilon^2 S.$$

Take L to be $L=L_0+\varepsilon^2 L_2$, where

$$L_0 u(\mathbf{x},\mathbf{y}, t)$$
$$=\int \eta\sigma_n^s(\mathbf{x}, \boldsymbol{\eta},\mathbf{y}, t)u(\mathbf{x}, \boldsymbol{\eta}, t)\, d\boldsymbol{\eta}-\sigma_\eta^T(\mathbf{x},\mathbf{y}, t)u(\mathbf{x},\mathbf{y}, t) \qquad (n=0,2).$$

Assume that 0 is an eigenvalue of L_0; that is, there exists a function $\phi=\phi(\mathbf{x},\mathbf{y}, t)$ such that $L_0\phi=0$, and $1=\int\phi\, d\mathbf{y}=4\pi\int|\mathbf{y}|^2\phi\, d|\mathbf{y}|$. (Note that in the Smoluchowski approximation ϕ is Maxwellian.) Assume that $\phi=\phi(|\mathbf{y}|)$, L_0 is particle conserving; that is $\int(L_0 f)\, dy=0$ for all f in the domain of L_0 (hence $\Phi^*=1$ is an eigenfunction of L_0^*). Conclude that 0 is an eigenvalue of the operator $T=L_0-\mathbf{y}\times\mathbf{B}\cdot\nabla_{\mathbf{v}}$ with ϕ and $\phi^*=1$ as eigenfunctions of T and T^*, respectively. Assume that 0 is an isolated eigenvalue of T and spec $(T)\subset[\mathrm{Re}\lambda<\gamma<0\}$ for all **B**. If $g(y)\perp\phi^*$, then $T^{-1}g$ exists and is unique if $\int T^{-1}g\, dy=0$ is required. Assume that

$$\sigma_0^s=\sum_{n=0}^{\infty}\frac{2n+1}{4\pi}P_n\left(\frac{\mathbf{y}}{|\mathbf{y}|}\cdot\frac{\boldsymbol{\eta}}{|\boldsymbol{\eta}|}\right)f_n(\mathbf{x}, \boldsymbol{\eta},\mathbf{y}, t),$$

where P_n are the Legendre polynomials. Conclude that L_0 leaves invariant the space of scalar functions of the form

$$\frac{\mathbf{y}}{|\mathbf{y}|}\cdot\mathbf{H}(\mathbf{x},|\mathbf{y}|,t).$$

Set $u=\sum_{n=0}^{\infty}u_n\varepsilon^n$ and obtain

$$u_0=A_0(\mathbf{x},t)\phi(\mathbf{x},|\mathbf{y}|,t)$$

$$u_1=A_1(\mathbf{x},t)\phi(\mathbf{x},|\mathbf{y}|,t)+\frac{\mathbf{y}}{|\mathbf{y}|}\cdot\psi(\mathbf{x},|\mathbf{y}|,t),$$

where

$$\psi=(|\mathbf{B}|^2+K^2)^{-1}\left[K\mathbf{h}+\mathbf{B}(\mathbf{B}\cdot K^{-1}\mathbf{h})+\mathbf{B}\times\mathbf{h}\right] \tag{6.1.33}$$

$$\mathbf{h}=(|\mathbf{y}|\phi\nabla_{\mathbf{x}}A_0+\left[|\mathbf{y}|\nabla_{\mathbf{x}}\phi+(\mathbf{E}+\mathbf{w}\times\mathbf{B})\left(\frac{\partial}{\partial|\mathbf{y}|}\phi\right)\right]A_0$$

and K is defined by

$$L_0\left[\frac{\mathbf{y}}{|\mathbf{y}|}\cdot\mathbf{H}(x,|y|,t)\right]=\frac{\mathbf{y}}{|\mathbf{y}|}\cdot(K\mathbf{H})(\mathbf{x},|\mathbf{y}|,t)$$

(note that $\mathbf{K}\psi-\mathbf{B}\times\psi=\mathbf{h}$). Show that the solvability condition for u_2 is

$$\frac{\partial A_0}{\partial t}+\nabla_{\mathbf{x}}\cdot\left(\frac{4\pi}{3}\int|\mathbf{y}|^3\psi\,d|\mathbf{y}|+\mathbf{w}A_0\right)=\sigma A_0+S_0, \tag{6.1.34}$$

where

$$S_0(x,t)=\int S(\mathbf{x},\mathbf{y},t)\,d\mathbf{y}$$

$$\sigma(\mathbf{x},t)=\int(L_2\phi)(\mathbf{x},\mathbf{y},t)\,d\mathbf{y}.$$

Rewrite (6.1.34) in the form

$$\frac{\partial}{\partial t}A_0+\nabla_{\mathbf{x}}\cdot A_0\omega=\sigma A_0+S_0$$

where

$$\omega = \mathbf{M}(|\mathbf{y}|\phi)\cdot\nabla \ln A_0 + \mathbf{y}\cdot(\mathbf{M}\nabla_{\mathbf{x}}\phi)$$
$$+\mathbf{M}\left(\frac{\partial}{\partial|\mathbf{y}|}\phi\right)\cdot(\mathbf{E}+\mathbf{w}\times\mathbf{B})+\mathbf{w}$$

where $\mathbf{M}$ is a tensor operator:

$$\mathbf{M}(\cdot) = \frac{4\pi}{3}\int |\mathbf{y}|^2(|\mathbf{B}|^2+K^2)^{-1}(IK+\mathbf{BB}K^{-1}+\mathbf{B}\times\mathbf{I})(\cdot)\,d|\mathbf{y}|.$$

Here $\mathbf{B}\times\mathbf{I}\cdot\mathbf{f}=\mathbf{B}\times\mathbf{f}$.

Consider the following cases:

(i) $\sigma=0$ (conservation of particles)

(ii) Assume that ϕ is Maxwellian:

$$\phi = \left(\frac{m}{2\pi kT}\right)^{3/2} e^{-m|\mathbf{y}|^2/2kT}$$

and analyze the drift term due to ω.

(iii) Consider the case $\mathbf{B}=0$ [hence $\mathbf{M}(\cdot)=(4\pi/3)\int \mathbf{I}|\mathbf{y}|^3K^{-1}(\cdot)\,d|\mathbf{y}|$].

(iv) $\mathbf{B}\to\infty$, $\mathbf{B}=|\mathbf{B}|\mathbf{e}$ [hence $\mathbf{M}(\cdot)=(4\pi/3)\int \mathbf{ee}|\mathbf{y}|^3K^{-1}(\cdot)\,d|\mathbf{y}|$].

(v) Assume that scattering is isotropic; that is, $\sigma_0^s=\sigma_0^T(\mathbf{x},|\mathbf{y}|,t)$, and so on. Show that

$$K\mathbf{h} = -|\mathbf{y}|\sigma_0^T(\mathbf{x},|\mathbf{y}|,t)\mathbf{h}(\mathbf{x},\mathbf{y},t)$$

and construct K^{-1} and $(|\mathbf{B}|^2+K^2)^{-1}$ (Larsen and D'Arruda 1976).

EXERCISE 6.1.6

Consider the effect of gravity on the Brownian motion and explain the phenomenon of sedimentation, that is, of variations in concentration of particles as a function of altitude. Use the Smoluchowski approximation to derive the approximate Langevin equation for the displacement

$$\frac{d\mathbf{x}}{ds} = \mathbf{K}(\mathbf{x}) + \sqrt{\frac{2kT}{m}}\,\frac{dw(s)}{ds}, \qquad s=\beta t$$

$\mathbf{x}=(x, y, z)^T$. Set

$$\mathbf{K}=\left(0,0,-\left(1-\frac{\rho_0}{\rho}\right)\right)^T g,$$

where ρ_0 is the density of the surrounding fluid and ρ is the concentration of Brownian particles ($\rho \geqslant \rho_0$). Let w be the solution of the corresponding Fokker–Planck equation.

(i) Show that $\mathbf{j}=(w\mathbf{K}+D\nabla w)$ is the diffusion current ($D=kT/m\beta$).

(ii) Disregard diffusion in the x, y directions and obtain

$$\frac{\partial w}{\partial t}=D\frac{\partial^2 w}{\partial z^2}+c\frac{\partial w}{\partial z} \tag{6.1.35}$$

$$c=\left(1-\frac{\rho_0}{\rho}\right)g.$$

Show that the initial and boundary conditions for this problem are

$$D\frac{\partial w}{\partial z}+cw=0 \quad \text{and} \quad w\to\delta(z-z_0) \tag{6.1.36}$$

at $z=0$, $t>0$ (at the bottom!).

(iii) Solve (6.1.35) by setting

$$w=U(z,t)\exp\left\{-\left[\frac{c}{2D}(z-z_0)+\frac{c^2}{4D}t\right]\right\}.$$

Show that $\lim_{t\to\infty} w(z,t,z_0)=(c/D)e^{-cz/D}$; that is, the equilibrium solution represents the law of isothermal atmospheres (Lamb 1945). Illustrate graphically the nature of the solution, in particular plot $w(z,t,0)$.

(iv) Show that

$$Ez=\frac{D}{c}=\frac{kT}{mg}\frac{\rho}{\rho-\rho_0},$$

that is, the height of the equivalent homogeneous atmosphere.

(v) Why isn't the state $z=0$ of minimum energy achieved? Show that particles initially at $z=0$ will not stay there, specifically

$$w(z,t,0)=\frac{1}{\sqrt{\pi Dt}}e^{-(z+ct)^2/4Dt}$$

$$+\frac{c}{D\sqrt{\pi}}e^{-cz/D}\int_{(2-ct)/2\sqrt{Dt}}^{\infty}e^{-x^2}\,dx\to\frac{c}{D}e^{-cz/D}$$

Hence the particles do a certain amount of work at the expense of the internal energy of the surrounding fluid, contrary to the second law of thermodynamics!

(vi) Show that the average work done in this manner is given by

$$EA=\frac{kT}{\text{particle}}.$$

Hence on the average there is a *decrease* in entropy of amount $k/$particle:

$$ES=S_{\max}-Nk,$$

where N denotes the number of Brownian particles. Can this work be utilized to run a heat engine with an efficiency higher than that of the Carnot cycle?

Consider the entropy of particles at $z\leqslant D/c$ and $z\geqslant D/c$ and the total entropy of the particles.

EXERCISE 6.1.7

Show that the relative displacement between two particles describing Brownian motions independently of each other and with diffusion coefficients D_1 and D_2 also follows the laws of Brownian motion with the diffusion coefficient $D_{12}=D_1+D_2$.

EXERCISE 6.1.8

A particle, assumed fixed in space, is in a medium of infinite extent in which a number of similar Brownian particles are distributed uniformly at time $t=0$. More precisely, assume that $N/V\to\nu$ as the volume v of the vessel containing the gas increases to infinity and N is the number of

Brownian particles in the vessel. Assume that the stationary particle is surrounded by a sphere of influence of radius R; that is, any particle that comes within distance R of the stationary particle is absorbed into the sphere. What is the rate at which particles arrive on the sphere of radius R surrounding the fixed particle? Let ν be the average concentration of particles exterior to $|r|=R$ at time $t=0$. [*Answer:* $4\pi DR\nu(1+R/\sqrt{\pi Dt})$ (Smoluchowski 1916).]

6.2 DIFFUSION APPROXIMATION TO MARKOV CHAINS, APPLICATIONS TO GENETICS

The construction of the Brownian motion as a limit of a rescaled random walk can be generalized to a class of Markov chains. Assume that for each N a Markov chain $x_N(t_n^N)$ is given on the real line. More specifically, let $x_N^1, x_N^2, \ldots$ be the possible states of the Markov chain, and

$$p\left[X_n(t_n^N)=x_N^j \mid X_N(t_{n-1}^N)=x_N^i\right]=p_{ij}^N(n),$$

where

$$t_1^N<t_2^N<\cdots.$$

Assume that

$$\max_n \Delta_N t_n \equiv \max_n \left(t_{n+1}^N-t_n^N\right)\to 0 \qquad \text{as} \quad N\to\infty.$$

We define a piecewise constant interpolation of $x_N(t_n^N)$ by setting

$$x_N(t)=x_N\left(t_n^N\right) \qquad \text{for} \quad t_n^N\leqslant t<t_{n+1}^N.$$

Assume that

$$\frac{1}{\Delta t}E_{x,t}\left[x_N(t+\Delta t)-x_N(t)\right]\to a(x,t)$$

as $N\to\infty$, $\Delta t=\Delta t(N)\to 0$,

$$\frac{1}{\Delta t}E_{x,t}\left[x_N(t+\Delta t)-x_N(t)\right]^2\to b(x,t)$$

and

$$\frac{1}{\Delta t} e_{x,t}[x_N(t+\Delta t)-x_N(t)]^{2+\delta} \to 0$$

and furthermore that the convergence is sufficiently rapid. To be more specific, setting

$$E_{x,t}[x_N(t+\Delta t)-x_N(t)] = a_N(x,t)\,\Delta t$$

$$E_{x,t}[x_N(t+\Delta t)-x_N(t)]^2 = b_N(x,t)\,\Delta t,$$

we assume that

$$\int_0^T \Delta t^{(1/2)+\alpha}\left|\frac{x_N(t+\Delta t)-x_N(t)}{\Delta t} - a_N(x_N(t),t)\right|^2 dt \to 0$$

and

$$\int_0^T \left|a_N(x_N(t),t) - a(x_N(t),t) + \sqrt{b_N(x_N(t),t)} - \sqrt{b(x_N(t),t)}\right|^2 dt \to 0$$

as $N \to \infty$. Furthermore, we assume that $a(x,t)$ and $\sqrt{b(x,t)}$ are bounded and continuous functions in $R \times [0,T]$ and that $a_N(x,t)$ and $b_N(x,t)$ are bounded functions. If the stochastic differential equation

$$dx(t) = a(x(t),t)\,dt + \sqrt{b(x(t),t)}\,dw(t)$$

has a unique solution, then $x_N(t) \to x(t)$ as $N \to \infty$ in the wak sense; that is, for each continuous function

$$E_x f(x_N(t)) \to E_x f(x(t)) \qquad \text{as} \quad N \to \infty.$$

[see Kushner (1974) for proof]. Consider, for example, the Wright–Fisher model (Feller 1951). Let two genes, A and a, determine a genetic type in a population. The three possible genotypes are assumed to be AA, aa, and Aa. Furthermore, we assume that the reproductive cells, or *gametes*, have only one gene; thus the gametes of an organism of genotype AA or aa have the gene A or a, respectively, whereas the gametes of an organism of genotype Aa may have the gene A or a with equal probability. An offspring receives one gene from each parent under the conditions of the Bernoulli scheme. More specifically, the genotype structure of N offspring is a result of $2N$ independent drawings from a set of $2N$ genes of types A and a. Suppose, now, that the population under consideration consists of N individuals in each generation. This may be achieved by an appropriate selection of organisms in each generation. If in some generation i of the genes are of type A$(0 \leqslant i \leqslant 2N)$, we say that the generation is in state E_i. Such a genetic process is clearly a Markov chain with $2N+1$ states $E_0, E_1, \dots, E_{2N}$. The transition probability from E_i to E_j in one generation is given by

$$P_{ij}^N = \binom{2N}{j}\left(\frac{i}{2N}\right)^j\left(1-\frac{i}{2N}\right)^{2N-j}. \tag{6.2.1}$$

Alternatively, we may think of a population of N individuals consisting of $X_N(n)=i$ individuals of type A in the nth generation. The next generation consists of N individuals randomly selected from a practically infinite offspring of the previous generation. Such is the case, for example, in a fish pond. The selection process is binomial with probability $x=i/N$ for type A. We are justified in making this assumption because the proportion i/N of A types is equal to the probability in a large offspring population. The transition probability is given by

$$P_{ij}^N = P(X_N(n+1)=j \mid X_N(n)=i) = \binom{N}{j} x^j (1-x)^{N-j}.$$

If selection forces act on the population, let s denote the *fitness* of A relative to a; that is,

$$P_{ij}^N = \binom{N}{j} x^{*j}(1-x^*)^{N-j},$$

where

$$\frac{x^*}{1-x^*} = (1+s)\frac{x}{1-x} \qquad \text{or} \qquad x^* = \frac{(1+s)x}{1+sx}.$$

If $s = s_N(n)$ is a random variable, the quantities of interest, such as the *probability of extinction* of a genotype, the *time until extinction*, the *total A population*, and so on, become very hard to calculate. In such a case we approximate the Markov chains $\{X_N(n)\}$ by a diffusion process, or more precisely, by a solution of a stochastic differential equation. To this end we rescale the process by setting

$$x_N(t) = \frac{X_N([Nt])}{N},$$

where $[Nt]$ is the greatest integer not exceeding Nt, and t is any positive number. Thus $x_N(t)$ represents the *proportion* of A types in population. We shall assume that

$$NEs_N(n)\to\sigma(t) \quad \text{as} \quad N\to\infty \qquad n=[Nt]$$

$$NEs_N^2(n)\to v(t) \quad \text{and} \quad NEs_N^k(n)\to 0 \quad \text{for all} \quad k>2.$$

Setting $\Delta t=1/N$, we get

$$a_N\equiv\frac{1}{\Delta t}E_{x,t}[x_N(t+\Delta t)-x_N(t)]$$
$$=E[X_N([Nt]+1)-X_N([Nt])|X_N([Nt])=xN].$$

Since $X_N(n)$ is $B(N, x^*)$, we have (cf. Chapter 1)

(6.2.2) $a_N=NE(x^*-x)$

$$=NE\left(\frac{(1+s_N(n))x}{1+s_N(n)x}-x\right)\to[\sigma(t)-v(t)x]x(1-x)\equiv a(x,t)$$

and similarly,

(6.2.3)

$$\frac{1}{\Delta t}E_{x,t}[x_N(t+\Delta t)-x_N(f)]^2\to x(1-x)[1+v(t)x(1-x)]\equiv b(x,t).$$

The higher moments tend to zero as $N\to\infty$.

To show that the convergence is sufficiently rapid as to satisfy the conditions for convergence specified at the beginning of this section, we consider the Wright–Fisher model again. We have

$$E[x_N(t+\Delta t)-x_N(t)|x_N(t)]=0$$

and

$$E[|x_N(t+\Delta t)-x_N(t)|^2|x_N(t)]=\frac{x_N(t)[1-x_N(t)]}{N}$$

$$E[|x_N(t+\Delta t)-x_N(t)|^4|x_N(t)]\leqslant\frac{K}{N^2}$$

where K is a constant. It can be easily seen now that the conditions for convergence hold, provided that the stochastic differential equation

$$dx(t)=a(x,t)\,dt+\sqrt{b(x,t)}\,dw(t) \tag{6.2.4}$$
$$x(0)=x$$

has a unique solution (with absorption at $x=0$ and $x=1$). Note that the coefficient $\sqrt{b(x,t)}$ in (6.2.4) does not satisfy the conditions of the existence and uniqueness theorem. To show existence and uniqueness, we consider (6.2.4) in the interval $I_\varepsilon=[\varepsilon,1-\varepsilon]$ with absorption at the boundary ∂I of I_ε. In I_ε the conditions of the existence and uniqueness theorem are satisfied; therefore, a unique solution $x_\varepsilon(t)$ in I_ε up to the time

$$\tau_\varepsilon=\inf\{t\,|\,x_\varepsilon(t)\in\partial I_\varepsilon\}.$$

If $\varepsilon_1<\varepsilon_2$, then $\tau_{\varepsilon_1}\geqslant\tau_{\varepsilon_2}$ and

$$x_{\varepsilon_1}(t)=x_{\varepsilon_2}(t), \qquad 0\leqslant t\leqslant\tau_{\varepsilon_2},$$

by the localization principle (cf. Exercise 4.2.3*). Taking the limit as $\varepsilon\to0$, we see that $x_\varepsilon(t)$ converges to a limit $x(t)$ and $\tau_\varepsilon\to\tau$, where τ is the absorption time of $x(t)$.

It follows that $x_N(t)\to x(t)$, where $x(t)$ is the solution of the stochastic differential equation $dx(t)=a(x,t)\,dt+\sqrt{b(x,t)}\,dw(t)$ with absorbing boundaries at $x=0$ and $x=1$. It is obvious that once a genotype is extinct, it will stay extinct for all future generations unless mutation occurs. Thus the probability of extinction is the probability of exit of $x(t)$ from the interval $(0,1)$.

EXERCISE 6.2.1

Prove (6.2.2) and (6.2.3) by expanding in Taylor's series for $|s_N(n)|<1$ and applying Chebyshev's inequality for $|s_N(n)|\geqslant1$ (Levikson 1974).

EXERCISE 6.2.2

Show that the assumption that the relative selection intensities of types AA, Aa, and aa, respectively, are given by $[1+s_N^1(n)]:1:[1+s_N^2(n)]$ leads to the drift coefficient

$$a(x,t)=x(1-x)\left[\sigma_1(t)-\sigma_2(t)-v_1(t)x+v_2(t)(1-x)+(2x-1)r(t)\right],$$

and diffusion coefficient

$$b(x,t)=x(1-x)[1+x(1-x)(v_1+v_2-2r)],$$

where

$$Es_N^i(n)\sim\frac{\sigma_i(t)}{N},\qquad E[s_N^i(n)]^2\sim\frac{v_i(t)}{N},$$

$$Es_N^1(n)s_N^2(n)\sim\frac{r(t)}{N},\qquad E|s_N^i(n)|^k=o\left(\frac{1}{N}\right)$$

for all $k>2$ and $n=[Nt]$.

EXERCISE 6.2.3

Assume that mutation occurs in the Wright–Fisher models with rate α_1 from A into a and α_2 from a into A. More precisely, x should be replaced by

$$x'=-x\alpha_1+(1-x)\alpha_2 \qquad \text{and} \qquad x^*=\frac{(1+s)x'}{1+sx'}.$$

Assume for simplicity that s_N is deterministic and

$$Ns_N\to\sigma,\qquad N\alpha_i\to\gamma_i\qquad (i=1,2)$$

as $N\to\infty$. Show $a(x)=\sigma x(1-x)-\gamma_1 x+\gamma_2(1-x), b(x)=x(1-x)$.

EXERCISE 6.2.4

Show that the degenerate *backward* parabolic boundary value problem

$$\frac{\partial u}{\partial t}+a(x,t)\frac{\partial u}{\partial x}+\tfrac{1}{2}b(x,t)\frac{\partial^2 u}{\partial x^2}=-1 \tag{6.2.4}$$

$$u(0,t)=u(1,t)=0$$

has a unique bounded solution.

(*Hint.* Show that the solutions $u_\varepsilon(x,t)$ of (6.2.4) in the interval $[\varepsilon,1-\varepsilon]$ such that $u_\varepsilon(\varepsilon,t)=u(1-\varepsilon,t)=0$ converge to a limit $u(x,t)$ and estimate the rate of convergence) (Levikson and Schuss 1977; Schuss 1977).

EXERCISE 6.2.5

Conclude from Exercise 6.2.4 that extinction of a genotype will occur in finite time with probability 1. Write down the differential equation satisfied by the extinction probability $u(x,t)=P$ (a will be extinct|proportion of A is x at time t):

$$\frac{\partial u}{\partial s}+Lu=0, \qquad s\geqslant t, \qquad u(0,s)=0, \qquad u(1,s)=1. \tag{6.2.5}$$

Solve (6.2.4) and (6.2.5) in the time-homogenous case. Use the solutions to estimate the probability and expected time of extinction:

$$E_{i,n_0}\tilde{n}_N=E(\tilde{n}\,|\,X_N(n_0)=i)\approx Nu(x,t),$$

where $u(x,t)$ is the solution of (6.2.4), $\tilde{n}_N$ is the number of generations until extinction, n_0 is the number of the initial generation, $t=n_0/N$, $x=i/N$. How good is this approximation? Compare to the Wright–Fisher model without selection.

EXERCISE 6.2.6

Show that the total A population may be approximated by

$$v(x,t)=N^2E_{x,t}\int_t^{\tau_x}x(s)\,ds,$$

where $\tau_x=\inf\{s\geqslant t\,|\,x(s)=0 \text{ or } x(s)=1,\ x(t)=x\}$ and show that $v(x,t)$ is the solution of the equation

$$\frac{\partial v}{\partial t}+Lv=-N^2x, \qquad v(0,t)=v(1,t)=0.$$

EXERCISE 6.2.7

Assume that selection decays exponentially, that is, $\sigma(t)=\sigma e^{-\alpha t}$, and let $v(t)=0$, that is, $a(x,t)=\sigma e^{-\alpha t}x(1-x)$ and $b(x,t)=x(1-x)$. Derive the following expansion for the expected exit time $w(x,t)$[$w(x,t)$ is the solution of (6.2.4)]

$$w(x,t)=\sum_{n=0}^{\infty}w_n(x)e^{-n\alpha t}, \tag{6.2.6}$$

where

$$\tfrac{1}{2}x(1-x)\frac{\partial^2 w_0}{\partial x^2} = -1, \qquad w_0(0)=w_0(1)=0;$$

that is, $w_0(x) = -2[x\ln x + (1-x)\ln(1-x)]$ and

$$\tfrac{1}{2}x(1-x)\frac{\partial^2 w_n}{\partial x^2} - \alpha n w_n(x) = -x(1-x)\frac{\partial w_{n-1}(x)}{\partial x}$$

$$w_n(0)=w_n(1)=0.$$

Show that

$$\int_0^1 w_n^2(x)\,dx \leqslant \frac{C^n}{n!}$$

where

$$C = \text{constant};$$

hence the series (6.2.6) converges uniformly for $t \geqslant 0$. Show that if $\sigma(t) = \sigma/(1+t)$, then the asymptotic series

$$w(x,t) \sim \sum_{n=0}^{\infty} w_n(x) t^{-n} \qquad (t \gg 1)$$

is *not* convergent. Compute $w_n(x)$ and show that

$$\left| w(x,t) - \sum_{n=0}^{m} w_n(x) t^{-n} \right| = O\left(t^{-(m+1)}\right) \qquad \text{as} \quad t \to \infty.$$

EXERCISE 6.2.8

Show that if we condition the process on exit at the right end point of the interval, we obtain a diffusion process with the same coefficient $b(x)$ as above, but with drift term modified to be

$$a^*(x) = a(x) + \frac{b(x)u'(x)}{u(x)},$$

where $u(x)$ is the probability of exit at the right end point (Ewens 1973).

EXERCISE 6.2.9*

Read Feller's papers (1952, 1954) and Mandl's book (1968).

EXERCISE 6.2.10

Consider the case of selective advantage due to fertility difference between the a and A types. Let α and β denote positive real numbers, fix the population size at n, suppose that in the current generation there are j gametes of type a and $n-j$ of type A, and define $p_j = j/k$. Let the transition probability matrix be defined by

$$p_{ij} = \binom{n}{j} \tilde{p}_i^j (1-\tilde{p}_i)^{n-j},$$

where we define

$$\tilde{p}_j = p_j + \frac{f_n(p_j)}{n}$$

and

$$f_n(p) = \frac{\beta pq(\alpha(p-q)+q)}{1-\beta q(2\alpha p+q)/n}, \qquad q = 1-p.$$

The parameters α and β are used to represent differences in fertility. Find the approximating stochastic differential equation and show that the conditions for convergence are satisfied (Crow and Kimura 1970).

CHAPTER 7

The Exit Problem and Singular Perturbations

7.1 SMALL DIFFUSION IN A FLOW

The cumulative effect on dynamical systems, of even very small random perturbations, may be considerable after sufficiently long times. For example, even if the corresponding deterministic system has an asymptotically stable equilibrium point, random effects can cause the trajectories of the system to leave any bounded domain with probability 1 (see Section 5.4). Consider the effect of a small random perturbation of white noise type, on a (deterministic) dynamical system

$$\frac{d\mathbf{x}}{dt}=\mathbf{b}(\mathbf{x}). \tag{7.1.1}$$

The vector $\mathbf{x}=(x_1,\ldots,x_n)^T$ then becomes a stochastic process $\mathbf{x}_\varepsilon(t)$ which satisfies the Itô stochastic differential equation

$$d\mathbf{x}_\varepsilon=\mathbf{b}(\mathbf{x}_\varepsilon)\,dt+\varepsilon\boldsymbol{\sigma}\,d\mathbf{w}(t), \tag{7.1.2}$$

where $\mathbf{w}(t)$ is the n-dimensional Brownian motion, $\mathbf{b}(\mathbf{x})$ is a vector field, $\boldsymbol{\sigma}(\mathbf{x})$ is the diffusion matrix, and $\varepsilon\neq 0$ is a small real parameter. In this chapter we shall study the following problem, originally posed by Kolmogorov: find the asymptotic expansion in ε of (i) the probability distribution of the points on the boundary of a domain, where trajectories of the perturbed system first exit, and (ii) of the expected exit time. To illustrate the nature of such problems, it is convenient to think of the random perturbations as a slow diffusion of particles in a deterministic flow field. Naturally, the results will depend on the nature of the underly-

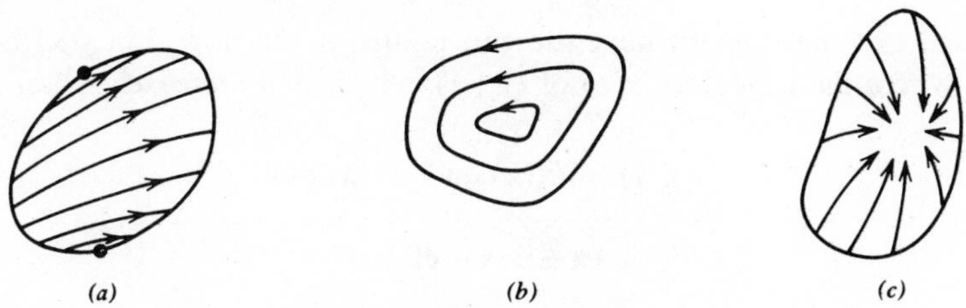

Figure 7.1.1 Flow fields

ing flow. We distinguish three cases according as the particles are diffusing (a) with a flow, (b) across a flow, or (c) against the flow (see Fig. 7.1.1). We shall relate this problem to the solution of certain singularly perturbed elliptic boundary value problems as described in Chapter 4.

7.2 DIFFUSION WITH A FLOW AND DIFFUSION WITH REFLECTION

Let Ω be a domain in R^n and assume that the flow lines of the vector field $\mathbf{b}(\mathbf{x})$ enter and leave Ω (Fig. 7.1.1*a*). More precisely, let $\mathbf{x}(t)$ be a solution of (7.1.1) such that $\mathbf{x}(0)=\mathbf{x}\in\Omega$. Assume that for some finite $t_1>0$ and $t_2\geqslant 0$, $x(t_i)\in\partial\Omega$ $(i=1,2)$. Let $\boldsymbol{\sigma}(\mathbf{x})$ be a nonsingular diffusion matrix in $\overline{\Omega}$ and set

$$a_{ij}(x)=\tfrac{1}{2}(\sigma\sigma^*)_{ij}=\tfrac{1}{2}\sum_{k=1}^{n}\sigma_{ik}\sigma_{kj}.$$

We shall assume for simplicity that $a_{ij}=\delta_{ij}$. Let $u_\varepsilon(\mathbf{x},t)$ be the solution of the boundary value problem

$$L_\varepsilon u_\varepsilon=\varepsilon^2\Delta u_\varepsilon+b(\mathbf{x})\cdot\nabla u_\varepsilon=0 \qquad \text{in } \Omega \tag{7.2.1}$$

$$u_\varepsilon(\mathbf{x})=f(\mathbf{x}) \qquad \text{on } \partial\Omega,$$

where $f(\mathbf{x})$ is a smooth function on $\partial\Omega$ [cf. (5.4.8)]. Then, by (5.4.9),

$$u_\varepsilon(\mathbf{x})=E_{\mathbf{x}}f(\mathbf{x}_\varepsilon(\tau))=\int_{\partial\Omega}p_\varepsilon(\mathbf{x},\mathbf{y})f(\mathbf{y})\,dS_y.$$

It is intuitively clear that the deviation of $\mathbf{x}_\varepsilon(t)$ from the deterministic trajectory $\mathbf{x}(t)$, if $\mathbf{x}_\varepsilon(0)=\mathbf{x}(0)=\mathbf{x}\in\Omega$, will be small if the random fluctuations are small. The diffusing particles will therefore be carried by the flow and will exit the domain near the exit points of the flow. Let $\mathbf{y}(\mathbf{x})$ be the point where the trajectory $\mathbf{x}(t)$ of (7.1.1) exits Ω as t increases. Then

$$p_\varepsilon(\mathbf{x},\mathbf{y}) \to \delta(\mathbf{y}(\mathbf{x})-\mathbf{y}) \qquad \text{as } \varepsilon \to 0$$

$$(\mathbf{x}\in\Omega,\ \mathbf{y}\in\partial\Omega).$$

Hence the solution $u_\varepsilon(\mathbf{x})$ of (7.2.1) will converge to

(7.2.2) $$u(\mathbf{x}) = \int_{\partial\Omega} \delta(\mathbf{y}(\mathbf{x})-\mathbf{y}) f(\mathbf{y})\, dS_{\mathbf{y}} = f(\mathbf{y}(\mathbf{x})).$$

Formula (7.2.2) implies that $u(\mathbf{x})$ is the solution of the *reduced equation*

(7.2.3) $$\mathbf{b}(\mathbf{x})\cdot\nabla u = 0 \qquad \text{in } \Omega$$

$$u(\mathbf{x}) = f(\mathbf{x}) \qquad \text{on } \partial\Omega_1,$$

where $\partial\Omega_1$ is that portion of $\partial\Omega$ where the trajectories of (7.1.1) leave Ω. Indeed, (7.2.3) can be written in the form

(7.2.4) $$\frac{d}{dt} u(\mathbf{x}(t)) = 0,$$

where $\mathbf{x}(t)$ is a solution of (7.1.1). Hence $u(\mathbf{x}(t))$ is constant on the *characteristics* $\mathbf{x}(t)$ and the values of the constant is the value $f(\mathbf{y}(\mathbf{x}))$. A more precise mathematical approach is based on the method of boundary layers and asymptotic expansions for (7.2.1) (the ray method). Since the reduced equation (7.2.3) is of first order, boundary conditions can be imposed at one end of each characteristic only. The solution $u_\varepsilon(\mathbf{x})$ of the Dirichlet problem (7.2.1) satisfies boundary conditions on *all* of $\partial\Omega$; therefore, a boundary layer will develop near $\partial\Omega_2 = \partial\Omega - \partial\Omega_1$, that is, at points where the field $\mathbf{b}(\mathbf{x})$ points *into* Ω. Such points are characterized by the condition

(7.2.5) $$b(\mathbf{x})\cdot\boldsymbol{\nu}(\mathbf{x}) < 0, \qquad \mathbf{x}\in\partial\Omega_2,$$

where $\boldsymbol{\nu}(\mathbf{x})$ is the outer normal to $\partial\Omega$ at $\mathbf{x}$. We shall therefore seek a solution of the form

$$u_\varepsilon(\mathbf{x}) = u(\mathbf{x}) + e^{-g/\varepsilon^2}\left(h + \varepsilon^2 h_1 + \cdots\right).$$

We get from (7.2.1)

$$\frac{e^{-g/\varepsilon^2}h}{\varepsilon^2}\left[|\nabla g|^2-\mathbf{b}\cdot\nabla g\right]+e^{-g/\varepsilon^2}(b-2\nabla g)\cdot\nabla h+O(\varepsilon^2). \tag{7.2.6}$$

The function g is found by setting the first bracket in (7.2.6) equal to zero, giving

$$|\nabla g|^2=\mathbf{b}\cdot\nabla g. \tag{7.2.7}$$

The boundary layer effect occurs if $g(\mathbf{x})=0$ for $\mathbf{x}\in\partial\Omega_2$ and $g(\mathbf{x})>0$ for all $\mathbf{x}\in\Omega$. The function $h(\mathbf{x})$ is the solution of

$$(\mathbf{b}-2\nabla g)\cdot\nabla h-\Delta g h=0 \quad \text{in} \quad \Omega \tag{7.2.8}$$

and h satisfied the boundary condition

$$h(\mathbf{x})=f(\mathbf{x})=u(\mathbf{x}) \quad \text{on} \quad \partial\Omega_2. \tag{7.2.8$'$}$$

It is necessary to construct g and h in a neighborhood of $\partial\Omega_2$ as $e^{-g/\varepsilon^2}h(\mathbf{x})$ is transcendentally small outside the boundary layer $\{\mathbf{x}|\text{dist}(\mathbf{x},\partial\Omega_2)<\varepsilon\}$; that is, $e^{-g/\varepsilon}h(\mathbf{x})\varepsilon^{-n}\to 0$ as $\varepsilon\to 0$ for all $n>0$. Equation (7.2.7) is solved by the method of characteristics, that is, by solving the system (cf. (10.2.7), (10.2.8))

$$\frac{d\mathbf{x}}{dt}=2\mathbf{p}-\mathbf{b}; \qquad \frac{d\mathbf{p}}{dt}=\nabla\mathbf{b}\cdot\mathbf{p}; \qquad \frac{dg}{dt}=|\mathbf{p}|^2 \tag{7.2.9}$$

where

$$\mathbf{p}=\nabla g \quad \text{and} \quad (\nabla\mathbf{b}\cdot\mathbf{p})_i=\frac{\partial\mathbf{b}}{\partial x_i}\cdot\mathbf{p}.$$

The variable $\mathbf{x}(t)$ is determined on $\partial\Omega_2$ as the coordinate of a boundary point; the variable $\mathbf{p}$ is determined in such a way that the characteristic curves $\mathbf{x}(t)$ of (7.2.8) enter Ω at $\partial\Omega_2$, for example, $\mathbf{p}=(\mathbf{b}\cdot\boldsymbol{\nu})\boldsymbol{\nu}$ on $\partial\Omega_2$ (Levinson 1950).

The function h is found by integration of (7.2.8) along the characteristics $\mathbf{x}(t)$; since (7.2.8) can be written in the form

$$\frac{dh}{dt}(\mathbf{x}(t))+g(\mathbf{x}(t))h=0,$$

we have

$$h(\mathbf{x}(t)) = h(\mathbf{x}(0))\exp\left[-\int_0^t \Delta g(\mathbf{x}(s))\,ds\right],$$

where $h(\mathbf{x}(0)) = f(\mathbf{x}) = u(\mathbf{x})$ by (7.2.8) [here $\mathbf{x}(0) = \mathbf{x} \in \partial\Omega_2$]. This construction shows that the intuitive description given above is correct.

EXERCISE 7.2.1

Show that the expected time of exit $E_{\mathbf{x}}\tau^{\varepsilon} \to \tau_{\mathbf{x}}$, where $\tau_{\mathbf{x}}$ is the time that the trajectory $\mathbf{x}(t)$ that starts at $\mathbf{x}$ needs to leave the domain [at $\mathbf{y}(\mathbf{x})$]. Derive an asymptotic expansion for the corresponding boundary value problem (5.4.5) in the time-homogeneous case.

EXERCISE 7.2.2

Consider the case $\mathbf{b}\cdot\boldsymbol{\nu} > 0$ on $\partial\Omega$, that is, the vector field $\mathbf{b}$ points outside Ω at all points of $\partial\Omega$. In this case there is at least one critical point $\mathbf{x}_0 \in \Omega$, where $\mathbf{b}(\mathbf{x}_0) = 0$ (without loss of generality, assume that only one such point $\mathbf{x}_0$ exists and $\mathbf{x}_0 = 0$). No boundary layers develop in this case, but since all characteristics $\mathbf{x}(t)$ converge to $\mathbf{x}_0 = 0$ as $t \to -\infty$, the solution $u(\mathbf{x})$ of the reduced problem (7.2.3) must be discontinuous at $\mathbf{x}_0 = 0$. Construct an *inner layer* expansion $u_\varepsilon(\mathbf{x}) = l(\varepsilon, \mathbf{x}) + O(\varepsilon^2)$ near $\mathbf{x} = 0$, where $l(\varepsilon, \mathbf{x})$ is the inner layer; that is, $l(\varepsilon, \mathbf{x})$ is smooth and $l(\varepsilon, \mathbf{x}) \to u(\mathbf{x})$ as $\varepsilon \to 0$, $l(\varepsilon, x) = e^{g/\varepsilon^2}(h + \cdots)$ (Mizel 1957; Friedman 1976).

EXERCISE 7.2.3*

Consider the problem when *killing* affects the system (see Section 5.1) (Kamienomostskaya 1955; Livne and Schuss 1973; Oleinik 1952; Ladyzhenskaya 1957; Holland 1976; Levinson 1950).

EXERCISE 7.2.4*

Assume that *absorption* occurs on $\partial\Omega_1$, *reflection* on $\partial\Omega_2$; and killing may affect the system. Find the expansions of the quantities of interest (Livne and Schuss 1973; Kamienomostskaya 1952; Lions 1973).

EXERCISE 7.2.5*

Develop an analogous asymptotic theory for the Fokker–Planck equation and for the distribution of the exit time (5.4.6) (Aronson 1956, 1959; Kamienomostskaya 1955).

EXERCISE 7.2.6

(i) Consider the case of *reflection* at $\partial\Omega$. Show that the deterministic process $\mathbf{x}(t)$ defined by (7.2.1) with reflection at $\partial\Omega$ satisfies the differential equation for all times t such that $\mathbf{x}(t)\in\Omega$ and for all t

$$\mathbf{x}(t)=\mathbf{x}(t_0)+\int_{t_0}^{t}\chi_{\bar{\Omega}}(\mathbf{x}(s))\mathbf{b}(\mathbf{x}(s))\,ds+\boldsymbol{\xi}(\mathbf{x},t),$$

where $\mathbf{x}(0)=\mathbf{x}$, $t_0=\inf\{t\,|\,\mathbf{x}(t_0)\in\partial\Omega\}$, and

$$\boldsymbol{\xi}(\mathbf{x},t)=-\int_0^t\chi_{\partial\Omega}(\mathbf{x}(s))\boldsymbol{\nu}(\mathbf{x}(s))\left[\mathbf{b}(\mathbf{x}(s))\cdot\boldsymbol{\nu}(\mathbf{x}(s))\vee 0\right]ds,$$

$$\boldsymbol{\xi}(\mathbf{x},0)=0$$

that is,

$$\frac{d\mathbf{x}(t)}{dt}=\mathbf{b}(\mathbf{x})-\chi_{\partial\Omega}(\mathbf{x})\boldsymbol{\nu}(\mathbf{x})\left[\mathbf{b}(\mathbf{x})\cdot\boldsymbol{\nu}(\mathbf{x})\vee 0\right].$$

Thus $\dot{\mathbf{x}}(t)$ is tangent to the boundary at points $\mathbf{x}\in\partial\Omega$ such that $\mathbf{b}\cdot\boldsymbol{\nu}>0$ and is equal to the tangential component of $\mathbf{b}(\mathbf{x})$; $\dot{\mathbf{x}}(t)=\mathbf{b}(\mathbf{x})$ at boundary points $\mathbf{x}$ such that

$$\mathbf{b}(\mathbf{x})\cdot\boldsymbol{\nu}(\mathbf{x})\leqslant 0.$$

(ii*) The analogous construction for (7.1.2) determines a random reflection process $\boldsymbol{\xi}_\varepsilon(\mathbf{x},t)$ such that $\boldsymbol{\xi}_\varepsilon(\mathbf{x},t)$ is measurable with respect to $\mathscr{F}_t$ (Gihman and Skorokhod 1972), continuous a.s., $\boldsymbol{\xi}_\varepsilon(\mathbf{x},0)=\mathbf{0}$, $\boldsymbol{\xi}_\varepsilon(\mathbf{x},t)$ is constant if $\mathbf{x}_\varepsilon(t)\in\Omega$ $\boldsymbol{\xi}_\varepsilon(\mathbf{x},t)\cdot\boldsymbol{\nu}(\mathbf{x}_\varepsilon(t))$ decreases on the set $\{t\,|\,\mathbf{x}_\varepsilon(t)\in\partial\Omega\}$. Equation (7.1.2) takes the form

(7.2.10) $$d\mathbf{x}_\varepsilon(t)=\mathbf{b}(\mathbf{x}_\varepsilon)\,dt+\varepsilon\boldsymbol{\sigma}(\mathbf{x}_\varepsilon)\,d\mathbf{w}+\chi_{\partial\Omega}(\mathbf{x}_\varepsilon)\boldsymbol{\nu}(\mathbf{x}_\varepsilon)\,d\xi_\varepsilon(\mathbf{x},t).$$

Show that

$$\mathbf{x}_\varepsilon(t)\to\mathbf{x}(t) \qquad \text{and} \qquad \xi_\varepsilon(\mathbf{x},t)\to\xi(\mathbf{x},t)$$

as $\varepsilon\to 0$ (Gihman and Skorokhod 1972; Anderson and Orey 1976).

(iii) Use Itô's formula for (7.2.10) to show that the solution u_ε of the Neumann problem for $\lambda>0$,

$$L_\varepsilon u-\lambda u=0 \quad \text{in} \quad \Omega$$

$$\frac{\partial u}{\partial \boldsymbol{\nu}}=g \quad \text{on} \quad \partial\Omega$$

can be represented by

$$u_\varepsilon(\mathbf{x})=E_{\mathbf{x}}\left[\int_0^\infty e^{-\lambda t}g(\mathbf{x}_\varepsilon(t))\,d_t\xi_\varepsilon(\mathbf{x},t)\right].$$

Conclude that

$$u_\varepsilon(\mathbf{x})\to u(\mathbf{x})=\int_0^\infty e^{-\lambda t}g(\mathbf{x}(t))\,d_t\xi(\mathbf{x},t),$$

where $\mathbf{x}(t)$ and $\xi(\mathbf{x},t)$ are defined in (ii*).

(iv) Consider the mixed boundary value problem

$$L_\varepsilon u=0 \quad \text{in} \quad \Omega \tag{7.2.11}$$

$$\lambda u+\frac{\partial u}{\partial \boldsymbol{\nu}}=h \quad \text{on} \quad \partial\Omega, \quad \lambda>0.$$

Show that the solution u_ε of (7.2.11) can be represented by

$$u_\varepsilon(\mathbf{x})=E_{\mathbf{x}}\left[\int_0^\infty h(\mathbf{x}(t))e^{-\lambda\xi_\varepsilon(\mathbf{x},t)}d_t\xi_\varepsilon(\mathbf{x},t)\right]$$

and

$$u_\varepsilon(\mathbf{x})\to u(\mathbf{x})=\int_0^\infty h(\mathbf{x}(t))e^{-\lambda\xi(x,t)}d_t\xi(\mathbf{x},t)$$

[note that $\xi_\varepsilon(x,t)\to\infty$ as $t\to\infty$ (Anderson and Orey 1976)].

7.3 SMALL DIFFUSION ACROSS THE FLOW

Consider (7.1.2) in case (7.1.1) has a center at $\mathbf{x}=0$ and $\partial\Omega$ is a closed characteristic (integral curve) of (7.1.1). Then (7.1.2) describes a small diffusion *across* the flow (Figure 7.1.1*b*). A simple example of the singular perturbation problem (7.2.1) is given by

$$\varepsilon^2 \Delta u+f(r)\frac{\partial u}{\partial \theta}=0 \qquad \text{in } R_1<r<R_2 \tag{7.3.1}$$

$$u(R_i,\theta)=\phi_i(\theta) \qquad (i=1,2)$$

in polar coordinates. The solution of the reduced problem is $u_0=u(r)$. Integrating (7.3.1) with respect to θ from 0 to 2π, we obtain

$$\varepsilon^2\Delta\hat{u}=\varepsilon^2\left(\hat{u}_{rr}+\frac{1}{r}\hat{u}_r\right)=0,$$

where

$$\hat{u}(r)=\hat{u}_\varepsilon(r)=\int_0^{2\pi}u(r,\theta)\,d\theta.$$

Thus

$$\hat{u}_\varepsilon(r)=c_1\ln r+c_2;$$

hence

$$u_\varepsilon(r,\theta)\to u_0(r)=a\ln r+b,$$

where

$$a=\frac{1}{2\pi}\int_0^{2\pi}\frac{\phi_2(\theta)-\phi_1(\theta)}{\ln R_2-\ln R_1}\,d\theta$$

and

$$b=\frac{1}{2\pi}\int_0^{2\pi}\frac{\phi_1\ln R_2-\phi_2\ln R_1}{\ln R_2-\ln R_1}\,d\theta.$$

It follows that

$$p_\varepsilon(R_1,\theta,r,\theta_0)\to\frac{1}{2\pi}\,\frac{\ln R_2-\ln r}{\ln R_2-\ln R_1}$$

and

$$p_\varepsilon(R_2,\theta,r,\theta_0)\to\frac{1}{2\pi}\frac{\ln r-\ln R_1}{\ln R_2-\ln R_1}\qquad\text{as } \varepsilon\to 0.$$

If $R_1=0$ and $\Omega=\{r<R\}$, then $c_1=0$; hence

$$u_0(r)=\text{constant}=\frac{1}{2\pi}\int_0^{2\pi}\phi(\theta)\,d\theta$$

and

$$p_\varepsilon(R,\theta,r,\theta_0)\to\frac{1}{2\pi}\qquad\text{as } \varepsilon\to 0.$$

The expected exit time $E_x\tau_\varepsilon$ must obviously become infinite as $\varepsilon\to 0$. It is the solution of Dynkin's equation (5.4.5): namely,

$$\text{(7.3.2)}\qquad L_\varepsilon u=\varepsilon^2\,\Delta u+f(r)\frac{\partial u}{\partial\theta}=-1,\qquad R_1<r<R_2$$

$$u(R_i,\theta)=0\qquad (i=1,2).$$

It can be readily seen that $u=u_\varepsilon$ is independent of θ; hence

$$E_{r,\theta}\tau_\varepsilon=u_\varepsilon(r)=\frac{\pi}{2\varepsilon^2}\left\{\frac{R_1^2\ln R_2-R_2^2\ln R_1}{\ln R_2-\ln R_1}+\frac{R_2^2-R_1^2}{\ln R_2-\ln R_1}\ln r-r^2\right\}.$$

If, in particular $\Omega=\{r<R\}$, then $c_1=0$ and

$$E_{r,\theta}\tau_\varepsilon=\frac{\pi}{2\varepsilon^2}(R^2-r^2),$$

so

$$E_0\tau_\varepsilon=\frac{\pi}{2\varepsilon^2}R^2$$

(Khasminskii 1963).

EXERCISE 7.3.1

(i). Write down the equation (7.1.2) that corresponds to the operator L_ε in (7.3.2) and write down the Fokker–Planck equation for this case.

(ii). Use separation of variables to construct

$$p_\varepsilon(r,\theta,r_0,\theta_0,t)=\sum_{n=1}^{\infty} c_n J_0\left(\frac{\sqrt{\lambda_n}}{\varepsilon}r\right)J_0\left(\frac{\sqrt{\lambda_n}}{\varepsilon}r_0\right)r_0 e^{-\lambda_n t}$$

where λ_n are the *eigenvalues* of L_ε [i.e., $L_\varepsilon u_n+\lambda_n u_n=0$, $u_n(R_i)=0$, $i=1,2$]. Here $J_0(x)$ is Bessel's function of order zero. Assume that $p_\varepsilon(r,\theta,r_0,\theta_0,0)=\delta(r-r_0)$. Show that $\lambda_n=\varepsilon^2 x_n^2/R^2$, where x_n is the nth zero of $J_0(x)$, and show that

$$E_0\tau_\varepsilon=\sum_{n=1}^{\infty}\frac{-J_0'(x_n)}{x_n J_1^2(x_n)}\frac{1}{\lambda_n}$$

$$=\frac{R^2}{\varepsilon^2}\sum_{n=1}^{\infty}\frac{-J_0'(x_n)}{x_n^3 J_1^2(x_n)}=\frac{\pi R^2}{2\varepsilon^2}.$$

7.4 SMALL DIFFUSION AGAINST THE FLOW

Consider (7.1.2) in case (7.1.1) has a stable equilibrium point in Ω. Without loss of generality we assume that $\mathbf{b}(\mathbf{0})=\mathbf{0}\in\Omega$ and $\mathbf{b}\cdot\boldsymbol{\nu}<0$ on $\partial\Omega$; that is, $\mathbf{b}$ points into Ω at $\partial\Omega$. The characteristics $\mathbf{x}(t)$ converge to the origin so that (7.1.2) describes small diffusion against the flow (cf. Fig. 7.1.1*c*).

(7.4.1) $$L_\varepsilon u=\varepsilon^2\sum_{k,j=1}^{n} a_{ij}(\mathbf{x})\frac{\partial^2 u}{\partial x_i\partial x_i}+\sum_{i=1}^{n} b_i(\mathbf{x})\frac{\partial u}{\partial x_i}=0 \qquad \text{in } \Omega$$

$$u(\mathbf{x})=f(\mathbf{x}) \qquad \text{on } \partial\Omega$$

and the reduced equation is

(7.4.2) $$Lu=\sum b_i(\mathbf{x})\frac{\partial u}{\partial x_i}\equiv\mathbf{b}\cdot\nabla u=0.$$

The solution $u_0(\mathbf{x})$ of the reduced equation (7.4.2) in this case must be constant along characteristics. If $u_\varepsilon(\mathbf{x})\to u_0(\mathbf{x})$ and the derivatives of $u(\mathbf{x})$ remain bounded in a neighborhood of the origin, then no discontinuities develop in the solution $u_\varepsilon(\mathbf{x})$ as $\varepsilon\to 0$, so that $u_0(\mathbf{x})$ must be continuous near the origin $\mathbf{0}$. Since all characteristics meet at $\mathbf{0}$ and $u_0(\mathbf{x})$ is constant on characteristics and continuous at $\mathbf{0}$, $u_0(\mathbf{x})$ must be constant in Ω, $u_0(\mathbf{x})=C_0$, say. The following example shows that this is the case. Let $\mathbf{b}(\mathbf{x})=$

$-(x_1, x_2)$ in the plane; then the Dirichlet problem (7.4.1), (7.4.2) is

$$\varepsilon\Delta u_\varepsilon - r\frac{\partial u_\varepsilon}{\partial r} = 0 \quad \text{in } \Omega \tag{7.4.3}$$

$$u = f \quad \text{on } \partial\Omega,$$

where Ω is a starlike domain containing the origin (i.e., any line through the origin intersects $\partial\Omega$ at a single point). Equation (7.4.3) can now be written in the form

$$\nabla\cdot\left(e^{-r^2/2\varepsilon}\nabla u_\varepsilon\right) = 0;$$

hence, multiplying by u_ε and integrating over the circle $\{r \leqslant R\}$, we obtain

$$\tfrac{1}{2}e^{-R^2/2\varepsilon}\int_0^{2\pi}\frac{\partial u^2}{\partial R}(R,\phi)\,d\phi = \int_0^{2\pi}\int_0^R e^{-r^2/2\varepsilon}|\nabla u|^2 r\,dr\,d\phi.$$

Since the last integrand is nonnegative, the value of the integral decreases if the domain of integration is decreased. So

$$\int_0^{2\pi}\frac{\partial u^2}{\partial R}(R,\phi)\,d\phi \geqslant 2e^{R^2/2\varepsilon}\int_0^{R/2}\int_0^{2\pi} e^{-r^2/2\varepsilon}|\nabla u|^2 r\,dr\,d\phi \tag{7.4.4}$$

$$\geqslant 2e^{3R^2/8\varepsilon}\int_0^{2\pi}\int_0^{R/2}|\nabla u|^2 r\,dr\,d\phi.$$

Next, we multiply inequality (7.4.4) by R, integrate with respect to R from 0 to δ, and change order of integration. We obtain

$$\int_0^{2\pi}\int_0^{\delta} R\frac{\partial u^2}{\partial R}(R,\phi)\,dR\,d\phi \geqslant 2\int_0^{2\pi}\int_0^{\delta/2}\int_{2r}^{\delta} e^{3R^2/8\varepsilon}R\,dR\,|\nabla u|^2 r\,dr\,d\phi$$

$$\geqslant \frac{16\varepsilon}{3}\int_0^{2\pi}\int_0^{\delta/3}\left(e^{3\delta^2/8\varepsilon} - e^{3r^2/2\varepsilon}\right)|\nabla u|^2 r\,dr\,d\phi$$

$$\geqslant \frac{16}{3}\varepsilon\int_0^{2\pi}\int_0^{\delta/3}\left(1 - e^{-5\delta^2/24\varepsilon}\right)|\nabla u|^2 r\,dr\,d\phi.$$

If ε is sufficiently small, we obtain

$$\delta\int_0^{2\pi}u^2(\delta,\phi)\,d\phi - \int_0^{2\pi}\int_0^{\delta}u^2(r,\phi)r\,dr\,d\phi \geqslant \tfrac{8}{3}\varepsilon e^{3\delta^2/8\varepsilon}\int_0^{2\pi}\int_0^{\delta/3}|\nabla u|^2 r\,dr\,d\phi.$$

The maximum principle for (7.4.3) implies that $\max_{\bar{\Omega}}|u_\varepsilon| \leqslant \max_{\partial\Omega}|f(\mathbf{x})|$; hence the last inequality leads to

$$\int_0^{2\pi}\int_0^{\delta/3}|\nabla u_\varepsilon|^2 r\,dr\,d\phi \leqslant e^{-\delta^2/4\varepsilon}\max_{\partial\Omega}|f|^2, \tag{7.4.5}$$

where C is a positive constant independent of ε and $f(\boldsymbol{x})$. Similar inequalities can be obtained for derivatives of u of all orders; hence, by Sobolev's inequality (Friedman 1964),

$$\nabla u_\varepsilon \to 0 \qquad \text{as } \varepsilon \to 0.$$

EXERCISE 7.4.1

Derive inequality (7.4.5) in n dimensions. [For the proof in the general case, see Kamin (1978).]

The leading term in the outer expansion of $u_\varepsilon(\mathbf{x})$ is therefore the constant $u_0(\mathbf{x}) = C_0$. The outer expansion is to be valid in the interior of Ω, but not near $\partial\Omega$, since it cannot in general satisfy the given boundary condition (7.4.2). Therefore, we construct a boundary layer expansion by introducing local coordinates in a neighborhood of the boundary. Specifically, let $\rho(\mathbf{x})$ be a smooth function in $\bar{\Omega}$ such that $\rho = 0$ on $\partial\Omega$, $\rho < 0$ in Ω, and $\rho(\mathbf{x}) = -\text{dist}(\mathbf{x}, \partial(\Omega)$ for any $\mathbf{x}$ near $\partial\Omega$. Then we introduce the coordinate system $\mathbf{y} = (y_1, y_2, \ldots, y_{n-1}, z)^T$ (in n dimensions), which we denote by $(\mathbf{y}', z)$, where, with no loss of generality, we take $y_i = x_i$ $(i = 1, \ldots, n-1)$ and $y_n = z = \rho(\mathbf{x})$.

Thus $y_i (i \leqslant n-1)$ are tangential variables in the plane $z = 0$, while z represent a normal variable defined so that $\nabla\rho = \boldsymbol{\nu}$ on $\partial\Omega$, where the outer normal to $\partial\Omega$ is $\boldsymbol{\nu} = (0, \ldots, 0, 1)$ in the new coordinate system. We now employ the stretching transformation $\eta = z/\varepsilon^2$ and seek solutions of the form

$$u_\varepsilon \sim \sum_j U_j(\mathbf{y}', \eta)\varepsilon^j. \tag{7.4.6}$$

Inserting (7.4.6) into (7.4.1) written in the new coordinate system, and equating the coefficient of each power of ε separately to zero, we find that the leading term U_0 satisfies the ordinary differential equation

$$a\frac{\partial^2 U_0}{\partial \eta^2} + b^0\frac{\partial U_0}{\partial \eta} = 0, \qquad -\infty < \eta < 0, \tag{7.4.7}$$

where

$$a=a(y')=\sum_{i,j=1}^{n} a_{ij}(\mathbf{x})\nu_i\nu_j>0$$

and

$$b^0=b^0(\mathbf{y}')=\sum_{i=1}^{n} b_i(\mathbf{x})\nu_i<0 \qquad (\mathbf{x}=(\mathbf{y}',0)).$$

The solution of (7.4.7) that satisfies the boundary condition $U_0(\mathbf{y}',0)=f(\mathbf{y}')$ is given by

$$U_0=C_1(\mathbf{y}')+(f(\mathbf{y}')-C_1(\mathbf{y}'))e^{-b^0\eta/a}.$$

Matching the leading terms in the outer and boundary layer expansions then implies that $C_1(\mathbf{y}')=C_0$. Therefore, the leading term in the uniform expansion of $u_\varepsilon(\mathbf{x})$ is given by

$$u_\varepsilon(\mathbf{x})\sim C_0+(f(\mathbf{x})-C_0)e^{\zeta(\mathbf{x})/\varepsilon^2}, \tag{7.4.7$'$}$$

where $\zeta(\mathbf{x})\equiv -b^0\rho/a$ and the constant C_0 is undetermined. To determine C_0, and thus to determine the unique asymptotic form of $u_\varepsilon(\mathbf{x})$, we employ a condition that is obtained by taking the inner product of (7.4.1) with the solution Z_ε of the adjoint equation

$$L_\varepsilon^*Z=\varepsilon^2\sum_{i,j=1}^{n}\frac{\partial(a_{ij}Z)}{\partial x_i\partial x_j}-\sum_{i=1}^{n}\frac{\partial(b_iZ)}{\partial x_i}=0 \tag{7.4.8}$$

normalized so that $Z_\varepsilon(0)=1$. The constant C_0 is then determined by employing the asymptotic forms of u_ε and Z_ε in the identity

$$\int_\Omega Z_\varepsilon L_\varepsilon u_\varepsilon\,dx=0. \tag{7.4.9}$$

We now construct the asymptotic expansion of Z_ε by employing the *ray method* (Ludwig 1975; Cohen and Lewis 1967). Thus we seek Z_ε in the form

$$Z_\varepsilon=e^{\phi/\varepsilon^2}w_\varepsilon(\mathbf{x}), \tag{7.4.10}$$

where

$$w_\varepsilon(\mathbf{x}) \sim \sum_{j=0}^{\infty} w_j(\mathbf{x})\varepsilon^j \tag{7.4.11}$$

with $\phi(\mathbf{0})=0$, $w_0(\mathbf{0})=1$.

Inserting (7.4.10) and (7.4.11) into the adjoint equation (7.4.8) and equating the coefficient of each power of ε separately to zero, we obtain equations for ϕ and w_j. In particular, ϕ satisfies

$$\sum_{i,j=1}^{n} a_{ij}\frac{\partial\phi}{\partial x_i}\frac{\partial\phi}{\partial x_i} = \sum_{i=1}^{n} b_i\frac{\partial\phi}{\partial x_i}, \qquad \mathbf{x}\in\bar{\Omega}, \tag{7.4.12}$$

while the leading term w_0, in the expansion of w, satisfies the equation

$$\sum_{i=1}^{n}\sum_{j=1}^{n}\left(2a_{ij}\frac{\partial\phi}{\partial x_j}-b_i\right)\frac{\partial w_0}{\partial x_i} + \sum_{i=1}^{n}\left[\sum_{j=1}^{n}\left(a_{ij}\frac{\partial^2\phi}{\partial x_i\partial x_j}+2\frac{\partial(a_{ij})}{\partial x_i}\frac{\partial\phi}{\partial x_j}\right)-\frac{\partial b_i}{\partial x_i}\right]w_0=0. \tag{7.4.13}$$

The leading term in the expansion of u_ε, given by (7.4.7′), and the leading term in the expansion of Z_ε, given by the solutions of (7.4.12) and (7.4.13), are now employed in (7.4.9). After integrating by parts and retaining only the leading terms, we obtain

$$\int_{\partial\Omega} e^{\phi/\varepsilon^2} w_0\left[(f(\mathbf{x})-C_0)\mathbf{b}\cdot\boldsymbol{\nu} - f(x)\left(\mathbf{b}\cdot\boldsymbol{\nu}-\frac{\partial\phi}{\partial n}\right)\right]dS \sim 0, \tag{7.4.14}$$

where the *conormal* derivative $\partial\phi/\partial n$ is defined by

$$\frac{\partial\phi}{\partial n} = \sum_{i,j} a_{ij}\nu_j\frac{\partial\phi}{\partial x_i}. \tag{7.4.14′}$$

Then, solving for C_0, we obtain

$$C_0 = \lim_{\varepsilon\to 0}\frac{\displaystyle\int_{\partial\Omega} e^{\phi/\varepsilon^2} w_0 f(\mathbf{x})\frac{\partial\phi}{\partial n}\,dS}{\displaystyle\int_{\partial\Omega} e^{\phi/\varepsilon^2} w_0\mathbf{b}\cdot\boldsymbol{\nu}\,dS} \tag{7.4.15}$$

The main contributions to the integrals in (7.4.15) come from points on $\partial\Omega$ where ϕ achieves its maximum. At such points $\nabla\phi/|\nabla\phi|=\boldsymbol{\nu}$, which together with (7.4.12) implies that $\partial\phi/\partial n=\mathbf{b}\cdot\boldsymbol{\nu}$ at points of maximum. Therefore,

$$C_0=\lim_{\varepsilon\to 0}\frac{\int_{\partial\Omega}e^{\phi/\varepsilon^2}w_0 f\,\mathbf{b}\cdot\boldsymbol{\nu}\,dS}{\int_{\partial\Omega}e^{\phi/\varepsilon^2}w_0\mathbf{b}\cdot\boldsymbol{\nu}\,dS}. \tag{7.4.16}$$

We note that the function w_0 is nonnegative on the boundary, since (7.4.13) may be written as a first-order ordinary differential equation along the characteristic directions. Thus

$$w_0(\mathbf{x}(t))=w_0(\mathbf{x}(0))\exp\left\{-\int_0^t\sum_{i=1}^n\left[\sum_{j=1}^n\left(a_{ij}\frac{\partial^2\phi}{\partial x_i\,\partial x_j}+2\frac{\partial(a_{ij})}{\partial x_i}\frac{\partial\phi}{\partial x_j}-\frac{\partial b_i}{\partial x_i}\right]ds\right\}, \tag{7.4.17}$$

where $\mathbf{x}(t)$ in (7.4.17) is the solution of the characteristic equations

$$\frac{dx_i}{dt}=\sum_{j=1}^n 2a_{ij}\frac{\partial\phi}{\partial x_j}-b_i \qquad (i=1,\dots,n) \tag{7.4.18}$$

(see Chapter 10), corresponding to (7.4.12). For any point $\mathbf{y}\in\partial\Omega$, let $\mathbf{x}(t,\mathbf{y})$ be the solution of (7.4.18) satisfying $\mathbf{x}(0,\mathbf{y})=\mathbf{y}$. We choose $\nabla\phi$ on $\partial\Omega$ so that $\mathbf{x}(t,\mathbf{y})\to 0$ as $t\to\infty$, and (7.4.12) is satisfied. This can be done by prescribing $\nabla\phi$ on the boundary $\partial\Omega'$ of a slightly bigger domain $\Omega'\supset\Omega$ so that $\mathbf{x}(t,\mathbf{y})\to 0$ as $t\to\infty$ and (7.4.12) is satisfied in Ω' and thus on $\partial\Omega$. Therefore, since $w_0(\mathbf{0})=1$, we have

$$w_0(\mathbf{y})=\exp\int_0^\infty\sum_{i=1}^n\left[\sum_{j=1}^n a_{ij}\frac{\partial^2\phi}{\partial x_i\,\partial x_j}+2\frac{\partial(a_{ij})}{\partial x_i}\frac{\partial\phi}{\partial x_j}-\frac{\partial b_i}{\partial x_i}\right]ds>0.$$

The constant C_0, and therefore the leading term in the uniform expansion of $u_\varepsilon(\mathbf{x})$, is now uniquely determined. By employing (5.4.9) or (5.4.9′), we can compute

$$\lim_{\varepsilon\to 0}P(\mathbf{x}_\varepsilon(\tau)=\mathbf{y}\,|\,\mathbf{x}_\varepsilon(0)=\mathbf{x})\equiv\lim_{\varepsilon\to 0}p_\varepsilon(\mathbf{x},\mathbf{y}) \qquad (\mathbf{x}\in\Omega,\ \mathbf{y}\in\partial\Omega) \tag{7.4.19}$$

from

$$\lim_{\varepsilon\to 0}\int_{\partial\Omega} f(\mathbf{y})p(\mathbf{x},\mathbf{y})\,dS_{\mathbf{y}}=C_0, \tag{7.4.20}$$

that is, from

$$\lim_{\varepsilon\to 0}\left[\int_{\partial\Omega} f(\mathbf{y})p(\mathbf{x},\mathbf{y})\,dS_{\mathbf{y}}-\frac{\int_{\partial\Omega} f(\mathbf{y})e^{\phi/\varepsilon^2}w_0\mathbf{b}\cdot\boldsymbol{\nu}\,dS_{\mathbf{y}}}{\int_{\partial\Omega} e^{\phi/\varepsilon^2}w_0\mathbf{b}\cdot\boldsymbol{\nu}\,dS_{\mathbf{y}}}\right]=0. \tag{7.4.21}$$

We note first that (7.4.20) implies that $\lim_{\varepsilon\to 0}p_\varepsilon(\mathbf{x},\mathbf{y})$ is independent of $\mathbf{x}$. We will compute this limit (in the sense of distributions) by asymptotically evaluating the integrals in (7.4.21), and as we shall see below, the limit can in fact be a distribution. Since the integrals in (7.4.21) are Laplace-type integrals, the major contributions come from points where $\phi(\mathbf{x})$ is maximum on $\partial\Omega$.

(i) If the maximum $\hat{\phi}$ of ϕ on $\partial\Omega$ is achieved on a set $U\subset\partial\Omega$, with nonempty interior U^0 (e.g., on an arc or a union of arcs in two dimensions), and the measure of $\partial\Omega=U-U^0$ in $\partial\Omega$ is zero, then

$$p(\mathbf{y})=\lim_{\varepsilon\to 0}p_\varepsilon(\mathbf{x},\mathbf{y})=\frac{\chi_U(\mathbf{y})w_0\mathbf{b}\cdot\boldsymbol{\nu}(\mathbf{y})}{\int_U w_0\mathbf{b}\cdot\boldsymbol{\nu}(\mathbf{y})\,dS_{\mathbf{y}}}, \tag{7.4.22}$$

where $\chi_U(\mathbf{y})$ is the characteristic function of the set U.

(ii) If the maximum $\hat{\phi}$ of ϕ on $\partial\Omega$ is achieved at the points $\mathbf{Q}_1,\ldots,\mathbf{Q}_m$ on $\partial\Omega$, and if

$$H_k=H(\mathbf{Q}_k)\equiv\det\left.\frac{\partial^2\phi}{\partial y_i\partial y_j}\right|_{\mathbf{Q}_k}\neq 0 \qquad (i,j=1,\ldots,n-1)$$

for $k=1,\ldots,m$, then

$$p(\mathbf{y})=\frac{\sum_{k=1}^{m}H^{-1/2}(\mathbf{Q}_k)w_0\mathbf{b}\cdot\boldsymbol{\nu}(\mathbf{Q}_k)\delta(\mathbf{y}-\mathbf{Q}_k)}{\sum_{k=1}^{m}H^{-1/2}(\mathbf{Q}_k)w_0\mathbf{b}\cdot\boldsymbol{\nu}(\mathbf{Q}_k)}, \tag{7.4.23}$$

where $\delta(\mathbf{y})$ is the Dirac delta function on $\partial\Omega$.

(iii) In the two-dimensional case ($n=2$), let $(x, y)=(x(s), y(s))$ be the parametric representation of $\partial\Omega$, where s denotes arc length on $\partial\Omega$. If the maximum $\hat{\phi}$ of ϕ on $\partial\Omega$ is achieved at the points $s_1,\ldots, s_m$ and

$$\phi(x(s_i), y(s_i))-\phi(x(s), y(s))=d_i^{-2k_i}(s-s_i)^{2k_i}(1+o(1)) \text{ as } s-s_i\to 0$$

with $d_i>0$ $(i=1,2,\ldots,m)$, and if $k\equiv k_1=k_2=\ldots=k_l=\max_i k_i (l\leqslant m)$, then

$$(7.4.24)\qquad p(x(s), y(s))=\frac{\sum_{j=1}^{l} d_j w_0 \mathbf{b}\cdot\boldsymbol{\nu}(s_j)\,\delta(s-s_j)}{\sum_{j=1}^{l} d_j w_0 \mathbf{b}\cdot\boldsymbol{\nu}(s_j)}.$$

The proof of convergence, outlined in Exercises 7.4.2*–7.4.5*, is due to Kamin (1978).

EXERCISE 7.4.2*

Let $u_\varepsilon(\mathbf{x})$ be the solution of

$$(7.4.25)\qquad L_\varepsilon u=\varepsilon\Delta u-\nabla\phi\cdot\nabla u=0 \quad \text{in } \Omega$$

$$u=f \quad \text{on } \partial\Omega.$$

Assume that $\phi(\mathbf{0})=0$, $\phi(\mathbf{x})>0$ for $\mathbf{x}\in\Omega-\{\mathbf{0}\}$, $\nabla\phi(\mathbf{0})=\mathbf{0}$,

$$\det\frac{\partial^2\phi}{\partial x_i\partial x_j}(\mathbf{0})\neq 0, \qquad \nabla\phi(\mathbf{x})\cdot\mathbf{x}\geqslant 0 \quad \text{in } \Omega,$$

$$\nabla\phi(\mathbf{x})\cdot\mathbf{x}\geqslant\gamma|\mathbf{x}|^2 \qquad (\gamma>0) \text{ on } \partial\Omega,$$

and $\partial\phi/\partial\nu>0$ on $\partial\Omega$. Introduce the variable $\mathbf{y}=\mathbf{x}/\sqrt{\varepsilon}$ and note that $|\nabla\phi(\mathbf{y}\sqrt{\varepsilon})|\leqslant C\beta\sqrt{\varepsilon}\,|\mathbf{y}|$, $(|\mathbf{y}|\leqslant\delta)$. Use interior Schauder estimates (Friedman 1964) to show that $|\partial u/\partial\mathbf{y}|\leqslant C$ $(|\mathbf{y}|\leqslant\delta)$. Conclude that

$$(7.4.26)\qquad |u_\varepsilon(\mathbf{x})-u_\varepsilon(\mathbf{0})|\leqslant C\sqrt{\varepsilon} \qquad (|\mathbf{x}|\leqslant\varepsilon).$$

EXERCISE 7.4.3*

Construct a function

$$v_\varepsilon(\mathbf{x}) = h_\varepsilon(\mathbf{x}) e^{-g(\mathbf{x})/\varepsilon} \qquad \text{where } h_\varepsilon(\mathbf{x})|_{\partial\Omega} = f(\mathbf{x}),$$

$h_\varepsilon(\mathbf{x}) = 0$ outside some neighborhood Ω' of $\partial\Omega$,

$$h_\varepsilon(\mathbf{x}) \in C^\infty(R^n)$$

$$g(\mathbf{x})|_{\partial\Omega} = 0, \qquad \left.\frac{\partial g}{\partial \boldsymbol{\nu}}\right|_{\partial\Omega} = -\left.\frac{\partial \phi}{\partial \boldsymbol{\nu}}\right|_{\partial\Omega},$$

$$g(\mathbf{x}) \geqslant 0 \text{ in } \Omega', \qquad L_\varepsilon v_\varepsilon(\mathbf{x}) = \varepsilon^m \rho(\mathbf{x}, \varepsilon), \qquad |\rho| \leqslant B,$$

$\partial h_\varepsilon / \partial \boldsymbol{\nu} \leqslant C$, where B and C are some constants independent of ε, and m is an integer. Similarly, construct a function $v_\varepsilon^{(1)}(\mathbf{x}) = h_\varepsilon^{(1)}(\mathbf{x}) e^{-g(\mathbf{x})/\varepsilon}$ satisfying similar conditions with f replaced by 1 (Vishik and Liusternik 1962).

EXERCISE 7.4.4*

Show that the solution $u_\varepsilon(\mathbf{x})$ admits the following asymptotic expansion in Ω:

$$u_\varepsilon(\mathbf{x}) = u_\varepsilon(\mathbf{0}) + v_\varepsilon(\mathbf{x}) - u_\varepsilon(\mathbf{0}) v^{(1)}(\mathbf{x}) + o(1), \tag{7.4.27}$$

where $o(1)$ is uniform in Ω as $\varepsilon \to 0$. Use (7.4.7) for $|\mathbf{x}| \leqslant \delta$ and proceed as follows in $\Omega_\varepsilon = \{\mathbf{x} \in \Omega \,|\, |\mathbf{x}| > \varepsilon\}$. Set

$$w_\varepsilon(\mathbf{x}) = \varepsilon\left(1 - \frac{\varepsilon^{n-2}}{|\mathbf{x}|^{n-2}}\right) \qquad \text{for } n \geqslant 3,$$

$$w_\varepsilon(\mathbf{x}) = \varepsilon\left(1 - \frac{\ln|\mathbf{x}|}{\ln \varepsilon}\right) \qquad \text{for } n = 2,$$

and note that $w_\varepsilon(\mathbf{x}) = 0$ for $|\mathbf{x}| = \varepsilon$ and $w_\varepsilon(\mathbf{x}) = O(\varepsilon)$. Show that

$$L_\varepsilon w = -\frac{1}{|\mathbf{x}|} \frac{\partial w_\varepsilon}{\partial |\mathbf{x}|} \nabla\phi(\mathbf{x}) \cdot \mathbf{x} \leqslant 0 \qquad \text{in } \Omega_\varepsilon$$

and

$$L_\varepsilon w_\varepsilon = -\lambda(n-2)\varepsilon^{n-1}|\mathbf{x}|^{2-n} \qquad \text{near } \partial\Omega$$

if $n\geqslant 3$, $L_\varepsilon w_\varepsilon \leqslant \lambda\varepsilon/\ln\varepsilon$ if $n=2$. Set $z_+(\mathbf{x})=v_\varepsilon(\mathbf{x})-u_\varepsilon(\mathbf{0})v_\varepsilon^{(1)}(\mathbf{x})+w_\varepsilon(\mathbf{x})$ and show that $z_+(\mathbf{x})=0$ for $|\mathbf{x}|=\varepsilon$, $z_+(\mathbf{x})=f(\mathbf{x})-u_\varepsilon(\mathbf{0})+w_\varepsilon(\mathbf{x})=f(\mathbf{x})-u_\varepsilon(\mathbf{0})+O(\varepsilon)$ for $\mathbf{x}\in\partial\Omega$ and $\varepsilon\to 0$. Show that $L_\varepsilon z_+(\mathbf{x})\leqslant 0$ for $\varepsilon\leqslant\varepsilon_0$ in Ω_ε; hence $L_\varepsilon[u_\varepsilon(\mathbf{x})-u_\varepsilon(\mathbf{0})-z_+(\mathbf{x})]\geqslant 0$ in Ω_ε. Use the maximum principle to show that $u_\varepsilon(\mathbf{x})-u_\varepsilon(\mathbf{0})-z_+(\mathbf{x})=O(\varepsilon)$. Similarly, construct $z_-(\mathbf{x})$ and show that the assertion holds.

EXERCISE 7.4.5*

Prove $u_\varepsilon(\mathbf{x})\to C_0$ as $\varepsilon\to 0$, uniformly on compact subsets of Ω, where

$$C_0=\lim_{\varepsilon\to 0}\frac{\displaystyle\int_{\partial\Omega} e^{-\phi/\varepsilon}\frac{\partial\phi}{\partial\boldsymbol{\nu}}f(\mathbf{x})\,ds}{\displaystyle\int_{\partial\Omega} e^{-\phi/\varepsilon}\frac{\partial\phi}{\partial\boldsymbol{\nu}}\,ds}.$$

Write (7.4.35) in the form $\nabla\cdot e^{-\phi/\varepsilon}\nabla u=0$; hence

$$\int_{\partial\Omega} e^{-\phi/\varepsilon}\frac{\partial u_\varepsilon}{\partial\boldsymbol{\nu}}\,ds=0.$$

Use the asymptotic expansion (7.4.27) where $o(1)=\psi_\varepsilon(\mathbf{x})$ and

$$L_\varepsilon\psi_\varepsilon(x)=O(\varepsilon). \tag{7.4.28}$$

Obtain

$$\varepsilon\int_{\partial\Omega} e^{-\phi/\varepsilon}\frac{\partial\psi\varepsilon}{\partial\boldsymbol{\nu}}\,ds=\int_{\Omega} e^{-\phi/\varepsilon}L_\varepsilon\psi_\varepsilon(\mathbf{x})\,d\mathbf{x}=O(\varepsilon^{1+\frac{n}{2}});$$

hence

$$\int_{\partial\Omega} e^{-\phi/\varepsilon}\frac{\partial\phi}{\partial\boldsymbol{\nu}}[u_\varepsilon(\mathbf{0})-f(\mathbf{x})]\,ds$$

$$=\varepsilon\int_{\partial\Omega} e^{-\phi/\varepsilon}\left[\frac{\partial h_\varepsilon}{\partial\boldsymbol{\nu}}-u_\varepsilon(\mathbf{0})\frac{\partial h_\varepsilon(1)}{\partial\boldsymbol{\nu}}\right]+O(\varepsilon^{1+\frac{n}{2}}).$$

Conclude that $u_\varepsilon(\mathbf{0})\to C_0$ and using (7.4.37), $u_\varepsilon(\mathbf{x})\to C_0$ (Kamin 1978).

7.5 EXAMPLES

In certain important cases, the results simplify and the formulas become explicit in that they involve only the known coefficients of the given problem. This is the case, for example, if the field $\mathbf{b}(\mathbf{x})$ can be decomposed into a field that is essentially a gradient of a potential and a field that is both orthogonal to the gradient field and whose divergence satisfies a certain condition. Thus, if there exists a smooth function $\psi(\mathbf{x})$ and a smooth vector field $\mathbf{l}(\mathbf{x})$ such that

$$b_j = \sum_{i=1}^{n} a_{ij}\frac{\partial\psi}{\partial x_i} + l_j \qquad (j=1,2,\ldots,n) \tag{7.5.1}$$

$$\mathbf{l}\cdot\nabla\psi = \sum_{i=1}^{n} l_i\frac{\partial\psi}{\partial x_i} = 0, \tag{7.5.2}$$

and

$$\nabla\cdot\mathbf{l} \equiv \sum_{i=1}^{n}\frac{\partial l_i}{\partial x_i} = \sum_{i,j=1}^{n}\frac{\partial a_{ij}}{\partial x_j}\frac{\partial\phi}{\partial x_i} = 0, \tag{7.5.3}$$

then the functions $\phi(\mathbf{x})$ and $w_0(\mathbf{x})$ are given explicitly by

$$\psi(\mathbf{x}) = \phi(\mathbf{x}) \tag{7.5.4}$$

and

$$w_0(\mathbf{x}) = \mathbf{1}. \tag{7.5.5}$$

In this case

$$C_0 = \lim \varepsilon\to 0 \frac{\int_{\partial\Omega}\exp(\psi/\varepsilon^2)f(\mathbf{x})\mathbf{b}\cdot\boldsymbol{\nu}(\mathbf{x})\,dS}{\int_{\partial\Omega}\exp(\psi/\varepsilon^2)\mathbf{b}\cdot\boldsymbol{\nu}(\mathbf{x})\,dS.} \tag{7.5.6}$$

Let us consider a problem (7.1.2) with $\sigma_{ij}=\delta_{ij}$ so that the principal part of the elliptic operator L_ε is the Laplacean, and with $\mathbf{b}(\mathbf{x})=\nabla\psi$ for some ψ. In this case at boundary points where ψ is maximum, we have $\partial\psi/\partial n = \partial\psi/\partial\boldsymbol{\nu} = -|\mathbf{b}|$ [see (7.4.14′)], where $|\mathbf{b}|$ denotes the length of $\mathbf{b}(\mathbf{x})$. It follows that

$$p(\mathbf{y}) = \frac{|\mathbf{b}(\mathbf{y})|\chi_U(\mathbf{y})}{\int_U |\mathbf{b}(\mathbf{x})|\,dS_{\mathbf{x}}} \tag{7.5.7}$$

in case (i) (see Section 7.4),

$$p(\mathbf{y}) = \frac{\sum_{K=1}^{m} H^{-1/2}(\mathbf{Q}_k)|\mathbf{b}(\mathbf{Q}_k)|\,\delta(\mathbf{y}-\mathbf{Q}_k)}{\sum_{k=1}^{m} H^{-1/2}(\mathbf{Q}_k)|\mathbf{b}(\mathbf{Q}_k)|} \tag{7.5.7$'$}$$

in case (ii), and

$$p(x(s), y(s)) = \frac{\sum_{i=1}^{l} d_i|\mathbf{b}(s_i)|\,\delta(s-s_i)}{\sum_{i=1}^{l} d_i|\mathbf{b}(s_i)|} \tag{7.5.7$''$}$$

in case (iii).

EXERCISE 7.5.1

Again taking $\sigma_{ij} = \delta_{ij}$, consider the two-dimensional problem with $\mathbf{b}(\mathbf{x}) = -(ax, by)$, where a and b are positive constants (this is the two-dimensional *Ornstein–Uhlenbeck process*). The boundary $\partial\Omega$ of the domain Ω is smooth and such that $\mathbf{b}\cdot\boldsymbol{\nu} < 0$ on $\partial\Omega$ but is otherwise arbitrary. Show:

(i) If the contact between $\partial\Omega$ and the largest ellipse $ax^2 + by^2 = \text{constant}$, inscribed in Ω, occurs along an arc (or union of arcs) S of $\partial\Omega$, then

$$\lim_{\varepsilon\to 0} u_\varepsilon(x, y) = C_0 = \frac{\int_S \left[a^2x^2(s) + b^2y^2(s)\right]^{1/2} f(\mathbf{x}(s))\,ds}{\int_S \left[a^2x^2(s) + b^2y^2(s)\right]^{1/2} ds}.$$

(ii) If the contact occurs at distinct points (x_i, y_i) $(i = 1, \ldots, m)$ on $\partial\Omega$,

then

$$C_0 = \frac{\sum_{i=1}^{l} d_i \left(a^2 x_i^2 + b^2 y_i^2\right)^{1/2} f(x_i, y_i)}{\sum_{i=1}^{l} d_i \left(a^2 x_i^2 + b^2 y_i^2\right)^{1/2},}$$

where (x_i, y_i) $(i = 1, \ldots, l)$ are the points of highest contact.

(iii) Consider the particular case $a = b$.

We now consider problems where the flow $\mathbf{b}(\mathbf{x})$ is a linear function of $\mathbf{x}$ as in the Ornstein–Uhlenbeck process. Thus we consider

(7.5.8) $$\mathbf{b}(\mathbf{x}) = \mathbf{B}\mathbf{x},$$

where $\mathbf{B}$ is a real matrix whose eigenvalues satisfy $\mathrm{Re}\,\lambda_i < 0$, so that the corresponding dynamical system has an asymptotically stable equilibrium point at the origin. To construct the functions ϕ and w_0 (see Section 7.4), we first note that $\mathbf{B}$ can be decomposed as follows:

(7.5.9) $$\mathbf{B} = \mathbf{A} + (\mathbf{B} - \mathbf{A})$$

with $\mathbf{A}$ a symmetric matrix, $\mathbf{A}(\mathbf{B} - \mathbf{A})$ antisymmetric, and trace $\mathbf{B}$ = trace $\mathbf{A}$.

EXERCISE 7.5.2

Define $\mathbf{Y} = \int_0^\infty e^{\mathbf{B}t} e^{\mathbf{B}^* t}\, dt$ and show that $\mathbf{Y}$ is a symmetric positive definite matrix (i.e., $\mathbf{x}^T \mathbf{Y} \mathbf{x} \geqslant c^2 |x|^2$ for all $\mathbf{x} \in R^n$), which satisfies

(7.5.10) $$\mathbf{B}\mathbf{Y} + \mathbf{Y}\mathbf{B}^* = -\mathbf{I}$$

$[\mathbf{I} = (\delta_{ij})]$. Show that the matrix $\mathbf{A}$, given by $\mathbf{A} = -\frac{1}{2}\mathbf{Y}^{-1}$, satisfies the requirements in the decomposition (7.5.9).

Hint. To show that trace $\mathbf{A}$ = trace $\mathbf{B}$, diagonalize $\mathbf{Y}$ by an orthogonal matrix $\mathbf{G}$, then use (7.5.10).

Now we define

(7.5.11) $$\phi(\mathbf{x}) = -\tfrac{1}{2}\mathbf{x}^T \mathbf{A} \mathbf{x} = -\tfrac{1}{2} \sum_{i,j=1}^{n} A_{ij} x_i x_j,$$

and note that $|\nabla\phi|^2=\mathbf{b}\cdot\nabla\phi=\mathbf{Bx}\cdot\nabla\phi$, $\Delta\phi=-\text{trace }\mathbf{A}=-\nabla\cdot\mathbf{Bx}=-\text{trace }\mathbf{B}$, so that $w_0\equiv 1$. Thus

$$p(\mathbf{y})=|\mathbf{Ay}|\chi_U(\mathbf{y}) \tag{7.5.12}$$

in case (i),

$$p(\mathbf{y})=\frac{\sum_{k=1}^{m} H_k^{-1/2}|\mathbf{Ay}_k|\,\delta(\mathbf{y}-\mathbf{y}_k)}{\sum_{k=1}^{m} H_k^{-1/2}|\mathbf{Ay}_k|}$$

in case (ii), and

$$p(\mathbf{y})=\frac{\sum_{i=1}^{l} d_i|\mathbf{Ay}_i|\,\delta(\mathbf{y}-\mathbf{y}_i)}{\sum_{i=1}^{l} d_i|\mathbf{Ay}_i|}$$

in case (iii).

EXERCISE 7.5.3

Consider the system

$$\begin{cases} \dfrac{dx}{dt}=-x+y^2+\varepsilon dw, \\ \dfrac{dy}{dt}=y-xy+\varepsilon dw_2, \end{cases} \tag{7.5.13}$$

where (w_1, w_2) is the two-dimensional Brownian motion. Derive the following Dirichlet problem for the probability distribution of the exit points (see Section 5.4).

$$\tfrac{1}{2}\varepsilon^2\Delta u-r\frac{\partial u}{\partial r}-r\sin\theta\frac{\partial u}{\partial\theta}=0 \quad \text{in } \Omega \tag{7.5.14}$$

$$u=f \quad \text{on } \partial\Omega,$$

where (r,θ) are polar coordinates in the plane. Show that $\phi=-r^2/2$ and

$w_0=(\cosh r+\sinh r\ \cos\theta)^{-1}$. Let r_0 be the radius of the maximal circle about the origin inscribed in Ω. Show that

$$p(x,y)=\frac{[\cosh r_0+\sinh r_0\cos\theta]^{-1}\chi_U}{\int_U[\cosh r_0+\sinh r_0\cos\theta]^{-1}r_0\,d\theta} \tag{7.5.15}$$

in case (i), and

$$p(x,y)=\frac{\sum_{i=1}^{l} d_i[\cosh r_0+\sinh r_0\cos\theta_i]^{-1}\delta(r_0,(\theta-\theta_i))}{\sum_{i=1}^{l} d_i[\cosh r_0+\sinh r_0\cos\theta_i]^{-1}}$$

in case (iii) where $(x,y)-r_0(\cos\theta,\sin\theta)$ (Matkowsky and Schuss 1977).

EXERCISE 7.5.4

Discuss the theory of exit in case the boundary $\partial\Omega$ is piecewise smooth (e.g., $\partial\Omega$ contains corners, etc.)

EXERCISE 7.5.5

Interpret the formulas for $p(x,y)$ in terms of diffusion currents across potential barriers (Chandrasekhar 1954).

7.6 THE EXPECTED EXIT TIME AND THE FIRST EIGENVALUE IN SINGULAR PERTURBATIONS

Let $\mathbf{b}(\mathbf{x})$ be a flow that is essentially a gradient of a potential. Specifically, we consider flows $\mathbf{b}(\mathbf{x})$ for which (7.5.1) holds with $\mathbf{l}=\mathbf{0}$. Thus $-\partial\psi/\partial\mathbf{n}=\mathbf{b}\cdot\boldsymbol{\nu}>0$ on $\partial\Omega$ [see (7.4.14′)]. We normalize ψ so that $\psi(\mathbf{0})=0$. Consider the case of a single stable equilibrium point in the flow, $\mathbf{b}(\mathbf{0})=\mathbf{0}$, say. Thus $\psi(\mathbf{x})$ attains its unique minimum at $\mathbf{x}=\mathbf{0}$, and $\psi(\mathbf{x})\geqslant 0$ in Ω. Let λ_1^ε be the principal eigenvalue of L_ε [given by (7.4.1)], that is, λ_1^ε is the smallest positive number λ such that the problem

$$L_\varepsilon u+\lambda u=0 \quad \text{in } \Omega \tag{7.6.1}$$

$$u=0 \quad \text{on } \partial\Omega$$

has a nontrivial solution (see Sect. 10.3). The transition probability density function $p_\varepsilon(\mathbf{x},\mathbf{y},t)$ of $\mathbf{x}_\varepsilon(t)$ (with absorption at $\partial\Omega$) is the Green's function of the first initial and boundary value problem for the backward Kolmogorov's equation, which for the case of time-independent coefficients is given by

$$L_\varepsilon u_\varepsilon-\frac{\partial u_\varepsilon}{\partial t}=0 \quad \text{in } Q=\Omega\times(0,\infty) \tag{7.6.2}$$

$$u_\varepsilon(\mathbf{x},t)=0 \quad \text{on } \partial\Omega\times(0,\infty)$$

$$u_\varepsilon(\mathbf{x},0)=f(\mathbf{x}) \quad \text{in } \Omega\times\{0\}.$$

The operator

$$\tilde{L}_\varepsilon=\varepsilon\sum_{i,j=1}^{n}\frac{\partial^2 a_{ij}}{\partial x_i\partial x_j}+\sum_{i=1}^{n}b_i\frac{\partial}{\partial x_i}$$

differs from L_ε by the first-order term

$$\varepsilon\sum_{i,j=1}^{n}\left(\frac{\partial a_{ij}}{\partial x_j}\right)\frac{\partial}{\partial x_j},$$

so (7.4.13) will change if L_ε is replaced by $\tilde{L}_\varepsilon$. The operator $\tilde{L}_\varepsilon$ is self-adjoint in the space $L^2(\Omega)$ with the inner product

$$(u,v)\equiv\int_\Omega e^{-\psi(\mathbf{x})/\varepsilon^2}u(\mathbf{x})v(\mathbf{x})\,d\mathbf{x}, \tag{7.6.3}$$

where $\psi(\mathbf{x})$ is defined by (7.4.14′). It follows that L_ε has a complete orthonormal set of real eigenfunctions $\{\phi_j\}$ that satisfy

$$\tilde{L}_\varepsilon\phi_j=-\lambda^\varepsilon_j\phi_j \quad \text{in } \Omega \tag{7.6.4}$$

$$\phi_j=0 \quad \text{on } \partial\Omega \tag{7.6.5}$$

and

$$\int_\Omega \phi_j \phi_k e^{-\psi/\varepsilon^2} d\mathbf{x} = \delta_{jk}, \tag{7.6.6}$$

where $\lambda_j^\varepsilon > 0$ are the eigenvalues of L_ε. Thus

$$p_\varepsilon(\mathbf{x},\mathbf{y},t) \sim \sum_{j=1}^{\infty} e^{-\psi(\mathbf{y})/\varepsilon^2} \phi_j(\mathbf{y},\varepsilon)\phi_j(\mathbf{x},\varepsilon) e^{-\lambda_j^\varepsilon t}. \tag{7.6.7}$$

The eigenfunctions $\phi_j(\mathbf{x},\varepsilon)$ normalized by (7.6.6) are unbounded in ε. Therefore, we write them as

$$\phi_j(\mathbf{x},\varepsilon) = c_j(\varepsilon) g_j(\mathbf{x},\varepsilon) \tag{7.6.8}$$

with the bounded function $g_j(\mathbf{x},\varepsilon)$ normalized by

$$\int_\Omega g_j^2(\mathbf{x},\varepsilon)\,d\mathbf{x} = 1 \qquad j = 1,2\ldots. \tag{7.6.9}$$

Now, the reduced equation (7.6.1) (with $\varepsilon = 0$) for each g_j is given by

$$\mathbf{b}\cdot\nabla g_j^0 = \lambda_j^0 g_j^0, \tag{7.6.10}$$

where the superscript 0 denotes the quantity evaluated at $\varepsilon = 0$. It can be shown (Friedman 1973) that $\lambda_1^\varepsilon \to 0$ exponentially in K/ε^2 for some negative constant K. This fact implies that the reduced equation for g_1^0 is given by

$$\mathbf{b}\cdot\nabla g_1 = 0. \tag{7.6.11}$$

Thus as $\varepsilon \to 0$,

$$g_1(\mathbf{x},\varepsilon) \to C, \tag{7.6.12}$$

where by (7.6.9) C is a nonzero constant. Employing (7.6.8) in (7.6.7) with $u = 1$, we have that

$$\int_\Omega g_1(\mathbf{x},\varepsilon) g_j(\mathbf{x},\varepsilon) e^{-\psi(\mathbf{x})/\varepsilon^2} d\mathbf{x} = 0 \qquad (j \geqslant 2). \tag{7.6.13}$$

We now compute the asymptotic expansion of the expected exit time $E_x \tau_\varepsilon$ by noting that integration of the backward Kolmogorov's equation for

$p_\varepsilon(\mathbf{x},\mathbf{y},t)$

(7.6.14)
$$L_\varepsilon p_\varepsilon = \frac{\partial p_\varepsilon}{\partial t} \quad \text{in } \Omega$$
$$p_\varepsilon(\mathbf{x},\mathbf{y},t)=0, \quad \text{on } \partial\Omega$$
$$p_\varepsilon(\mathbf{x},\mathbf{y},0)=\delta(\mathbf{x}-\mathbf{y})$$

with respect to y over Ω and with respect to $\mathbf{t}$ over $[0,\infty]$ leads to

(7.6.15)
$$L_\varepsilon v_\varepsilon(\mathbf{x}) = -1 \quad \text{in } \Omega$$
$$v_\varepsilon(\mathbf{x})=0 \quad \text{on } \partial\Omega.$$

where $v_\varepsilon(\mathbf{x}) = \int_0^\infty \int_\Omega p_\varepsilon(\mathbf{x},\mathbf{y},t)\,d\mathbf{y}\,dt$. Equation (7.6.15) is identical to (5.4.5); hence $v_\varepsilon(\mathbf{x}) = E_\mathbf{x}\tau_\varepsilon$. Therefore, (7.6.7) implies that

$$E_\mathbf{x}\tau^\varepsilon = \int_0^\infty \int_\Omega p_\varepsilon(\mathbf{x},\mathbf{y},t)\,d\mathbf{y}\,dt$$
$$\sim \sum_{j=1}^\infty \frac{\phi_j(\mathbf{x})}{\lambda_j^\varepsilon} \int_\Omega \phi_j(\mathbf{y}) e^{-\psi(\mathbf{y})/\varepsilon^2} d\mathbf{y}.$$

In particular,

(7.6.16)
$$E_\mathbf{0}\tau^\varepsilon = \frac{\phi_1(\mathbf{0})}{\lambda_1^\varepsilon} \int_\Omega \phi_1(\mathbf{y}) e^{-\psi(\mathbf{y})/\varepsilon^2} d\mathbf{y}$$
$$+ \sum_{j=2}^\infty \frac{\phi_j(\mathbf{0})}{\lambda_1^\varepsilon} \int_\Omega \phi_j(\mathbf{y}) e^{-\psi(\mathbf{y})/\varepsilon^2} d\mathbf{y}.$$

Now, since $g_j(\mathbf{0}) \to 0$ for $j \geq 2$ by (7.6.14), $0 < \lambda_j^\varepsilon$ for $j \geq 2$ and

(7.6.17)
$$\int_\Omega C_j^2(\varepsilon) e^{-\psi(\mathbf{y})/\varepsilon^2} d\mathbf{y} = 1,$$

which is implied by (7.6.6), and (7.6.8) and (7.6.9), we have that

(7.6.18)
$$E_\mathbf{0}\tau^\varepsilon \sim (\lambda_1^\varepsilon)^{-1} \phi_1(\mathbf{0}) \int_\Omega \phi_1(\mathbf{y}) e^{-\psi(\mathbf{y})/\varepsilon^2} d\mathbf{y}.$$

Finally, employing (7.6.6), (7.6.8), and (7.6.12) in (7.6.16), we obtain

(7.6.19)
$$E_\mathbf{0}\tau^\varepsilon \sim (\lambda_1^\varepsilon)^{-1}.$$

To find the asymptotic behavior of $E_0\tau^\varepsilon$, and therefore of λ_1^ε, we derive the leading term in the asymptotic expansion of the boundary value problem

$$L_\varepsilon v = \varepsilon^2 \sum_{i,j=1}^{n} a_{ij}(\mathbf{x}) \frac{\partial^2 v}{\partial x_i \partial x_j} + \sum_{i=1}^{n} b_i(\mathbf{x}) \frac{\partial v}{\partial x_i} = g(\mathbf{x}) \quad \text{in } \Omega \tag{7.6.20}$$

$$v = 0 \quad \text{on } \partial\Omega, \tag{7.6.21}$$

which reduces to (7.6.15) for $g \equiv -1$. We know by the maximum principle that the solution of (7.6.20) and (7.6.21) is bounded by $e^{K/\varepsilon^2}\max_\Omega|g(\mathbf{x})|$ with $K=$ a positive constant. In addition, simple examples of the form (7.6.20) and (7.6.21) show that the solution can indeed grow as large as e^{K/ε^2}. Therefore, we scale the solution $v_\varepsilon(x)$ of (7.6.20) and (7.6.21) by defining

$$V_\varepsilon(\mathbf{x}) = e^{-K/\varepsilon^2} v_\varepsilon(\mathbf{x}) \tag{7.6.22}$$

with K to be chosen later. We note that $v_\varepsilon(\mathbf{x})$ is $O(1)$ in ε^2, and satisfies (7.6.20) and (7.6.21) with $g(\mathbf{x})$ replaced by $e^{-K/\varepsilon^2}g(\mathbf{x})$. Then the leading term in the expansion of $V_\varepsilon(\mathbf{x})$ is derived as in Section 7.4,

$$v_\varepsilon \sim C_0(1 - e^{\zeta/\varepsilon^2}) \tag{7.6.23}$$

with the constant C_0 to be determined as above.

Thus the leading term in the uniform expansion of $v_\varepsilon(\mathbf{x})$ is given by

$$v_\varepsilon(\mathbf{x}) \sim C_0(\varepsilon) e^{K/\varepsilon^2}(1 - e^{\zeta/\varepsilon^2}). \tag{7.6.24}$$

As above, we integrate (7.6.20) and (7.6.21) over Ω and employ the divergence theorem to obtain

$$\varepsilon \int_{\partial\Omega} e^{-\psi/\varepsilon^2} w_0 \frac{\partial v_\varepsilon}{\partial \mathbf{n}}\, dS_{\mathbf{y}} = \int_\Omega e^{-\psi/\varepsilon^2} w_0 g(\mathbf{x})\, d\mathbf{x} \tag{7.6.25}$$

(see Section 7.4). Then employing (7.6.24) in (7.6.25), we arrive at

$$\int_{\partial\Omega} e^{-\psi/\varepsilon^2} w_0 \mathbf{b}\cdot\boldsymbol{\nu}\, dS_{\mathbf{y}} = \frac{e^{-K/\varepsilon^2}}{C_0(\varepsilon)} \int_\Omega e^{-\psi/\varepsilon^2} w_0 g(\mathbf{x})\, d\mathbf{x}. \tag{7.6.26}$$

We evaluate these integrals asymptotically and observe that the main contribution to the left (right)-hand side of (7.6.26) comes from the points where $\psi=\hat{\psi}$ [from the origin where $\psi(\mathbf{0})=0$ is the minimum of ψ in Ω]. Therefore, we choose $K=\hat{\psi}$ and find that

$$C_0(\varepsilon)=\frac{(\sqrt{2\pi}\,\varepsilon)^n g(\mathbf{0})}{H^{1/2}(\mathbf{0})\int_{\partial\Omega} e^{-(\psi-\hat{\psi})/\varepsilon^2} w_0 \mathbf{b}\cdot\boldsymbol{\nu}\, dS_{\mathbf{y}}}, \tag{7.6.27}$$

where

$$H(\mathbf{0})=\det\left.\frac{\partial^2\psi}{\partial x_i \partial x_j}\right|_{\mathbf{x}=\mathbf{0}}.$$

We now calculate $C_0(\varepsilon)$ and subsequently the asymptotic behavior of $v_\varepsilon(\mathbf{x})$ as $\varepsilon\to 0$ by evaluating the Laplace-type integral in (7.6.27) in each of cases (i)–(iii) of Section 7.4. Thus we obtain

$$C_0(\varepsilon)=\frac{(\sqrt{2\pi}\,\varepsilon)^n g(\mathbf{0})}{H^{1/2}(\mathbf{0})\int_U w_0 \mathbf{b}\cdot\boldsymbol{\nu}\, dS_{\mathbf{y}}} \tag{7.6.28}$$

in case (i),

$$C_0(\varepsilon)=\frac{\sqrt{2\pi}\,\varepsilon g(\mathbf{0})}{H^{1/2}\sum_{i=1}^{m} H_k^{-1/2} w_0 \mathbf{b}\cdot\boldsymbol{\nu}(\mathbf{0}_k)} \tag{7.6.29}$$

in case (ii), and

$$C_0(\varepsilon)=\frac{2\pi\varepsilon^{(2k-1)/k} g(\mathbf{0})}{\Gamma\left(\frac{1}{2k}\right) H^{1/2}(\mathbf{0}) \sum_{i=1}^{l} d_i w_0 \mathbf{b}\cdot\boldsymbol{\nu}(s_i)} \tag{7.6.30}$$

in case (iii), where $\Gamma(z)$ is the gamma function.

Now that the leading term in the expansion of $v_\varepsilon(\mathbf{x})$ as (7.6.24) with C_0 given by one of (7.6.28)–(7.6.30) has been found, we employ the identity $v_\varepsilon(\mathbf{x})=E_{\mathbf{x}}\tau^\varepsilon$ in case $g\equiv -1$ to compute the leading term of the expected exit time. Thus

$$
(7.6.31)\quad E_{\mathbf{x}}\tau^{\varepsilon} \sim \begin{cases} -\dfrac{(\sqrt{2\pi}\,\varepsilon)^{n} e^{\hat{\psi}/\varepsilon^{2}}(1-e^{\zeta/\varepsilon^{2}})}{H^{1/2}(\mathbf{0})\displaystyle\int_{U} w_{o}\mathbf{b}\cdot\boldsymbol{\nu}\,dS_{\mathbf{y}}} & \text{for case (i),} \\[2ex] -\dfrac{\sqrt{2\pi}\,\varepsilon e^{\hat{\psi}/\varepsilon^{2}}(1-e^{\zeta/\varepsilon^{2}})}{H^{1/2}(\mathbf{0})\displaystyle\sum_{k=1}^{m} H_{k}^{-1/2} w_{0}\mathbf{b}\cdot\boldsymbol{\nu}|_{\mathbf{Q}_{k}}} & \text{for case (ii),} \\[2ex] -\dfrac{2\pi\varepsilon^{(2k-1)/k} e^{\hat{\psi}/\varepsilon^{2}}(1-e^{\zeta/\varepsilon^{2}})}{\Gamma\left(\dfrac{1}{2k}\right)H^{1/2}(\mathbf{0})\displaystyle\sum_{i=1}^{l} d_{i}w_{0}\mathbf{b}\cdot\boldsymbol{\nu}|_{s_{i}}} & \text{for case (iii).} \end{cases}
$$

Finally, employing (7.6.19), the asymptotic form of λ_1^{ε} is obtained by setting $\mathbf{x}=\mathbf{0}$ and taking the inverse of (7.6.31). We note that the expected exit time (principal eigenvalue) increases (decreases) exponentially as $\varepsilon \to 0$. Further eigenvalue problems, arising in the context of chemical reactions rates, are discussed in Section 8.6.

EXERCISE 7.6.1

Find the asymptotic expansion of the solution $y_{\varepsilon}(x)$ of the problem

$$\varepsilon y'' - xy' = g(x) \qquad \text{in } (a,b),$$

$$y(a)=y(b)=0 \qquad (a<0<b).$$

EXERCISE 7.6.2

Derive (7.6.23).

EXERCISE 7.6.3

Consider the diffusion approximation to the Wright–Fisher model in genetics in the presence of random large selection. Take $s=O(1)$ as $N\to\infty$

and write the approximating Fokker–Planck equation in the form

$$\frac{\partial^2}{\partial x^2}[x(1-x)p]+4N\sigma\frac{\partial}{\partial x}[(2x-1-a)(1-x)p]=4N\frac{\partial p}{\partial t} \qquad (0\leqslant x\leqslant 1),$$

$$t\geqslant 0, \quad -1\leqslant a\leqslant 1, \quad \sigma\geqslant 0$$

(see Section 5.2). Expand

$$p(x,t)=\sum_{n=1}^{\infty} e^{-\lambda_n t} f_n(2x-1)$$

or

$$p(x,t)\sim e^{-\lambda_1 t} f_1(2x-1)$$

Obtain the asymptotic form of λ_1 as $N\to\infty$ (Miller 1962).

EXERCISE 7.6.4

Consider a predator–prey interaction based on the following assumptions:

(i) The birth rate of each species is proportional to the size of the population.
(ii) The death rate due to food shortage is proportional to the square of the population.
(iii) The death rate of the prey is proportional to the sizes of the prey and predator population and the same is true for the birthrate of the predator.

Construct a *deterministic* model for the evolution of the predator–prey population and find the asymptotic behavior of the population as $t\to\infty$. Specifically, if $x(t)$ and $y(t)$ are the sizes of the prey and predator populations at time t, respectively, then

$$(7.6.32) \qquad \begin{cases} \dfrac{dx}{dt}=ax-bx^2-cxy \\ \dfrac{dy}{dt}=Ay-By^2+Cxy \end{cases}$$

Add "noise" to (7.6.32) and assume that the constants can be chosen so that

$$
(7.6.33)\qquad \begin{cases} \dfrac{dx}{dt} = x(1+\alpha-x-\alpha y) + \sqrt{\varepsilon}\,\dfrac{x\,dw_1}{dt} \\ \dfrac{dy}{dt} = y(1-\alpha-y+x) + \sqrt{\varepsilon}\,\dfrac{y\,dw_2}{dt}. \end{cases}
$$

Write down the Fokker–Planck and the backward Kolmogorov's equations for (7.6.33). Change variables $\xi = \log x$, $\eta = \log y$ to avoid degeneration. Show that the "quasiponential" ϕ is given by $\phi\,(\xi,\eta) = 2(e^{\xi} - \xi - 1 + e^{\eta} - \eta - 1)$. Define Ω by $\xi > \eta_1$, $\eta > \eta_1$, and compute the expected exit time and exit distribution for small ε. Use numerical computations to find w_0 (Ludwig 1974, 1975).

EXERCISE 7.6.5

Use similar analysis for the model of two competing species

$$
\frac{dx}{dt} = x(1+\alpha-x-\alpha y)
$$

$$
\frac{dy}{dt} = y(1+\alpha-y-\alpha x).
$$

Use numerical integration to find ϕ and w_0 (Ludwig 1974, 1975).

EXERCISE 7.6.6

Use (5.4.6), (7.6.19), and separation of variables to show that τ_ε becomes an exponentially distributed random variable as $\varepsilon \to 0$.

CHAPTER 8

Diffusion Across Potential Barriers

8.1 A DIFFUSION MODEL OF CHEMICAL REACTIONS

A particle inside a molecule is held by molecular bonds and may be considered resting or performing small oscillations about a stable equilibrium. In a chemical reaction, external forces (e.g., molecular collisions) may activate the particle to such a degree that it overcomes the molecular bonds and leaves the molecule. The random forces due to such random collisions, as described in Chapter 2, are described mathematically as "white noise." This is, of course, a mathematical idealization of the physical process (see Section 4.1). Their "intensity" determines the temperature of the reactants. Once outside the molecule such a particle either forms new molecular bonds and remains in a more stable equilibrium or is removed permanently from the molecule by other means. This situation is depicted in Fig. 8.1.1, where the particle B forms a new, more stable chemical bond after overcoming a potential barrier of height E. The rate at which the external forces push such particles over the edge of the potential well in which they initially rest determines the kinetics of the reaction.

If the particle has n degrees of freedom, its motion can be described as that of a particle in the $2n$-dimensional phase space, that is, by n independent displacement coordinates $\mathbf{x}=(x_1,\dots,x_n)^T$ and by n velocity coordinates $\mathbf{y}=d\mathbf{x}/dt$. The potential well confining the particle may consist of a succession of holes and barriers through which the particle passes before it escapes (see Fig. 8.1.2). This is the case of successive chemical reactions. Here the reaction may be considered complete only after the highest potential barrier has been surmounted. The energy that the initial reactants must acquire before they can surmount the highest barrier separating them

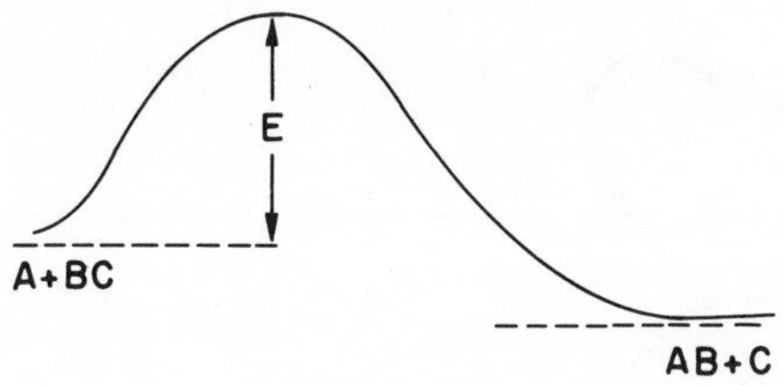

Figure 8.1.1 Potential barrier for a chemical reaction.

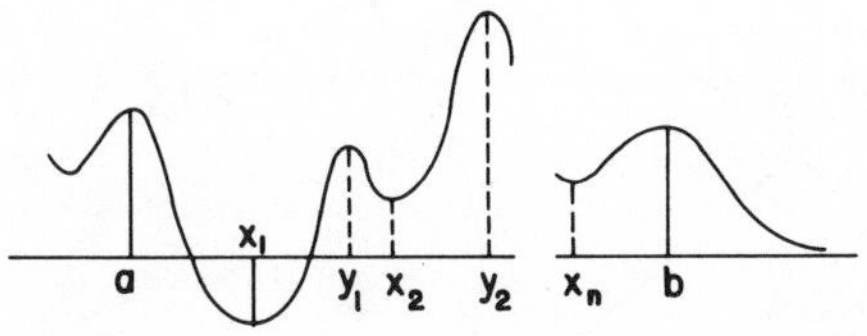

Figure 8.1.2 Potential for successive reactions.

from the final products is called the *activation energy*. The presence of intermediate states is inconsequential except insofar as such a state contains an appreciable fraction of the total molecules, which makes them really part of the initial state. For multiatomic molecules the potential energy is a function of the n independent internuclear distances measured from some arbitrary origin. In this case there are many directions of escape over the potential barrier in R^n. The different directions correspond to different chemical reactions; for example, the same activation energy is required for the splitting of any of the three equivalent H atoms in CH_3F, resulting in three different reactions (Benson 1960; Glasstone et al. 1941).

Finally, if the potential barrier is very sharp, as in Fig. 8.1.3, we may consider the forces to be discontinuous; that is, once the bound is broken, the force changes abruptly from attraction to repulsion. We shall compute for such models the rate of escape of particles over the barrier, thus giving the activation energy as a function of the potential well and the temperature (i.e., the "intensity" of the white noise). In particular, we shall find the effect of the spatial geometry of the potential well on the reaction rate and on the composition of the final products (Figs. 8.1.4–8.1.7).

We assume that the medium surrounding the molecule is in thermal equilibrium, so that the velocity distribution of the ions, due to random

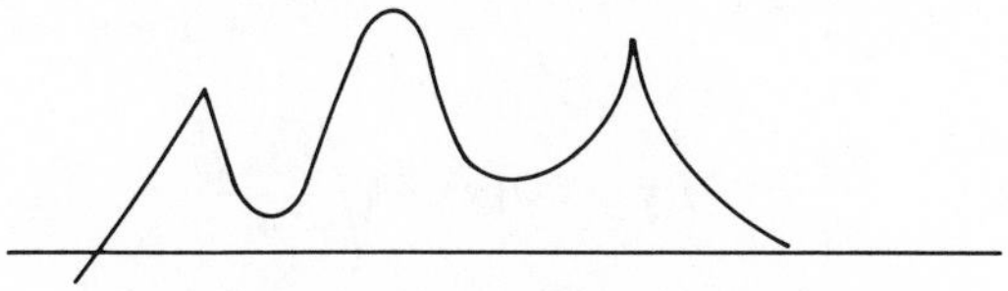

Figure 8.1.3 Potential barrier with sharp edges.

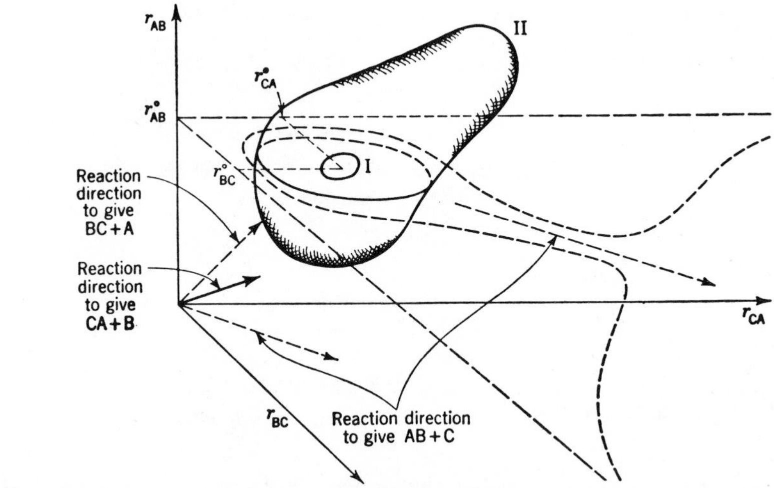

Figure 8.1.4 Two surfaces of constant potential energy (I, II) for ABC, a triatomic molecule as a function of the internuclear distances. (from Benson 1960)

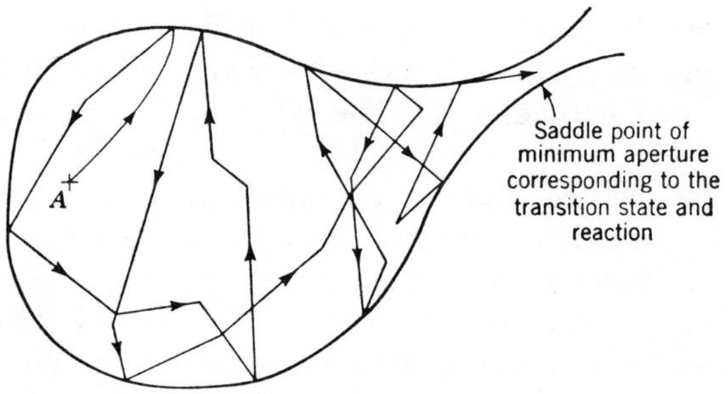

Figure 8.1.5 Schematic representation of the motion in phase space of the internal coordinates of a critically energized molecule with a single possible mode of decomposition. Bounding surface is one of constant potential energy. (from Benson 1960)

collisions, is Maxwellian in the n components $\mathbf{y}$. To be more specific, we describe the motion of such a particle by the Langevin equation.

$$\frac{d\mathbf{x}}{dt}=\mathbf{y}$$

$$\frac{d\mathbf{y}}{dt}=-\beta\mathbf{y}-\nabla\Phi(\mathbf{x})+\sqrt{\frac{2\beta kT}{m}}\;\dot{\mathbf{w}}.$$

Here the term $-\beta\mathbf{y}$ is the dynamic viscosity per unit mass, which expresses the slowing-down rate of particles by random collisions with the particles

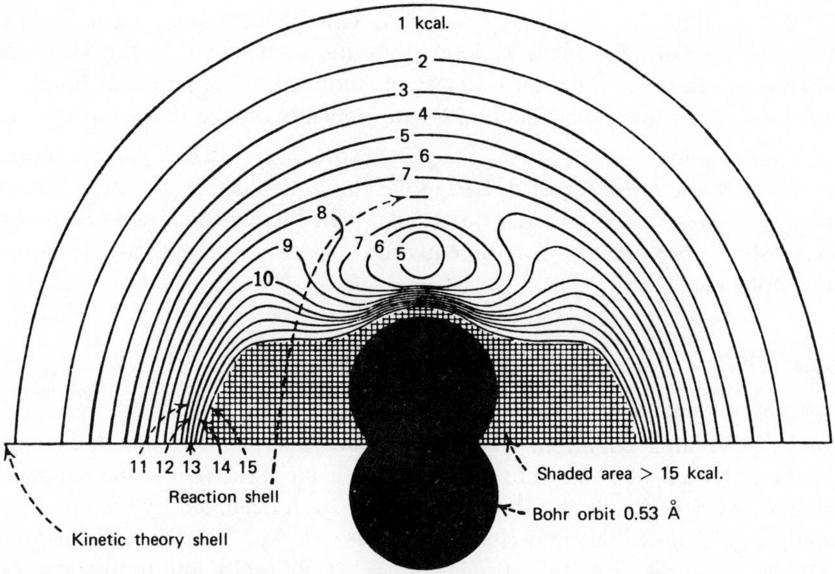

Figure 8.1.6 Potential-energy contours for approach of a hydrogen atom to a rigid hydrogen molecule. (From Hirschfelder, Eyring, and Topley.)

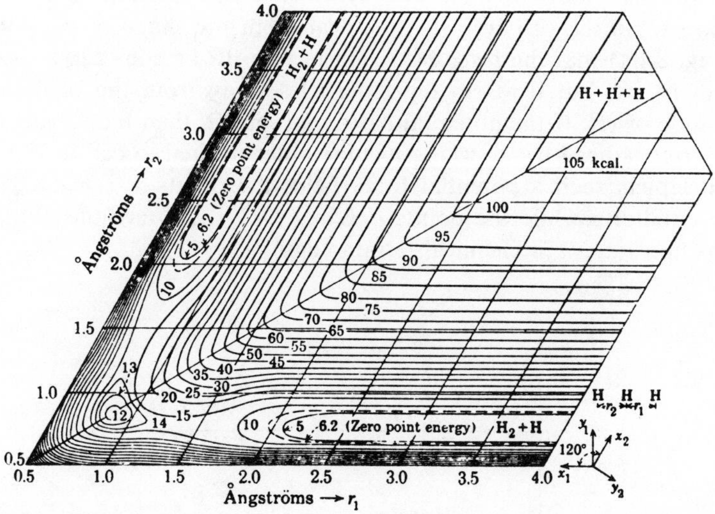

Figure 8.1.7 Potential-energy surface for the system of three hydrogen atoms based on 14 % Coulombic energy. (From Eyring, Gershinowitz, and Sun.)

of the medium in which they move. It can be computed from Stokes' formula $\beta = 6\pi a\eta/m$ for a sphere of radius a, where η is the viscosity coefficient and m is the reduced mass of a collision. The potential function $\Phi(\mathbf{x})$ describes the molecular bonds and depends on the relative distances of atomic nuclei only (Fig. 8.1.8). The expression $\sqrt{2\beta kT/m}\,dw$ represents the white noise forces due to molecular collisions. If β is large (i.e., if many collisions occur in each time unit), then the Smoluchowski equation is a valid approximation for the equation of motion of the particle inside the molecule. Thus

$$d\mathbf{x} = -\nabla\Phi(\mathbf{x})\,ds + \sqrt{\frac{2kT}{m}}\,d\mathbf{w}(s) \tag{8.1.1}$$

is the governing equation, where $\beta s = t$. We also assume that the depth of the potential well (the activation energy) is large relative to the temperature. More specifically, let Ω represent the region occupied by the potential well, and assume that $\min_{\Omega}\Phi = 0$, $\min_{\partial\Omega}\Phi = Q$, where $\partial\Omega$ is the potential barrier. Then rescaling by setting $-\boldsymbol{\nabla}\Phi = -Q\boldsymbol{\nabla}\phi \equiv Q\mathbf{b}$, and letting $s' = Qs$, (i.e., changing the original time scale by $t \to \beta t/Q$) we obtain the equation

$$d\mathbf{x} = b(\mathbf{x})\,dt + \varepsilon\,d\mathbf{w},$$

where $\varepsilon = \sqrt{2kT/mQ} \ll 1$ by assumption. We have used the fact that if $\mathbf{w}$ is a Brownian motion, then $w(C^2t)/C$ is a Brownian motion for any constant $C > 0$ (cf. (2.1.16)(ii)). It is clear from the shape of the potential (see Fig. 8.1.1) that the force $\mathbf{b}(\mathbf{x})$ vanishes on $\partial\Omega$ or points in a direction tangent to $\partial\Omega$, but tends to repel particles away from the boundary on either side of $\partial\Omega$. If the potential has a cusp at $\partial\Omega$, then $\mathbf{b}$ does not vanish on $\partial\Omega$, but rather $\mathbf{b}\cdot\boldsymbol{\nu} < 0$, where $\boldsymbol{\nu}$ is the outer normal to $\partial\Omega$. In Fig. 8.1.2, which depicts such a potential force field, the points x_i $(i = 1, \ldots, n)$ are stable equilibria while the points a, b, y_j $(j = 1, \ldots, k)$ are unstable states ("transition states" in chemical terminology).

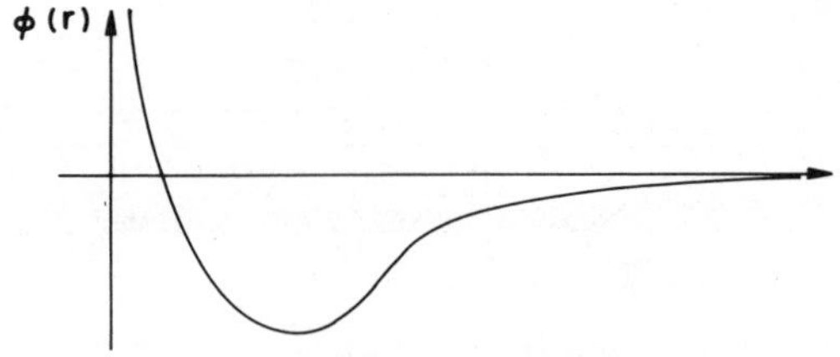

Figure 8.1.8 Potential of internuclear forces.

Let $\tau^\varepsilon=\tau(\mathbf{x},\varepsilon)$ be the first exit time of the solution $\mathbf{x}_\varepsilon(t)$ of (8.1.1) from Ω; that is, $\tau^\varepsilon=\inf\{t>0|\mathbf{x}_\varepsilon(t)\in\partial\Omega,\mathbf{x}_\varepsilon(0)=\mathbf{x}\in\Omega\}$. The average exit time $v_\varepsilon(\mathbf{x})=E_\mathbf{x}\tau^\varepsilon$ of a particle from the potential well determines the reaction rate

$$\kappa=\tfrac{1}{2}\int_\Omega\left[\frac{\rho(\mathbf{x})}{v_\varepsilon(\mathbf{x})}\right]d\mathbf{x}$$

where $\rho(\mathbf{x})$ is the relative concentration of particles in the well. We may assume that initially all particles are concentrated at the stable equilibria $\mathbf{x}_i$, so that

$$\kappa=\tfrac{1}{2}\sum\frac{\rho(\mathbf{x}_i)}{E_{\mathbf{x}_i}\tau^\varepsilon}.$$

If, as is generally the case, most of the particles are concentrated at the bottom of the deepest trough, located at $\mathbf{x}=\mathbf{0}$ say, then $E_{\mathbf{x}_i}\tau^\varepsilon\sim E_\mathbf{0}\tau^\varepsilon$ for small ε, so that

$$K=\tfrac{1}{2}E_\mathbf{0}\tau^\varepsilon.$$

That is, the reaction rate κ, which is the frequency at which particles escape over the potential barrier, is one-half of the inverse of the average escape time from the potential well. The factor $\frac{1}{2}$ is the probability that a particle reaching $\partial\Omega$ crosses over into the complement of Ω. The constant κ appears in the equations of chemical kinetics as follows. Let $C(t)$ be the concentration of a given reactant in a first-order reaction. Then the governing equation is given by (Benson 1960)

$$\frac{dC(t)}{dt}=-\kappa C(t).$$

The first formula for κ in terms of temperature and activation energy was given by Arrhenius in 1894 and has been considerably improved since that time. A computation of $E_\mathbf{0}\tau^\varepsilon$ is given in Section 8.3 for the case where the potential barrier is smooth. In Chapter 7, it was calculated for the case of a sharp barrier.

We will show that κ is given by a modified Arrhenius law. The modification accounts for the effects of directionality in that collisions in different directions may have different probabilities of activating the molecule. This constitutes a theory for the computation of the elusive "steric factor" in the Arrhenius law. This theory also gives the proportions of the various possible reactions occurring in different directions due to the steric effect. Finally, note that the first attempt to compute the reaction rate from a similar model in one dimension was due to Kramers (1940), who considered the Fokker–Planck (forward) equation. Of course, in one dimension there are no steric effects.

In a more realistic model of a chemical reaction one has to assume that the particle is confined to an infinite well and that the reaction consists of the transition of the particle from one stable equilibrium to another, more stable equilibrium. During the transition the particle may pass several other equilibrium states before a stationary regime is achieved. We describe this situation by the Smoluchowski–Kramers equation (8.1.1) with a potential Φ that has several minima, and $\Phi(\mathbf{x})\to\infty$, $|\mathbf{x}|\to\infty$. The transition probability density $p_\varepsilon(x, x_0, t)$ of the solution $\mathbf{x}_\varepsilon(t)$ of the Smoluchowski equation (8.1.1) represents the density of the reacting particles in space relative to the potential $\Phi(\mathbf{x})$. Thus

$$P(\mathbf{x}, t)=\int p_\varepsilon(\mathbf{x},\mathbf{y}, t)f(\mathbf{y})\,d\mathbf{y}$$

represents the fraction of particles in different reaction stages, characterized by the coordinate $\mathbf{x}$, at time t, given an initial distribution $f(\mathbf{x})$. The density $p_\varepsilon(\mathbf{x},\mathbf{x}_0, t)$ is the solution of the Fokker–Planck equation corresponding to (8.1.1):

$$Lu=\varepsilon\Delta u+\nabla\cdot(\nabla\phi u)=u_t \qquad \mathbf{x}\in R^n, \quad t>0 \tag{8.1.2}$$

$$u(\mathbf{x},\mathbf{x}_0, t)\to\delta(\mathbf{x}-\mathbf{x}_0) \qquad \text{as} \quad t\to\infty$$

The stathonary soluthon is given by $u^0(\mathbf{x})=e^{-\phi/\varepsilon}/\int e^{-\phi/\varepsilon}d\mathbf{x}$ and it describes the final distribution of the reaction products. The full solution $u(\mathbf{x},\mathbf{x}_0, t)$ can be written as an eigenfunction expansion

$$u(\mathbf{x},\mathbf{x}_0, t)=u^0(\mathbf{x})+e^{-\phi(\mathbf{x}_0)/\varepsilon}\sum_{n=1}^{\infty} u_n(x)u_n(x_0)e^{\lambda_n t},$$

where λ_n are the eigenvalues and $u_n(\mathbf{x})$ are the eigenfunctions of the problem

$$Lu=\lambda u. \tag{8.1.3}$$

Here L is the operator defined in (8.1.2). The stationary solution $u^0(\mathbf{x})$ is the eigenfunction corresponding to the eigenvalue $\lambda_0=0$. Thus the eigenvalue λ_1 determines the reaction rate, and therefore we must have $\kappa=\lambda_1$. It will be shown in Section 8.6 that $\tau=1/\lambda_1$ is the average time it takes a particle to surmount the highest potential barrier on its way to the deepest well.

8.2 ATOMIC MIGRATION IN CRYSTALS

A crystal is a lattice of atoms arranged in a periodic pattern. It consists of identical cells held together by interatomic forces (see Figs. 8.2.1 and 8.2.2). Very often impurities are introduced into the crystal. Impurity atoms are located at interstitial positions and move through the crystal by squeezing their way past the atoms surrounding them into new interstitial positions near the original ones (Fig. 8.2.3). Our purpose is to describe the migration of such impurities through the lattice. A similar phenomenon arises when defects are created in the lattice. In a pure crystal, the most important kind of point defect is a vacant site referred to as a vacancy (Fig. 8.2.4). Another type of point defect is an atom of the pure material that is not at a lattice site but at an interstitial position (Fig. 8.2.5). This occurs, for example, in the metallic lattice of nuclear reactor shields

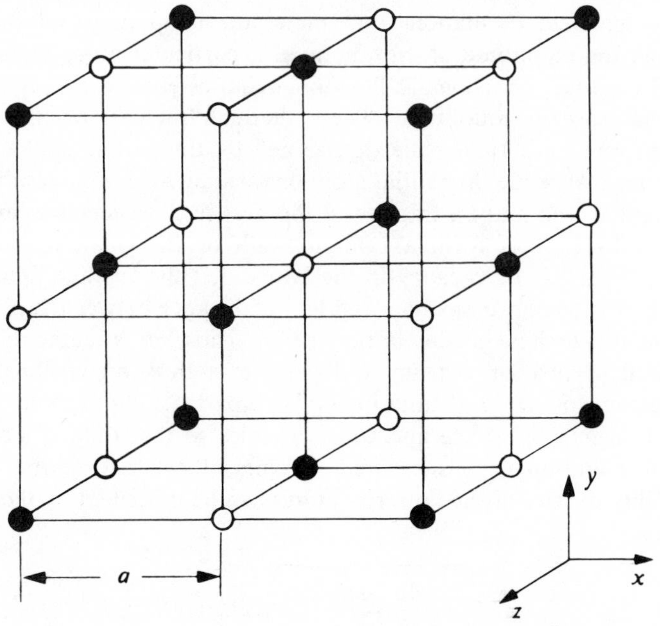

Figure 8.2.1 Sodium chloride structure. (From Girifalco, 1964.)

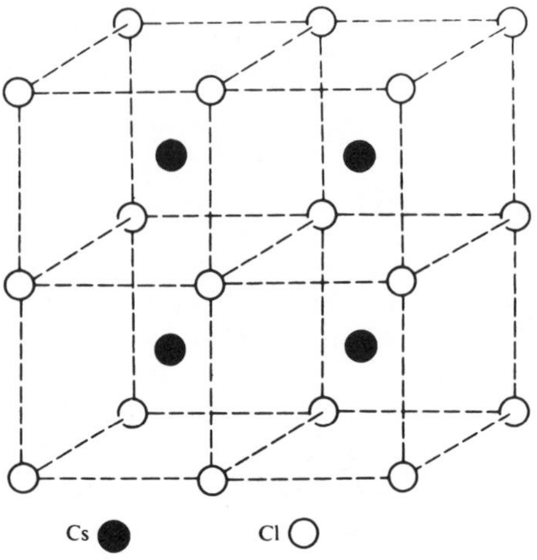

Figure 8.2.2 Cesium chloride structure. (From Girifalco, 1964.)

subjected to radiation. The creation and migration of such vacancies causes swelling and other changes in the physical properties of the metal.

If the lattice atoms were stationary, the interstitial particles or vacancies would be held at rest at the bottom of a potential well of the atomic forces (see Fig. 8.2.6). However, the atoms in a crystal are not stationary, but execute small, random oscillations about their average positions. The intensity of these oscillations determines the temperature of the crystal. Although the amplitude of vibrations of a particular atom remains small most of the time, occasionally the atom acquires enough energy to considerably increase its amplitude. These thermal fluctuations in amplitude, although rare on a time scale determined by the period of the vibrating atoms, may cause the interstitial atom or vacancy to climb over the edge of the potential well and be trapped in the next well. Successive jumps from one well to another form a random walk or Brownian motion of the interstitial particle or vacancy in the lattice. Let the average time between such fluctuations or jumps be $\bar{\tau}$ and let the distance between two neighboring interstitial sites in the lattice be a. Then, in a cubic lattice, the interstitial particle or vacancy will perform a three-dimensional random walk in steps of size a at time intervals $\bar{\tau}$ apart (i.e., with jump frequency $1/\bar{\tau}$). Rescaling the space and time variables so that time is measured in units of $\bar{\tau}$ and space is measured in units of cell size a, the transition probability density of the impurity atoms can be described by the diffusion equation

$$\frac{\partial u}{\partial t} = \frac{\alpha a^2}{\bar{\tau}}\Delta u \equiv D\Delta u \tag{8.2.1}$$

as in Section 2.2.

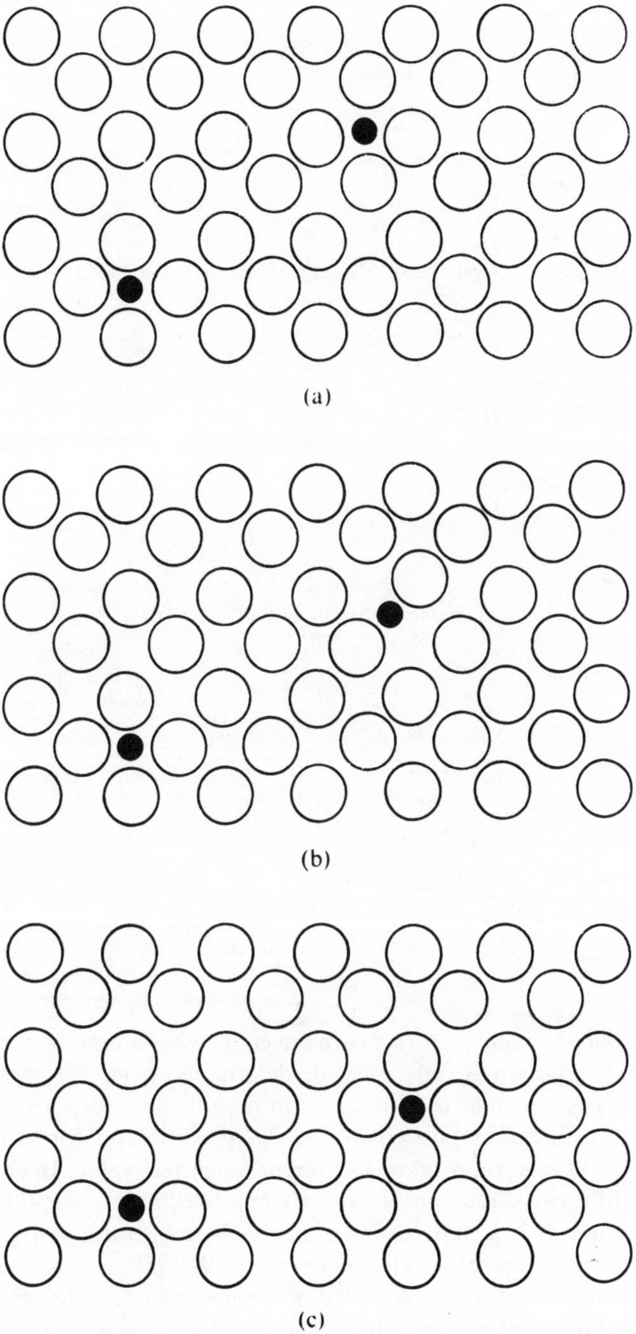

Figure 8.2.3 (*a*) Impurity atoms in a monatomic crystal before an atomic jump; (*b*) impurity atoms in activated state; (*c*) impurity atoms in a monatomic crystal after an atomic jump. (From Girifalco, 1964.)

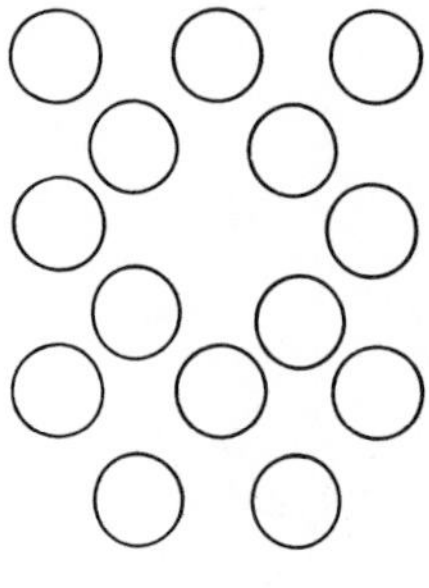

Figure 8.2.4 Vacancy. (From Girifalco, 1964.)

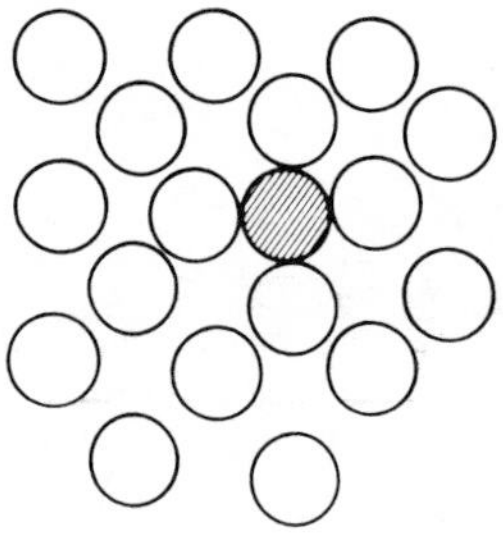

Figure 8.2.5 (From Girifalco, 1964.)

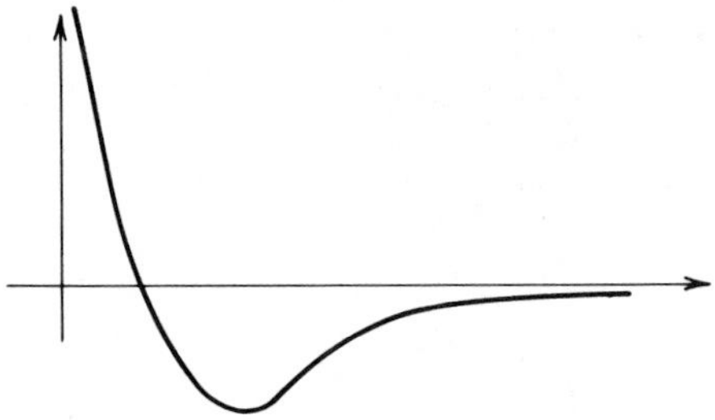

Figure 8.2.6 Potential of internuclear forces.

The quantity $\bar{\tau}$, which represents the average escape time of a Brownian particle from a potential well, must be determined from the properties of the crystal. The constant α equals $1/6$ in a cubic structure for which the jumps in all 6 directions are equally likely. In structures for which other than 6 possibilities exist, α takes a different numerical value. In addition, if jumps in different directions have different likelihoods, the diffusion is anisotropic and D is a matrix. Then the Laplace operator is replaced by a more general second-order elliptic operator. We will compute the diffusion matrix and show its dependence on temperature, on the structure of the lattice, and on other physical parameters. Knowledge of the diffusion coefficient enables us to determine certain important physical properties of crystals. For example, the electrical conductivity σ of ionic crystals is determined by the Nernst–Einstein formula

$$\sigma = \frac{Cq^2D}{kT},$$

where D is the diffusion coefficient, C the ratio of the concentration of interstitials to their atomic fraction, q the charge, T the temperature, and k is the Boltzmann constant (Girifalco 1964). In Section 8.7 a derivation of the Nernst–Einstein formula from the Smoluchowski equation is presented.

We now propose a model to describe the random motion of the interstitial particle or vacancy, in the cell, by computing the average escape time $\bar{\tau}$ and the probability distribution of the escape directions. This model is analogous to the model for chemical reaction rates where the random motion is due to the random collisions between the molecules of the reactants and the surrounding medium. In this model, the random motion is due to the thermal vibrations of the crystallic lattice. We consider each lattice atom as a Brownian harmonic oscillator, so that its velocity distribution is normal with zero mean and variance kT/m (see Section 2.1).

The motion of the interstitial atom is therefore governed by the Smoluchowski differential equation

$$d\mathbf{x} = -\nabla\Phi(\mathbf{x})\,dt + \sqrt{\frac{2kT}{m}}\,d\mathbf{w}(t), \tag{8.2.2}$$

where $\mathbf{x}$ is the displacement of the interstitial and Φ is the potential of the atomic forces binding it to its position. Another model, to describe the migration of vacancies, say, is to consider a lattice consisting of N atoms as a point in the $3N$-dimensional configuration space. Each equilibrium state of the lattice is a minimum of the potential Φ of the binding forces of the lattice. The potential Φ is therefore a function of the n independent coordinates (n degrees of freedom). Thus the migration of the vacancy corresponds to jumps of an n-dimensional Brownian particle from one potential well to another. Equation (8.2.2) for $\mathbf{x} = (x_1, \dots, x_n)^T$ describes this situation. The mathematical method we describe below is applicable to both models.

To determine the diffusion tensor for nonisotropic lattices [e.g., the zinc or graphite lattice (Fig. 8.2.7)] we note that the jump frequencies will be different in different directions and that the probability of passage through the saddle points connecting the potential wells may be different in

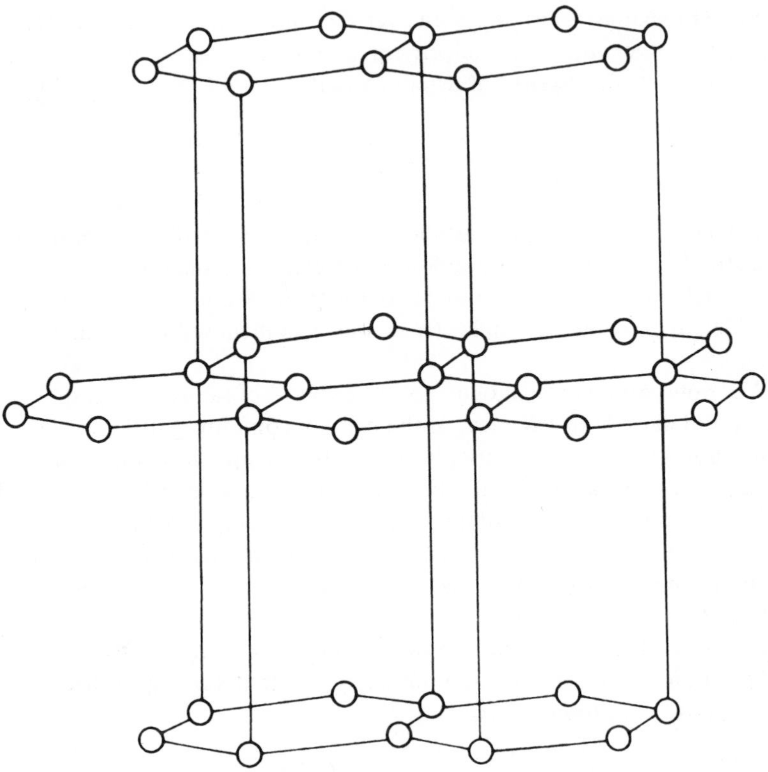

Figure 8.2.7 Graphite structure. (From Girifalco, 1964.)

different directions as well. As an example, the diffusion coefficient in the (x, y) plane may differ from that in the z-direction, so that the diffusion equation will take the form

$$D_1\left(\frac{\partial^2 u}{\partial x^2}+\frac{\partial^2 u}{\partial y^2}\right)+D_2\frac{\partial^2 u}{\partial z^2}=\frac{\partial u}{\partial t}. \tag{8.2.3}$$

The different values of D_1 and D_2 may be caused by jump frequencies in the (x, y) plane which differ from those in the z-direction. In addition, differences in the sizes of the jumps as well as differences in the probabilities of the jumps may cause the diffusion coefficients D_1 and D_2 to differ. This last situation is best illustrated by an example. Consider a lattice in which the interstitial atom may move from one cell to another along one of six possible paths, as shown in Fig. 8.2.8. Let the jump probabilities along lines with one, two, or three hatches have probabilities p, q, and r,

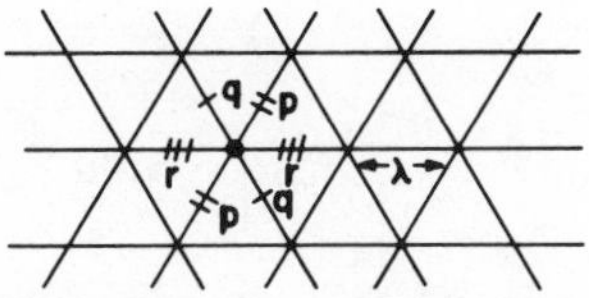

Figure 8.2.8 Nonisotropic lattice.

respectively. Thus $2(p+q+r)=1$. The backward Kolmogorov equation for the transition probability of this random walk is given by $P(x, y, a, b, (n+1)\bar{\tau})$ = probability of reaching the point (x, y) in the lattice in $(n+1)$ jumps $\bar{\tau}$ apart, given that initially the particle is at (a, b)

$$\begin{aligned}
&= r\left[P(x, y, a+\lambda, b, \bar{\tau}n) + P(x, y, a-\lambda, b, \bar{\tau}n)\right] \\
&\quad + p\left[P\left(x, y, a+\frac{1}{2}\lambda, b+\frac{\sqrt{3}}{2}\lambda, \bar{\tau}n\right)\right. \\
&\quad \left. + P\left(x, y, a-\frac{1}{2}\lambda, b-\frac{\sqrt{3}}{2}\lambda, \bar{\tau}n\right)\right] \\
&\quad + q\left[P\left(x, y, a-\frac{1}{2}\lambda, b+\frac{\sqrt{3}}{2}\lambda, \bar{\tau}n\right)\right. \\
&\quad \left. + P\left(x, y, a+\frac{1}{2}\lambda, b-\frac{\sqrt{3}}{2}\lambda, \bar{\tau}n\right)\right].
\end{aligned}$$

Thus

$$\begin{aligned}
&P(x, y, a, b, \bar{\tau}(n+1)) - P(x, y, a, b, \bar{\tau}n) \\
&\quad = P_t(x, y, a, b, \bar{\tau}n)\bar{\tau} + o(\bar{\tau}) \\
&\quad = \lambda^2\left[P_{aa}(x, y, a, b, \bar{\tau}n)\left(r+\frac{1}{2}p+\frac{1}{2}q\right) + P_{ab}(x, y, a, b, \bar{\tau}n)\right. \\
&\quad\quad \left. \times \frac{\sqrt{3}}{2}(p-q) + P_{bb}(x, y, a, b, \bar{\tau}n)\frac{3}{2}(p+q)\right] + o(\lambda^2).
\end{aligned}$$

Thus

$$\frac{\bar{\tau}}{\lambda^2}\frac{\partial p}{\partial t} = \frac{1}{2}\frac{\partial^2 p}{\partial a^2}\left(\frac{1}{2}+r\right) + \sqrt{3}\,(p-q)\frac{\partial^2 p}{\partial a \partial b} + \frac{3}{2}\left(\frac{1}{2}-r\right)\frac{\partial^2 p}{\partial b^2}$$

or

$$\frac{\partial p}{\partial t}=\frac{\lambda^2}{\bar{\tau}}\left[\frac{\frac{1}{2}+r}{2}\frac{\partial^2 p}{\partial a^2}+\sqrt{3}\,(p-q)\frac{\partial^2 p}{\partial a\partial b}+\frac{3}{2}\left(\frac{1}{2}-r\right)\frac{\partial^2 p}{\partial b^2}\right].$$

Therefore, the diffusion tensor has the form

$$D=\frac{\lambda^2}{\bar{\tau}}\begin{bmatrix}\frac{1}{2}\left(\frac{1}{2}+r\right) & \frac{\sqrt{3}}{2}(p-q)\\ \frac{\sqrt{3}}{2}(p-q) & \frac{3}{2}\left(\frac{1}{2}-r\right)\end{bmatrix}.$$

In principal-axis coordinates, D is given by

$$D=\begin{bmatrix}\lambda_1 & 0\\ 0 & \lambda_2\end{bmatrix},$$

where $\lambda_{1,2}$ are the distinct eigenvalues of the matrix above. This example illustrates how the diffusion coefficients may differ in different directions. In fact, however, in-plane anisotropic effects in hexagonal lattices are not observed experimentally. They are observed in other lattices (e.g., in orthorhombic lattices).

More generally, let the possible sites into which a particle, initially located at the origin, can jump, be given by the vectors $\mathbf{z}_k=(z_k^1, z_k^2, z_k^3)^T$ $(k=1,\ldots,2n)$. Furthermore, assume that the jumps occur at time intervals $\bar{\tau}$ apart, with probabilities $p_1,\ldots, p_{2n}$ respectively, where $\Sigma_i p_i=1$. Then the transition probability of the random walk satisfies the backward equation

(8.2.4) $$P(\mathbf{x},\mathbf{y},(n+1)\bar{\tau})=\sum_{j=1}^{2n} P(\mathbf{x}+\mathbf{z}_j,\mathbf{y}, n\bar{\tau})p_j,$$

where $\mathbf{x}=(x_1, x_2, x_3)^T, \mathbf{y}=(y_1, y_2, y_3)^T$. Hence, expanding the left- and right-hand sides of (8.2.11) in Taylor series about $n\bar{\tau}$ and $\mathbf{z}_j$, respectively, and retaining only the leading terms, we obtain

$$\frac{\partial p}{\partial t}=\sum_{i,j=1}^{3} D_{ij}\frac{\partial^2 p}{\partial x_i\,\partial x_j},$$

where $t=n\bar{\tau}$ and

(8.2.5) $$D_{ij}=\frac{\sum_{k=1}^{n} p_k z_k^i z_k^j}{\bar{\tau}}.$$

We have used the facts that $\mathbf{z}_{n+j}=-\mathbf{z}_j$ and $p_{n+j}=p_j$ $(j=1,\ldots,n)$, which follow from the periodicity of the lattice, in order to cancel the first-order spatial derivatives. Thus the distribution of the exit points, as well as the expected exit time, determines the diffusion matrix in the lattice. The computations of these quantities are given in Sections 8.6 and 8.7.

8.3 THE EXIT PROBLEM IN ONE DIMENSION AND MULTIPLE TRANSITION STATES

We consider the one-dimensional stochastic differential equation

$$dx=b(x)\,dt+\sqrt{2\varepsilon}\,dw$$

and we set

$$b(x)=-\frac{d}{dx}\Phi(x).$$

Our aim is to compute the exit distribution of the exit points. More specifically, we shall compute the proportion of particles leaving $\Omega=(a,b)$ at a and at b given their initial distribution. Equation (5.4.8) in this case is given by

$$\varepsilon u_\varepsilon''+b(x)u_\varepsilon'=0 \tag{8.3.1}$$

$$u_\varepsilon(a)=\alpha, \qquad u_\varepsilon(b)=\beta.$$

The solution of the reduced equation (8.3.1) with $\varepsilon=0$ must be constant on intervals where $b(x)\neq 0$. Therefore, as $\varepsilon\to 0$, internal layers, or regions where the solution of (8.3.1) changes rapidly from one constant to another, can only develop near zeros of $b(x)$. We first determine which of the zeros are sites of internal layers.

Let z_i be the zeros of $b(x)$. Assume that the Taylor series expansion of $b(x)$ near $x=z_i$ is given by $b(x)=A_i(x-z_i)^{k_i}+\cdots$. We introduce a stretching transformation in a neighborhood of $x=z_i$ by setting $\xi=(x-z_i)/\varepsilon^{\alpha_i}$. Then (8.3.1) becomes

$$\varepsilon^{1-2\alpha_i}u_{\xi\xi}+\left[A_i\varepsilon^{\alpha_i(k_i-1)}\xi^{k_i}+\cdots\right]u_\xi=0. \tag{8.3.2}$$

Hence, choosing $\alpha_i=1/(k_i+1)$, we see that the leading term u^0 in the internal layer expansion of u_ε about $\xi=0$ satisfies the equation

$$u_{\xi\xi}^0+A_i\xi^{k_i}u_\xi^0=0. \tag{8.3.3}$$

Since $u_\varepsilon(x)\to C_i$ if $z_i<x<z_{i+1}$, the matching conditions imply that $u^0(\xi)\to C_{i-1}$ as $\xi\to-\infty$ and $u^0(\xi)\to C_i$ as $\xi\to\infty$. The solution of (8.3.3) is given by

$$u^0(\xi)=\gamma_1\int_0^\xi \exp\left(-\frac{A_i\xi^{k_i+1}}{k_i+1}\right)d\xi+\gamma_2, \tag{8.3.4}$$

where the constants γ_1 and γ_2 are to be determined. If k_i is even, then $|u^0(\xi)|\to\infty$ either as $\xi\to\infty$ or as $\xi\to-\infty$, unless $\gamma_1=0$. Thus $\gamma_1=0$ and no layers occur, since $C_{i-1}=C_i=\gamma_2$. If k_i is odd and $A_i<0$, the same conclusion is drawn. Thus layers may occur only if k_i is odd and $A_i>0$, that is, at zeros of $b(x)$ at which $b(x)$ is increasing [transition points of $\Phi(x)$, namely at $z_i=y_i$ in fig. 8.1.2]. Then

$$\gamma_1=(C_i-C_{i-1})B_i \quad \text{and} \quad \gamma_2=\frac{C_i+C_{i-1}}{2} \tag{8.3.5}$$

with

$$B_i=\frac{\frac{1}{2}}{\int_0^\infty \exp\left[-A_i\xi^{k_i+1}/(k_i+1)\right]d\xi} =\frac{A_i^{(k_i+1)}(k_i+1)^{k_i/(k_i+1)}}{2\Gamma\left[1/(k_i+1)\right]}, \tag{8.3.6}$$

where $\Gamma(z)$ denotes the gamma function and the constants C_i will be determined below. Not all transition points will turn out to be internal layer sites, as will be seen below. Therefore, the expansion of $u_\varepsilon(x)$ near $x=y_i$ is given by

$$u_\varepsilon(x)\sim\frac{C_i+C_{i-1}}{2} +(C_i-C_{i-1})B_i\int_0^{(x-y_i)\varepsilon^{-1/(k_i+1)}}\exp\left(-\frac{A_i\xi^{k_i+1}}{k_i+1}\right)d\xi. \tag{8.3.7}$$

Near the end points a and b boundary layers may develop. The boundary layer expansion of $u_\varepsilon(x)$ near $x=a\equiv y_0$ is given by

$$u_\varepsilon(x)\sim\alpha+(C_0-\alpha)B_0\int_0^{(x-a)\varepsilon^{-1/(k_i+1)}}\exp\left(-\frac{A_0\xi^{k_0+1}}{k_0+1}\right)d\xi, \tag{8.3.8}$$

and near $x=b\equiv y_{n+1}$ by

(8.3.9) $$u_\varepsilon(x)\sim\beta+(C_n-\beta)B_n\int_0^{(b-x)\varepsilon^{-1/(k_{n+1}+1)}}\exp\left(-\frac{A_{n+1}\xi^{k_{n+1}+1}}{k_{n+1}+1}\right)d\xi.$$

To find the constants C_i, we multiply (8.3.1) by $e^{-\Phi(x)/\varepsilon}$ and integrate over each interval (y_i, y_j) $(i<j)$. We obtain

(8.3.10) $$e^{-\Phi(y_i)/\varepsilon}u'_\varepsilon(y_i)-e^{-\Phi(y_j)/\varepsilon}u'_\varepsilon(y_j)=0.$$

Hence, using the expansion (8.3.7), we obtain

(8.3.11) $$\begin{aligned} &e^{-\Phi(y_i)/\varepsilon}B_i(C_i-C_{i-1})\varepsilon^{-1/(k_i+1)}\\ &-e^{-\Phi(y_j)/\varepsilon}B_j(C_j-C_j-1)\varepsilon^{-1(k_j+1)}\sim 0.\end{aligned}$$

If $\Phi(y_i)<\Phi(y_j)$, then the first term in (8.3.11) dominates the second one, so that $C_i-C_{i-1}=0$, and no layer occurs at $x=y_i$. Clearly, therefore, internal layers can occur only at points y_i, for which $\Phi(y_i)=\max_j\Phi(y_j)$. Let $y'_1,\ldots, y'_{n'}$ denote the points where $\Phi(y_i)$ is maximal. Applying condition (8.3.11) at two such points of maximum, we see that $C_i=C_{i-1}$ if $k_i\neq\max_i k_i$. Let $y''_1,\ldots, y''_{n''}$ denote the points y'_i at which $k_i=\max_i k_i$. Henceforth we shall drop all primes for simplicity of notation. We have shown that internal layers occur only at transition points at which Φ achieves its absolute maximum in (a, b), and at which $b(x)$ vanishes to maximal order. Now we consider (8.3.10) with $i=0$. Then using the expansion (8.3.8), we obtain

(8.3.12) $$\begin{aligned} &e^{-\Phi(a)/\varepsilon}B_0(C_0-\alpha)\varepsilon^{-1/(k_0+1)}\\ &-e^{-\Phi(y_j)/\varepsilon}B_j(C_j-C_{j-1})\varepsilon^{-1/(k_j+1)}\sim 0.\end{aligned}$$

Therefore, if $\Phi(a)<\max_j\Phi(y_j)$, then $C_0=\alpha$. If both $\Phi(a)$ and $\Phi(b)$ are maximal, we obtain the following system of $n+1$ independent linear equations for the $n+1$ constants C_i:

(8.3.13)

$$\begin{bmatrix} B_0+B_1 & -B_1 & 0 & 0 & \cdots & \cdot \\ -B_1 & B_1+B_2 & -B_2 & 0 & & \cdot \\ 0 & -B_2 & B_2+B_3 & -B_3 & & \cdot \\ \vdots & & & \ddots & & 0 \\ \cdot & \cdot & 0 & -B_{n-1} & B_n & +B_{n+1} \end{bmatrix}\begin{bmatrix} C_0 \\ \cdot \\ \cdot \\ \cdot \\ C_n \end{bmatrix}=\begin{bmatrix} B_0\alpha \\ 0 \\ \vdots \\ 0 \\ B_{n+1}\beta \end{bmatrix}.$$

If $\Phi(a)<\max\Phi$, then (8.3.12) implies that $C_0=\alpha$. Similarly, if $\Phi(b)<\max\Phi$, then $C_n=\beta$. In either case the system reduces in the obvious way to a lower-order system. The solution of (8.3.13) is given by

$$C_j=(\alpha P_{n,j}+\beta Q_{n,j})P_n \tag{8.3.14}$$

$$P_n=\sum_{j=0}^{n+1}\left\{\frac{\prod_{i=0}^{n+1} B_i}{B_{n-j+1}}\right\}\equiv\sum_{j=0}^{n+1} P_j,$$

$$P_{n,j}\equiv\sum_{i=0}^{j} P_i \qquad \text{and} \qquad Q_{n,j}\equiv P_n-P_{n,j}.$$

We can now summarize this result as follows. *Let $u_\varepsilon(x)$ be the solution of the boundary value problem*

$$\varepsilon u''+b(x)u'=0 \quad \text{in} \quad (a,b)$$

$$u(a)=\alpha, \qquad u(b)=\beta,$$

where $b(x)=-\Phi'(x)$. *Let $y_1,\ldots,y_n$ be transition points of $\Phi(x)$ in (a,b). Let $s_1<s_2<\cdots<s_l$ be those points y_i where $\Phi(s_j)=\max_i\Phi(y_i)(j=1,\ldots,l)$. Let $t_1<t_2<\cdots<t_m$ be those points s_j where $b(x)$ vanishes to maximal order k (maximum with respect to j, $1\leqslant j\leqslant m$). Assume that the Taylor series expansion of $b(x)$ about t_i is given by*

$$b(x)=A_i(x-t_i)^k+\cdots,$$

where $A_i>0$. Set $B_i=(A_i)^{1/(k+1)}$; then

$$u_\varepsilon(x)\to\sum_{j=0}^{m} C_j\chi_{(t_j,t_{j+1})}(x) \qquad \textit{as} \quad \varepsilon\to 0, \quad x\in(t_j,t_{j+1})$$

where

$$C_j=\frac{\alpha P_{m,j}+\beta Q_{m,j}}{P_m} \qquad \text{and} \qquad \chi_{(t_j,t_{j+1})}(x)=\begin{cases}1 & \text{if} \quad x\in(t_j,t_{j+1})\\ 0 & \text{otherwise.}\end{cases}$$

The corresponding result for the exist problem can be stated as follows. *Let $x_\varepsilon(t)$ be the solution of the stochastic differential equation*

$$dx_\varepsilon(t)=b(x_\varepsilon)\,dt+\varepsilon\,dw(t),\ x_\varepsilon(0)=x\in(a,b)$$

and let $\tau=\inf\{t\,|\,x_\varepsilon(t)=a \text{ or } x_\varepsilon(t)=b\}$ *be the exist time of* $x(t)$ *from* (a, b). *If* $b(x)$ *and* $t_j(1\leqslant j\leqslant m)$ *are as above, then the probability of exiting at* $x=a$ *is given by*

$$(8.3.15)\quad P\{x_\varepsilon(\tau)=a\,|\,x_\varepsilon(0)=x\}\to\sum_{j=0}^{m} D_j\chi_{(t_j,t_{j+1})}(x)\quad \text{as}\quad \varepsilon\to 0,$$

where $D_j=P_{m,j}/P_m$, *and the probability of exiting at* $x=b$ *is given by*

$$P(x_\varepsilon(\tau)=b\,|\,x_\varepsilon(0)=x)=1-P(x_\varepsilon(\tau)=a\,|\,x_\varepsilon(0)=x).$$

Note. The exact solution of (8.3.1) in case $\alpha=0$, $\beta=1$, say, is given by

$$u_\varepsilon(x)=\frac{\int_a^x e^{\Phi(s)/\varepsilon}\,ds}{\int_a^b e^{\Phi(s)/\varepsilon}\,ds}.$$

Expansion about the absolute maxima of Φ in case of simple zeros of $b(x)$ yields

$$u_\varepsilon(x)\to\frac{\sum_{t_i<x}|\Phi''(t_i)|^{-1/2}}{\sum_i|\Phi''(t_i)|^{-1/2}}\qquad (x\neq t_i),$$

which agrees with (8.3.14).

EXERCISE 8.3.1

Discuss the cases $f(a)>0$, $f(b)<0$, and so on.

8.4 MULTIPLE TRANSITION STATES FOR ELLIPTIC PARTIAL DIFFERENTIAL EQUATIONS AND THE EXIT PROBLEM

As mentioned in Section 8.1, different directions of exit from the potential well may result in different reactions. The proportion of the different reactions are therefore determined by the exit density on $\partial\Omega$. In the context of the problem of atomic migration in crystals, the exit density determines

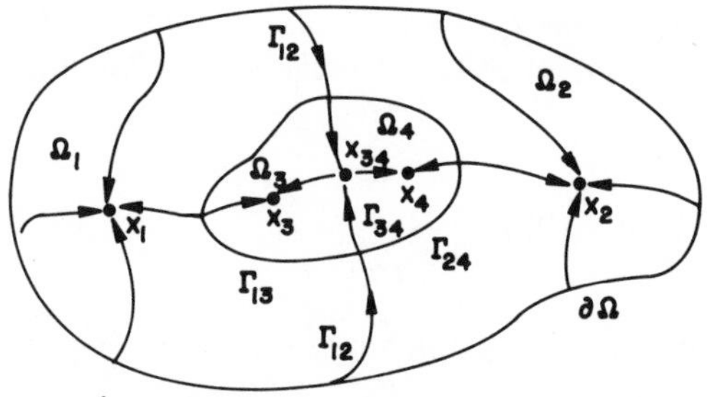

Figure 8.4.1 Potential well with multiple transition states.

the diffusion matrix in the anisotropic case. Let Ω be a multiple potential well (see Fig. 8.4.1). Let $\mathbf{x}_1,\ldots,\mathbf{x}_n$ be the isolated local minima of the potential $\mathbf{\Phi}$ in Ω.

These are the stable equilibria of the system

$$\text{(8.4.1)}\qquad \begin{cases} \dfrac{d\mathbf{x}}{dt} = \mathbf{b}(\mathbf{x}) \\ \mathbf{b}(\mathbf{x}) = -\nabla\Phi \end{cases}$$

in Ω. Since $\partial\Omega$ is a repulsive boundary on either side, all trajectories that begin in Ω will tend to equilibrium points. We assume that Ω is partitioned into domains of attraction of $\mathbf{x}_i$; that is, there are subdomains $\Omega_i \subset \Omega$ such that:

(i) $\mathbf{x}_j \in \Omega_j (j=1,\ldots,n)$, and if $\mathbf{x}\in\Omega_j$ and $\mathbf{x}(0)=\mathbf{x}$, then $\mathbf{x}(t)\to\mathbf{x}_j$ as $t\to\infty$.
(ii) $\cup_{j=1}^{n}\bar{\Omega}_j=\bar{\Omega}$.

We assume that $\partial\Omega_k$ consists of piecewise smooth surfaces Γ_{ij}. Assume that $\Omega_1,\ldots,\Omega_k$ are continguous to the boundary $\partial\Omega$, and $\Omega_{k+1},\ldots,\Omega_n$ are surrounded by the other $\Omega_i - s$ (see Fig. 8.4.1 and 8.6.4). Let $u_\varepsilon(\mathbf{x})$ be the solution of the Dirichlet problem

$$\text{(8.4.2)}\qquad \varepsilon\Delta u_\varepsilon + \mathbf{b}(\mathbf{x})\cdot\nabla u_\varepsilon = 0 \quad \text{in} \quad \Omega$$

$$\text{(8.4.3)}\qquad u_\varepsilon(\mathbf{x}) = f(\mathbf{x}) \quad \text{on} \quad \partial\Omega.$$

As $\varepsilon\to 0$ the solution u_ε converges to a solution u_0 of the reduced problem $\mathbf{b}(\mathbf{x})\cdot\nabla u = 0$. Thus, as shown before, $u_0(\mathbf{x})$ must be constant along the trajectories $\mathbf{x}(t)$ of (8.4.1). It follows that $u_0(\mathbf{x}) = C_j$ if $\mathbf{x}\in\Omega_j$. Therefore, boundary and internal layers will appear at $\partial\Omega$ and at Γ_{ij}, respectively.

Internal layers across Γ_{ij} are constructed as follows. Let (ξ,η) be local coordinates near Γ_{ij}, where ξ is a tangential coordinate in Γ_{ij} and η is the distance from Γ_{ij}. The surface Γ_{ij} is a "singular surface" of characteristics; that is, any solution $\mathbf{x}(t)$ of (8.4.1) such that $\mathbf{x}(0)=\mathbf{x}\in\Gamma_{ij}$ stays on Γ_{ij}. Therefore, $\mathbf{x}(t)$ must approach the ω-limit set on Γ_{ij}. We assume that this set consists of isolated stable equilibrium points only. This represents no loss of generality, since the other cases lead to the same results. Any equilibrium point on Γ_{ij} is unstable in Ω, although it may be stable on Γ_{ij}. For example, in Fig. 8.4.1, the point $\mathbf{x}_{34}$ is stable on Γ_{34} but is unstable in Ω. We will employ the characteristics as coordinates in Γ_{ij} since the solution of the internal layer equation is independent of the choice of the origin (Schuss 1972). Let $\boldsymbol{\xi}$ be the distance from a point $\mathbf{x}$ on Γ_{ij} to a stable equilibrium point $\mathbf{x}_{ij}$ on Γ_{ij} measured along a characteristic $\mathbf{x}(t)$ on Γ_{ij}, so that $\mathbf{x}(t_1)=\mathbf{x}$, and $\mathbf{x}(t)\to\mathbf{x}_{ij}$ as $t\to\infty$. Assume that in the neighborhood of Γ_{ij}, $\mathbf{b}\cdot\boldsymbol{\nu}=\eta b_1(\xi,\eta)+o(\eta)$. This assumption on the shape of the potential Φ at the transition states Γ_{ij} is usually made in the physical literature. We note that different assumptions may lead to different physical results. In the coordinate system (ξ,η), (8.4.2) then takes the form

$$u^0_{\omega\omega}+\omega B(s)u^0_{\omega}+C(s)u_s=0, \tag{8.4.4}$$

where d/ds is differentiation along the characteristic direction. Since $C(s)$ vanishes at all equilibrium points s on Γ_{ij}, (8.4.4) is a degenerate parabolic equation that must be solved for all s. There exists a unique solution to the boundary value problem for (8.4.4), determined from the matching conditions (Schuss 1972). The matching conditions are

$$u^0(\omega,s)\to C_j \quad \text{as} \quad \omega\to\infty \tag{8.4.5}$$

and

$$u^0(\omega,s)\to C_i \quad \text{as} \quad \omega\to-\infty. \tag{8.4.6}$$

The solution $u^0(\omega,s)$ is given by

$$u^0(\omega,s)=(C_j-C_i)\int_0^{\omega\gamma(s)}e^{-s^2/2}\,ds+\frac{C_j+C_i}{2}, \tag{8.4.7}$$

where $\gamma(s)$ is the solution of the Bernoulli equation

$$\frac{d\gamma}{ds}+\frac{B(s)}{C(s)}\gamma-\frac{1}{C(s)}\gamma^3=0 \tag{8.4.8}$$

with $\gamma(0)=\sqrt{B(0)}$, so that $u^0(\omega,0)$ is the solution of (8.4.4) at $s=0$. Hence

$$\gamma(s)=\left[w\int_0^s C(\sigma)^{-1}\exp\left(-2\int_\sigma^s \frac{B(t)}{C(t)}\,dt\right)d\sigma\right]^{-1/2}.$$

Therefore,

$$\frac{\partial u^0}{\partial \omega}(0,s)=\frac{C_j-C_i}{2\sqrt{\pi}}\gamma(s). \tag{8.4.9}$$

In a similar manner, the boundary layer expansion is constructed in terms of the local coordinates $\mathbf{x}=(\mathbf{x}', y)$, where $y=\rho(\mathbf{x})=\text{dist}(\mathbf{x},\partial\Omega)$, and $\mathbf{x}'$ is a tangential coordinate, as

$$u_\varepsilon(\mathbf{x})\sim\sqrt{\frac{2}{\pi}}\,(f(\mathbf{x}')-C_i)\int_{\rho(\mathbf{x})\gamma(\mathbf{x}')/\varepsilon^{1/2}}^{\infty} e^{-s^2/2}\,ds+C_i. \tag{8.4.10}$$

The only difference between the derivation of (8.4.7) and the derivation of (8.4.10) is that (8.4.6) is replaced by the condition $u^0(\omega,s)\to f(\mathbf{x}')$ as $\omega\to 0$. To find C_i, we multiply (8.4.2) by $e^{-\Phi/\varepsilon}$ and integrate over each Ω_i. We obtain

$$\int_{\partial\Omega\cap\partial\Omega_i} e^{-\Phi/\varepsilon}\frac{\partial u_\varepsilon}{\partial \nu}\,dS+\sum_j\int_{\Gamma_{ij}} e^{-\Phi/\varepsilon}\frac{\partial u_\varepsilon}{\partial \nu}\,dS=0. \tag{8.4.11}$$

The main contribution to the integrals comes from the points $\mathbf{x}'$, where Φ is minimal on $\partial\Omega\cap\partial\Omega_i$ and Γ_{ij}, respectively. At such points (8.4.4) reduces to

$$u^0_{\omega\omega}+\omega b_1(\omega,0)u^0_\omega=0; \tag{8.4.12}$$

hence (8.4.9) can be replaced by

$$\frac{\partial u^0}{\partial \nu}(\mathbf{x}',0)=\varepsilon^{-1/2}\frac{C_j-C_i}{2\sqrt{\pi}}[b_1(\mathbf{x}')]^{1/2} \quad \text{on} \quad \Gamma_{ij} \tag{8.4.13}$$

and the expansion (8.4.10) yields

$$\frac{\partial u^0(\mathbf{x}',0)}{\partial \nu}=\varepsilon^{-1/2}\frac{(f(\mathbf{x}')-C_i)[b_1(\mathbf{x}')]^{1.2}}{2\sqrt{\pi}}. \tag{8.4.14}$$

Employing (8.4.13) and (8.4.14) in (8.4.11), we obtain the equation

$$(8.4.15)\qquad \frac{\varepsilon^{-1/2}}{2\sqrt{\pi}}\int_{\partial\Omega\cap\partial\Omega_i}(f(\mathbf{x}')-C_i)[b_1(\mathbf{x}')]^{1/2}e^{-\Phi/\varepsilon}dS_{\mathbf{x}'}$$
$$+\sum_j \frac{C_j-C_i}{2\sqrt{\pi}}\varepsilon^{-1/2}\int_{\Gamma_{ij}} e^{-\Phi/\varepsilon}[b_1(\mathbf{x}')]^{1/2}\,dS_{\mathbf{x}'}\sim 0.$$

First we consider the case that the region Ω_i is contiguous to $\partial\Omega$. Then we must compare $\Phi_i=\min_{\partial\Omega\cap\partial\Omega_i}\Phi$ with $\Phi_{ij}=\min_{\Gamma_{ij}}\Phi$. If $\Phi_i>\Phi_{ij}$, then $C_i=C_j$ and no internal layer is formed, and we may disregard Γ_{ij}, across which no discontinuity appears. Thus we consider $\Phi_{ij}\geqslant\Phi_i$. If $\Phi_{ij}>\Phi_i$, then

$$(8.4.16)\qquad C_i=\lim_{\varepsilon\to 0}\frac{\displaystyle\int_{\partial\Omega\cap\partial\Omega_i} e^{-\Phi/\varepsilon}b_1^{1/2}(\mathbf{x}')f(\mathbf{x}')dS}{\displaystyle\int_{\partial\Omega\cap\partial\Omega_i} e^{-\Phi/\varepsilon}b_1^{1/2}(x')\,dS}$$
$$\equiv\int_{\partial\Omega\cap\partial\Omega_i} f(\mathbf{x}')\mu_i(\mathbf{x}')\,dS_{\mathbf{x}'}.$$

This case is analogous to the case $\Phi(a)<\max\Phi$ in the previous section, where C_0 was given explicitly in terms of the boundary data. The measure $\mu_i(\mathbf{x}')\,dS_{\mathbf{x}'}$ is determined by evaluating the integral in (8.4.16) asymptotically as in Section 7.4. If $\Phi_{ij}=\Phi_i$, we obtain a linear equation for the coefficient C_i.

Next we consider the case that the region Ω_i is not contiguous to $\partial\Omega$. Then integration over Ω_i yields

$$(8.4.17)\qquad C_i=\lim_{\varepsilon\to 0}\frac{\displaystyle\sum_j C_j\int_{\Gamma_{ij}} b_1^{1/2}(\mathbf{x}')e^{-\Phi/\varepsilon}dS_{\mathbf{x}'}}{\displaystyle\int_{\partial\Omega_i} b_1^{1/2}(\mathbf{x}')e^{-\Phi/\varepsilon}dS_{\mathbf{x}}},\qquad (i=k+1,\ldots,n),$$

where $\partial\Omega_i=\cup_j\Gamma_{ij}$. Evaluating the integrals in (8.4.16) and (8.4.17) asymptotically as in Section 7.4, we obtain

$$(8.4.18)\qquad C_i=\sum_j C_jD_{ij}$$

where

$$D_{ij}=\lim_{\varepsilon\to 0}\frac{\int_{\Gamma_{ij}} b_1^{1/2}(\mathbf{x}')e^{-\Phi/\varepsilon}\,ds_{\mathbf{x}'}}{\int_{\partial\Omega_i} b_1^{1/2}(\mathbf{x}')e^{-\Phi/\varepsilon}\,dS_{\mathbf{x}'}}$$

$$=\int_{\Gamma_{ij}} b_1^{1/2}(\mathbf{x}')\mu_{ij}(\mathbf{x}')\,dS_{\mathbf{x}'}\qquad (i=k+1,\ldots,n).$$

Note that $\Sigma_j D_{ij}=1$, $D_{ij}\geqslant 0$, so that the coefficients are probabilities. Equations (8.4.18) are the analogues of the second through $(n-1)$th equation in (8.3.10), which are homogeneous linear equations. Equations (8.4.16) are the analogues of the first and last equation in (8.3.10), which are inhomogeneous. They may be written in the form

(8.4.19) $$\int_{\partial\Omega\cap\partial\Omega_i} f(\mathbf{x}')b_1^{1/2}(\mathbf{x}')\mu_i(\mathbf{x}')\,dS_{\mathbf{x}'}$$

$$=C_i\int_{\partial\Omega\cap\partial\Omega_i} b_1^{1/2}(\mathbf{x}')\mu_i(\mathbf{x}')\,dS_{\mathbf{x}'}+\sum_j\frac{C_i-C_j}{2}\int_{\Gamma_{ij}} b_1^{1/2}(\mathbf{x}')\mu_{ij}(\mathbf{x}')\,dS_{\mathbf{x}'}.$$

Hence (8.4.18) and (8.4.19) imply that each coefficient C_i is some linear combination of the left-hand side of equation (8.4.19), that is,

$$C_i=\sum_{j=1}^{k} d_{ij}\int_{\partial\Omega\cap\partial\Omega_i} f(\mathbf{x}')b_1^{1/2}(\mathbf{x}')\mu_j(\mathbf{x}')\,dS_{\mathbf{x}'},$$

where the constants d_{ij} may be determined for any given system of equations. The exit density on $\partial\Omega$ is therefore given by

(8.4.20) $$p_\varepsilon(\mathbf{x},\mathbf{y})=P(\mathbf{x}_\varepsilon(\tau)=\mathbf{y}\in\partial\Omega\,|\,\mathbf{x}_\varepsilon(0)=\mathbf{x}\in\Omega)$$

$$\sim\sum_{i,j=1}^{n}\chi_{\Omega_i}(\mathbf{x})d_{ij}b_1^{1/2}(\mathbf{y})\mu_j(\mathbf{y}).$$

In case of a single stable equilibrium point in Ω,

$$\lim_{\varepsilon\to 0}\int_{\partial\Omega} f(\mathbf{y})p_\varepsilon(\mathbf{x},\mathbf{y})\,dS_{\mathbf{y}}=\lim_{\varepsilon\to 0}\frac{\int_{\partial\Omega} e^{-\Phi/\varepsilon}f(\mathbf{x}')b_1^{1/2}(\mathbf{x}')\,dS_{\mathbf{x}'}}{\int_{\partial\Omega} e^{-\Phi/\varepsilon}b_1^{1/2}(\mathbf{x}')\,ds_{\mathbf{x}'}},$$

so that (8.4.20) reduces to

$$(8.4.21)\qquad p(\mathbf{x},\mathbf{y}) \sim \frac{\chi_S(\mathbf{y}) b_1^{1/2}(\mathbf{y})}{\int_S b_1^{1/2}(\mathbf{y})\, dS_{\mathbf{y}}} \equiv p(\mathbf{y}), \qquad \mathbf{x}\in\Omega,\quad \mathbf{y}\in\partial\Omega$$

$$\equiv \frac{\chi_S(\mathbf{y})\left[\partial^2\Phi(\mathbf{y})/\partial\boldsymbol{\nu}^2\right]^{1/2}}{\int_S \left[\partial^2\Phi(\mathbf{y})/\partial\boldsymbol{\nu}^2\right]^{1/2} dS_{\mathbf{y}}}$$

where

$$S = \left\{\mathbf{y}\in\partial\Omega \,|\, \Phi = \min_{\partial\Omega}\Phi\right\} \qquad \text{and} \qquad \chi_S(\mathbf{y}) = \begin{cases} 1 & \text{if } \mathbf{y}\in S \\ 0 & \text{if } \mathbf{y}\notin S \end{cases},$$

provided that S has a nonempty interior. If $S=\{\mathbf{y}_1,\ldots,\mathbf{y}_m\}$ and

$$\det\frac{\partial^2\Phi}{\partial x_i'\partial x_j'}\bigg|_{\mathbf{y}_k} = H_k \neq 0 \qquad (i, j = 1,\ldots, n-1),$$

then

$$(8.4.22)\qquad p_\varepsilon(\mathbf{x},\mathbf{y}) \sim \frac{\sum_{j=1}^{m} (\partial^2\Phi/\partial\nu^2)^{1/2}(\mathbf{y}_j)\,\delta(\mathbf{y}-\mathbf{y}_j) H_j^{-1/2}}{\sum_{j=1}^{m} (\partial^2\Phi/\partial\boldsymbol{\nu}^2)^{1/2}(\mathbf{y}_j) H_j^{-1/2}} \equiv p(\mathbf{y}).$$

Note the difference between these results [(8.4.21) and (8.4.22)] for the case of smooth barriers, and the results for the case of sharp barriers as given in the previous pages.

Consider, for example, the case of two equilibrium points in Ω. Let u_ε be the solution of the Dirichlet problem

$$(8.4.23)\qquad \begin{aligned} \varepsilon\Delta u + x(1-x^2)u_x - yu_y &= 0 \quad \text{in } \Omega \\ u &= f \quad \text{on } \partial\Omega, \end{aligned}$$

where Ω is a bounded planar domain containing the three equilibrium points $(0,0)$ and $(\pm 1, 0)$. The boundary $\partial\Omega$ is smooth and $\mathbf{b}\cdot\boldsymbol{\nu} < 0$ on $\partial\Omega$, where $\mathbf{b} = (b_1, b_2)^T$ with $b_1 = x(1-x^2)$ and $b_2 = -y$. In this problem $\mathbf{b} = -\nabla\phi$ with $-\phi = x^2/2 - x^4/4 - y^2/2$. The points $(\pm 1, 0)$ are stable equilibria of the deterministic dynamical system

$$(8.4.24)\qquad \begin{cases} \dfrac{dx}{dt} = x(1-x^2) \\ \dfrac{dy}{dt} = -y; \end{cases}$$

thus there are no internal layers about the points $(\pm 1,0)$.

The solution of the reduced equation is a constant C_1 in $\Omega_1 = \Omega \cap \{x<0\}$, and is a (possibly) different constant C_2 in $\Omega_2 = \Omega \cap \{x>0\}$. It follows that an inner layer may be formed along the line $x=0$, containing the point $(0,0)$, which is a saddle point of the dynamical system (8.4.24). We introduce the stretched variable $\omega = x\varepsilon^{-1/2}$ near $\Gamma_{12} = \{x=0\}$ and find that the leading term of the inner layer expansion solves the degenerate parabolic boundary value problem

$$(8.4.25)\qquad u_{\omega\omega} + \omega u_\omega - y u_y = 0$$

for $-\infty < \omega < \infty$ and y in an open interval containing $y=0$. The matching conditions (8.4.6) and (8.4.7) imply that $u(\omega, y) \to C_{1,2}$ as $\omega \to \pm\infty$. The unique solution of (8.4.25) is given by

$$(8.4.26)\qquad \tilde{u}(\omega, y) = \frac{C_2 - C_1}{\sqrt{\pi}} \int_0^\omega e^{-s^2/2}\, ds + \frac{C_2 + C_1}{2}.$$

Finally, in the usual way we construct a boundary layer function in a neighborhood on $\partial\Omega$, so that the composite expansion is given by

$$(8.4.27)\qquad u_\varepsilon(x, y) \sim \tilde{u}(x\varepsilon^{-1}, y) + \left[f(x, y) - \tilde{u}(x\varepsilon^{-1}, y) \right] e^{\zeta(x,y)/\varepsilon}.$$

To calculate the constants C_1 and C_2, we multiply (8.4.23) by $e^{-\phi/\varepsilon}$ and integrate over each of the subdomains Ω_1 and Ω_2 to obtain the two equations

(8.4.28)

$$C_i \int_{\partial\Omega \cap \partial\Omega_i} e^{-\phi/\varepsilon} \mathbf{b}\cdot\boldsymbol{\nu}\, ds \sim \int_{\partial\Omega \cap \partial\Omega} e^{-\phi/\varepsilon} f\, \mathbf{b}\cdot\boldsymbol{\nu}\, ds$$
$$+ (-1)^i \frac{C_2 - C_1}{\sqrt{\pi}} \sqrt{\varepsilon} \int_{y_1}^{y_2} e^{-t^2/\varepsilon}\, dt, \qquad (i = 1,2)$$

where y_1 and y_2 are the points of intersection of $\partial\Omega$ with Γ_{12}. Now the contribution from the line segment Γ_{12} is $O(\varepsilon)$, while the contribution from the integrals over $\partial\Omega\cap\partial\Omega_i$ is $O(\sqrt{\varepsilon}\, e^{-\hat{\phi}/\varepsilon})$, where $\hat{\phi}_i$ is the minimum of ϕ on $\partial\Omega\cap\partial\Omega_i$. Therefore, the values of C_1 and C_2 depend on the signs of $\hat{\phi}_i(i=1,2)$. In particular, in case (A), if either of $\hat{\phi}_i(i=1,2)$ is positive, the contribution from the line segment Γ_{12} dominates and $C_1=C_2$. To calculate the value of $C_1=C_2$, we add the two equations (8.4.28) and asymptotically evaluate the resulting integral to obtain

(8.4.29)
$$C_1=C_2=\lim_{\varepsilon\to 0}\frac{\int_{\partial\Omega} e^{-\phi/\varepsilon} f\,\mathbf{b}\cdot\boldsymbol{\nu}\,ds}{\int_{\partial\Omega} e^{-\phi/\varepsilon}\mathbf{b}\cdot\boldsymbol{\nu}\,ds}.$$

However, in case (B), if both $\hat{\phi}_i(i=1,2)$ are nonpositive, the integrals over $\partial\Omega\cap\partial\Omega_i$ dominate and the C_i is given by

$$C_i=\lim_{\varepsilon\to 0}\frac{\int_{\partial\Omega\cap\partial\Omega_i} e^{-\phi/\varepsilon} f\,\mathbf{b}\cdot\boldsymbol{\nu}\,ds}{\int_{\partial\Omega\cap\partial\Omega_i} e^{-\phi/\varepsilon}\mathbf{b}\cdot\boldsymbol{\nu}\,ds}\qquad (i=1,2).$$

Therefore, in case (A), $p=\lim_{\varepsilon\to 0}p(x,y)$ is given by formulas (7.5.7) and (7.5.7′) in cases (i) and (ii), respectively. In case (B)

$$p(x,y,\xi,\eta)=\lim_{\varepsilon\to 0}\Pr\left[(x_\varepsilon(\tau),y_\varepsilon(\tau))=(\xi,\eta)\,|\,(x_\varepsilon(0),y_\varepsilon(0))=(x,y)\right]$$

is given by

(8.4.30)
$$p(x,y,\xi,\eta)=\sum_{i=1}^{2}\chi_{\Omega_i}(x,y)p_i(\xi,\eta),$$

where

(8.4.31)
$$p_i(\xi,\eta)=\frac{\chi_{U_i}(\xi,\eta)\mathbf{b}\cdot\boldsymbol{\nu}(\xi,\eta)}{\int_{U_i}\mathbf{b}\cdot\boldsymbol{\nu}\,ds}\qquad (i=1,2)$$

if case (i) holds on $\partial\Omega\cap\partial\Omega_i$ (see Section 7.4), and

$$p_i(\xi,\eta)=\frac{\sum_{k=1}^{l} d_k \mathbf{b}\cdot\boldsymbol{\nu}(s_k)\,\delta(s-s_k)}{\sum_{k=1}^{l} d_k \mathbf{b}\cdot\boldsymbol{\nu}(s_k)} \qquad (i=1,2)$$

if case (ii) holds on $\partial\Omega\cap\partial\Omega_i$. If $x=0$, then $p(0,y,\xi,\eta)=\frac{1}{2}(p_1+p_2)$. To illustrate the nature of these results in terms of stochastic processes, let Ω be the circle $x^2+y^2<R^2$ with $R>1$. We see that there exists a critical value of R, namely $R=\sqrt{2}$, such that the case (A) obtains if $R>\sqrt{2}$, while case (B) obtains if $R\leqslant\sqrt{2}$. Thus $\varepsilon\to 0$, if $R\leqslant\sqrt{2}$, the stochastic process $(x_\varepsilon(t), y_\varepsilon(t))$, which starts in Ω_i, will exit from $\partial\Omega\cap\partial\Omega_i$ with probability almost 1 and has a vanishingly small probability of crossing the line $x=0$. In contrast, if $R<\sqrt{2}$, the probability of crossing the line $x=0$ is in general not zero, and the process may leave from either $\partial\Omega\cap\partial\Omega_1$ or $\partial\Omega\cap\partial\Omega_2$, independent of the point where the process starts. In this case, the probability distribution of the exit points depends only on the distribution of the maxima of ϕ on $\partial\Omega$. We see, therefore, that if R is sufficiently small, the particle is most likely to hit the boundary of that part Ω_i where it starts, before crossing the line Γ_{12}. If R is large, the particle will cross Γ_{12}, so that its separating influence disappears.

8.5 THE EXIT TIME

The expected exit time $E_{\mathbf{x}}\tau^\varepsilon=v_\varepsilon(\mathbf{x})$ is a solution of the problem

$$\varepsilon\,\Delta v_\varepsilon(\mathbf{x})+\mathbf{b}(\mathbf{x})\cdot\nabla v_\varepsilon(\mathbf{x})=-1 \quad \text{in } \Omega \tag{8.5.1}$$

$$v_\varepsilon(\mathbf{x})=0 \quad \text{on } \partial\Omega. \tag{8.5.2}$$

The boundary value problem (8.5.1), (8.5.2) has a bounded solution. Therefore, the exit time is finite with probability 1. Since the solution $v_\varepsilon(\mathbf{x})$ increases exponentially as $\varepsilon\to 0$, we rescale by setting

$$v_\varepsilon(\mathbf{x})=e^{K/\varepsilon}C(\varepsilon)W_\varepsilon(\mathbf{x}), \tag{8.5.3}$$

so that $C(\varepsilon)=o(e^{K/\varepsilon})$ and $W_\varepsilon(\mathbf{x})=O(1)$, where the constant K is as yet undetermined. We have $W_\varepsilon(\mathbf{x})\sim 1$ in the interior of Ω. Then, following the procedure of Section 8.1, we construct a boundary layer expansion near

$\partial\Omega$. Thus we introduce local coordinates $\mathbf{x}=(\mathbf{x}', y)$ near $\partial\Omega$, where $y = -\operatorname{dist}(\mathbf{x}, \partial\Omega)$. We then introduce the stretching transformation $\eta = -\varepsilon^{-1/2}y$, in terms of which (8.5.1) takes the form

$$\frac{\partial^2 W_\varepsilon}{\partial\eta^2} + \eta b_1(\mathbf{x}')\frac{\partial W_\varepsilon}{\partial\eta} + \mathbf{b}_2\cdot\nabla_{\mathbf{x}'}W_\varepsilon + o(1) = O(e^{-K/\varepsilon}). \tag{8.5.4}$$

Here ηb_1 and $\mathbf{b}_2$ represent the normal and tangential components of $\mathbf{b}$ near $\partial\Omega$, respectively. The leading term W^0 in the boundary layer expansion thus satisfies the equation

$$W^0_{\eta\eta} + \eta b_1(\mathbf{x}')W^0_\eta + \mathbf{b}_2\cdot\nabla_{\mathbf{x}'}W^0 = 0. \tag{8.5.5}$$

This equation is exactly equation (8.4.4), which we will solve in the same manner, subject to the boundary condition $W^0|_{\eta=0}=0$, which follows from (8.5.2), and the condition $W^0\to 1$ as $\eta\to\infty$, which is obtained by matching the boundary layer and interior expansions. Then, using the procedure of the preceding section, we construct a solution which we call $W^0(\eta,\xi)$, where ξ is the function introduced in the preceding section. Next we employ $W^0(\eta,\xi)$ as well as the leading term in the interior expansion in the condition

$$\varepsilon\nabla\cdot\left(e^{-\Phi/\varepsilon}\nabla v_\varepsilon\right) = -e^{-\Phi/\varepsilon}, \tag{8.5.6}$$

which follows from (8.5.2). We integrate this equation and employ the divergence theorem to obtain

$$\varepsilon\int_{\partial\Omega} e^{-\Phi/\varepsilon}\frac{\partial v_\varepsilon}{\partial \boldsymbol{\nu}}\,dS = -\int_\Omega e^{-\Phi/\varepsilon}d\mathbf{x}. \tag{8.5.7}$$

We employ (8.5.3) and the leading term in the boundary layer expansion given in terms of $W^0(\eta,\xi)$, in (8.5.8), and evaluate the resulting Laplace integrals asymptotically as $\varepsilon\to 0$. The main contribution to the integral on the right-hand side comes from the absolute minima of Φ in Ω. The main contribution to the integral on the left-hand side comes from the neighborhood of the lowest saddle point of Φ on $\partial\Omega$. Thus it is only necessary to explicitly construct $W^0(\eta,\xi)$ in the neighborhood of such saddle points $\eta=0$, $\mathbf{x}'=\mathbf{x}'_i$. In the neighborhood of such points, the last term in the boundary layer equation (8.5.5) may be neglected, as in Section 8.1. The solution of the resulting equation is then given by

$$(8.5.8)\qquad W^0=\frac{\int_0^\eta e^{-(s^2/2)b_1(\mathbf{x}')}\,ds}{\int_0^\infty e^{-(s^2/2)b_1(\mathbf{x}')}\,ds};$$

that is,

$$(8.5.9)\qquad W_\varepsilon(\mathbf{x})\sim\frac{\int_0^{|\mathbf{y}|/\sqrt{\varepsilon}} e^{-(s^2/2)b_1(\mathbf{x}')}\,ds}{\int_0^\infty e^{-(s^2/2)b_1(\mathbf{x}')}\,ds}$$

for $\mathbf{x}=(\mathbf{x}', y)$ with $|y|\ll 1$. In particular,

$$(8.5.10)\qquad \left.\frac{\partial W_\varepsilon(\mathbf{x})}{\partial \boldsymbol{\nu}(\mathbf{x})}\right|_{\partial\Omega}=\left.\frac{\partial W_\varepsilon}{\partial y}\right|_{y=0}\sim-\varepsilon^{-1/2}\int_0^\infty e^{-(s^2/2)b_1(\mathbf{x}')}\,ds$$

$$=-2\varepsilon^{-1/2}\sqrt{\frac{2}{\pi}}\,[b_0(\mathbf{x}')]^{1/2}.$$

We employ (8.5.3) and (8.5.10) in (8.5.7) to obtain

$$(8.5.11)\qquad 2\sqrt{\frac{2}{\pi}}\,\varepsilon^{1/2}e^{K/\varepsilon}C(\varepsilon)\int_{\partial\Omega}e^{-\Phi/\varepsilon}b_1^{1/2}(\mathbf{x}')\,dS_{\mathbf{x}'}\sim\int_\Omega e^{-\Phi/\varepsilon}\,d\mathbf{x},$$

which must be evaluated asymptotically as $\varepsilon\to 0$.

If Φ achieves the minimum $\hat{\Phi}$ on $\partial\Omega$ along an $(n-1)$-dimensional "arc" S of $\partial\Omega$, we obtain

$$(8.5.12)\qquad 2\sqrt{\frac{2}{\pi}}\,\varepsilon^{1/2}e^{K/\varepsilon}e^{-\hat{\Phi}/\varepsilon}C(\varepsilon)\int_S b_1^{1/2}(\mathbf{x}')\,dS_{\mathbf{x}'}$$

$$\sim(2\pi\varepsilon)^{n/2}\sum_j \mathcal{H}^{-1/2}(\mathbf{y}_j),$$

where

$$\min_\Omega \phi=\Phi(\mathbf{y}_j)=0\qquad\text{and}\qquad \mathcal{H}(\mathbf{y}_j)=\det\frac{\partial^2\Phi}{\partial x_i\,\partial x_k}(\mathbf{y}_j).$$

Therefore, we choose $K=\hat{\Phi}$ and

$$(8.5.13)\qquad C(\varepsilon)=2^{(n-3)/2}\pi^{(n+1)/2}\varepsilon^{(n-1)/2}\frac{\sum_j \mathcal{H}^{-1/2}(\mathbf{y}_j)}{\int_S b_1^{1/2}(\mathbf{x}')\,dS_{\mathbf{x}'}}.$$

If the minimum of Φ is achieved at isolated points $\mathbf{x}'_i$ of $\partial\Omega$, and

$$\det\frac{\partial^2\Phi}{\partial x'_j\,\partial x'_k}\bigg|_{\mathbf{x}'_i}=H(\mathbf{x}'_i)\neq 0\qquad (j,k=1,\ldots,n),$$

K is chosen as above and

$$(8.5.14)\qquad C(\varepsilon)=\frac{(\pi/2)\sum_j \mathcal{H}^{-1/2}(\mathbf{y}_j)}{\sum_i b_1^{1/2}(\mathbf{x}'_i)H^{-1/2}(\mathbf{x}'_i)}.$$

Therefore, we have computed the expected exist time $E_{\mathbf{x}}\tau^{\varepsilon}=v_{\varepsilon}(\mathbf{x})$ asymptotically.

In the case of a single minimum of Φ, at $\mathbf{x}=\mathbf{0}$, say, $E_{\mathbf{0}}\tau^{\varepsilon}$ is given by

$$(8.5.15)\qquad E_{\mathbf{0}}\tau^{\varepsilon}=\begin{cases}\dfrac{e^{\hat{\Phi}/\varepsilon}}{2^{(1-n)/2}\varepsilon^{(1-n)/2}\mathcal{H}^{1/2}(\mathbf{0})\int_S b_1^{1/2}(\mathbf{x}')\,dS_{\mathbf{x}'}} & \text{in case (i)}\\[2ex] \dfrac{e^{\hat{\Phi}/\varepsilon}(1/\pi)}{\mathcal{H}^{1/2}(\mathbf{0})\sum_i b_i^{1/2}(\mathbf{x}'_i)H^{-1/2}(\mathbf{x}'_i)} & \text{in case (ii)}\end{cases}$$

with

$$\hat{\Phi}=\min_{\partial\Omega}\Phi\qquad\text{and}\qquad b_1(\mathbf{x}')=\frac{1}{2}\left|\frac{\partial^2\Phi}{\partial\boldsymbol{\nu}^2}(\mathbf{x}')\right|.$$

In the case of a number of minima, similar explicit expressions can be given.

EXERCISE 8.5.1

Show that in one dimension, in the case of a single well, at $x=x_0$, say, we have

$$(8.5.16)\quad E_{x_0}\tau^{\varepsilon}\cong\frac{2\pi}{\sqrt{-\Phi''(x_0)\Phi''(x_2)}}\exp\frac{[-(\Phi(x_2)-\Phi(x_0))]}{\varepsilon}$$

in (8.5.1). Here $b(x)=-\Phi(x)$, $\Omega=(x_1,x_2)$, where $x_1<x_0<x_2$, Φ has a single minimum in Ω at x_0 and maxima at x_1,x_2, $\Phi(x_1)=\Phi(x_2)$, $\Phi'(x_1)=\Phi'(x_2)=0$.

8.6 APPLICATIONS TO CHEMICAL REACTION RATES AND THE SECOND EIGENVALUE

We showed above that the chemical reaction rate is given by $\kappa=1/E_0\tau^{\varepsilon}$. Therefore, using formula (8.5.15), with $\varepsilon=k\mathrm{T}/mQ$, and the potential per unit mass Φ replaced by the potential $\Psi=m\Phi$, we obtain in the physical time scale, the modified Arrhenius law

$$(8.6.1)\quad \kappa=\frac{1}{2\pi m\beta}\mathcal{H}^{1/2}(\mathbf{0})\sum_i H^{-1/2}(\mathbf{x}_i')\left[\frac{\partial^2\Psi}{\partial\nu^2}(\mathbf{x}_i')\right]^{1/2}e^{-\hat{\Psi}/k\mathrm{T}}.$$

In case (i),

$$\kappa=\frac{1}{2\pi m\beta}(2k\mathrm{T})^{(n-1)/2}\mathcal{H}^{1/2}_{(\mathbf{0})}\int_S\left[\frac{\partial^2\Psi}{\partial\nu^2}(\mathbf{x})\right]^{1/2}ds_{\mathbf{x}}e^{-\hat{\Psi}/k\mathrm{T}},$$

the term $\mathcal{H}(\mathbf{0})$ is the Hessian of Ψ evaluated at the bottom of the well. It is known from classical mechanics that $\mathcal{H}^{1/2}(\mathbf{0})$ is the product of the principal frequencies of vibration of the particle at the bottom of the well. Thus the faster it oscillates, the more likely it is to escape. The term $\mathrm{H}(\mathbf{x}_i')$ denotes the Hessian of Ψ evaluated at the transition states $\mathbf{x}_i'$. $\mathrm{H}(\mathbf{x}_i)$ is the product of the frequencies of oscillation at the transition point, which are saddle points of Ψ. Thus the faster the particle oscillates, the longer it will stay at the transition point, thus lowering the reaction rate. The term $\partial^2\Psi/\partial\nu^2(\mathbf{x}_i')$ is proportional to the curvature of the transition state in the direction across the saddle point, which is the path the particle takes upon exiting. Thus the flatter the path through the transition point, the longer it takes to escape. Thus the greater the curvature of the path, the greater is the reaction rate. The expression $\Sigma_j\mathrm{H}^{-1/2}(\mathbf{x}_j')[\partial^2\Psi/\partial\nu^2]^{1/2}$ expresses the dependence of the preexponential factor in κ on the spatial structure of the potential and thus accounts for the steric factor in chemical reaction rates. This factor take into account the fact that collisions in different directions have different likelihoods of being chemically effective (i.e., of activations). The term m in the denominator appears as a result of change from Φ to Ψ. Here m is the reduced mass of the particle in a collision. The factor β in the denominator appears since the reaction rate is expressed in the physical time variable, whereas the exit time was expressed in the scaled time variable. β represents the viscous force, which slows the particles down and thus inhibits the reaction. The exponential factor is the Arrhenius law of chemical rates, in which the term $\hat{\Psi}$ is the activation energy given by the height of the lowest transition point above the bottom of the well. The expression for the steric factor in the preexponential factor seems to be new. In one dimension, this formula reduces to that of Kramers (1940). Of course, in one dimension, there are no steric effects.

In the more realistic model of a chemical reaction, as described in Section 8.1, the reaction rate constant κ was shown to be the eigenvalue λ_1 of the problem (8.1.3). Next we compute the eigenvalue λ_1. We assume first that $\Phi(x)$ has two local minima, at x_1 and x_3, respectively, and one local maximum at x_2, with $x_1<x_2<x_3$. We also assume that $\Phi(x_1)<\Phi(x_3)<\Phi(x_2)$ (see Fig. 8.6.1). Most of the particles initially located in the shallower well near x_3 will overcome the potential barrier at x_2 and eventually come to rest in the deeper well at x_1. A graph of the final equilibrium distribution $u^0(x)$ is given in Fig. 8.6.2. Thus the rate of approach to equilibrium is determined by the time $\bar{\tau}$ required to overcome the potential barrier at x_2. This time has been shown to be $O(e^{K/\varepsilon})$ with $K=\Phi(x_3)-\Phi(x_2)>0$. Therefore, we expect that λ_1 is exponentially small. We assume that $\lambda_1=O(e^{-K_1/\varepsilon})$ with $K_1>0$, and will in fact demonstrate that $\lambda_1=1/\bar{\tau}$, so that $K_1=K$.

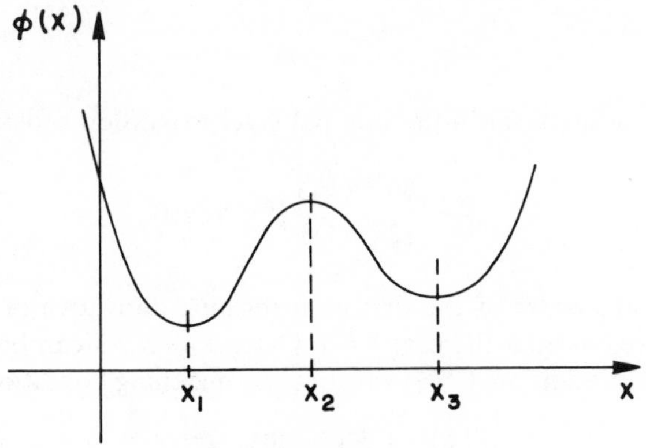

Figure 8.6.1

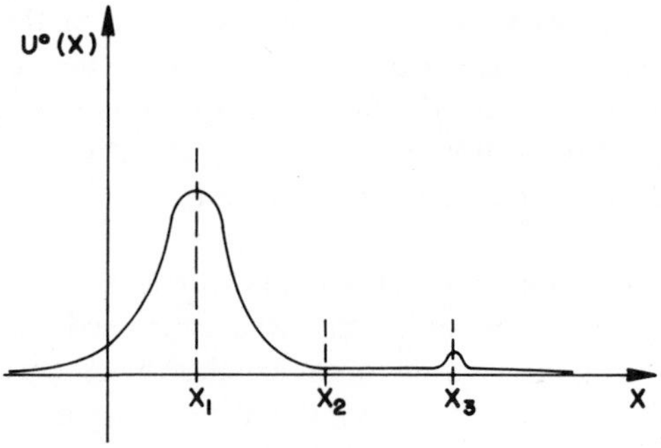

Figure 8.6.2

We set $u_1(x)=e^{\Phi/\varepsilon}v(x)$ with v assumed to be an $O(1)$ function, to obtain

(8.6.2) $$\varepsilon v_{xx}-\Phi_x v_x=\lambda_1 v.$$

Since λ_1 is exponentially small, the leading term V^0, in the expansion of v as $\varepsilon\to 0$, is a solution of the reduced equation

(8.6.3) $$\Phi_x v_x^0=0.$$

Thus

(8.6.4) $$v^0(x)=\begin{cases} C_1 & x<x_2 \\ C_3 & x>x_2 \end{cases}$$

We observe that, as shown above, the constants C_1 and C_3 are not necessarily equal, and an internal layer may exist in the neighborhood of the unstable equilibrium point x_2. The internal layer is characterized by the stretching transformation

(8.6.5) $$\xi=\frac{x-x_2}{\varepsilon^{1/2}},$$

so that the leading term in the internal layer expansion satisfies

(8.6.6) $$v_{\xi\xi}^0-\frac{\Phi^{(2k)}(x_2)}{(2k-1)!}\xi^{2k-1}v_\xi^0=0,$$

where $2k$ is the order of the first nonvanishing derivative of Φ at x_2. For simplicity, we consider the case $k=1$. Cases with $k>1$ can be treated as in Section 8.3. In addition $V^0(\xi)$ satisfies the matching conditions

(8.6.7) $$V^0(\xi)\to\begin{cases} C_1 & \text{as } \xi\to-\infty \\ C_3 & \text{as } \xi\to+\infty. \end{cases}$$

Hence

(8.6.8) $$V^0(\xi)=\frac{C_3-C_1}{2B}\int_0^\xi \exp\frac{\Phi''(x_2)s^2}{2}\,ds+\frac{C_1+C_2}{2}.$$

where

(8.6.9) $$B=\sqrt{\frac{\pi}{2|\Phi''(x_2)|}};$$

we observe that $\Phi''(x_2)<0$. Thus the uniform asymptotic expansion of v, for $\varepsilon \ll 1$, is given by

(8.6.10) $$v(x)=V^0\left(\frac{x-x_2}{\varepsilon^{1/2}}\right)+o(1) \quad \text{as} \quad \varepsilon\to 0.$$

One of the constants $C_{1,3}$ may be chosen arbitrarily, since $e^{-\Phi/\varepsilon}v$ is an eigenfunction. To determine the other constant, we multiply (8.6.2) by $e^{-\Phi/\varepsilon}$ and integrate from $-\infty$ to $+\infty$, to obtain

(8.6.11) $$\int_{-\infty}^{\infty}\varepsilon\left(e^{-\Phi/\varepsilon}v_x\right)_x dx=\lambda_1\int_{-\infty}^{\infty}e^{-\Phi/\varepsilon}v\,dx.$$

The integral on the left of (8.6.11) vanishes, since $e^{-\Phi/\varepsilon}v\to 0$ as $x\to\pm\infty$. Therefore, since $\lambda_1\neq 0$,

(8.6.12) $$\int_{-\infty}^{\infty}e^{-\Phi/\varepsilon}v\,dx=0.$$

We now employ (8.6.10) in (8.6.12) to find that

(8.6.13) $$\frac{C_1}{C_3}\sim\frac{\displaystyle\int_{-\infty}^{\infty}e^{-\Phi/\varepsilon}\left[1+(1/B)\int_0^\xi e^{\Phi''(x_2)s^2/2}\,ds\right]dx}{\displaystyle\int_{-\infty}^{\infty}e^{-\Phi/\varepsilon}\left[1-(1/B)\int_0^\xi e^{\Phi''(x_2)s^2/2}\,ds\right]dx}.$$

(8.6.14)

$$-\frac{C_1}{C_3}\sim\frac{1}{2}\left(1+\frac{1}{B}\int_0^{(x_1-x_2)/\varepsilon^{1/2}}e^{\Phi''(x_2)s^2/2}\,ds\right)$$

$$+\frac{1}{2}\sqrt{\frac{\Phi''(x_1)}{\Phi''(x_3)}}\,e^{-[\Phi(x_3)-\Phi(x_1)]/\varepsilon}\left(1+\frac{1}{B}\int_0^{(x_3-x_2)/\varepsilon^{1/2}}e^{\Phi''(x_2)s^2/2}\,ds\right),$$

so that C_1 approaches zero exponentially in $1/\varepsilon$. Then we may choose $C_3=1$, so that the leading term in the expansion for v is determined as

$$v \sim \frac{1}{2B}\int_0^{\xi} e^{\Phi''(x_2)s^2/2}\,ds + \frac{1}{2}. \tag{8.6.15}$$

We now calculate the eigenvalue λ_1 by multiplying (8.6.2) by $e^{-\Phi/\varepsilon}$ and integrating from x_2 to ∞. Then, after evaluating the resulting integrals asymptotically, we obtain

$$-\varepsilon v_x(x_2)e^{-\Phi(x_2)/\varepsilon} \sim \lambda_1 e^{-\Phi(x_3)/\varepsilon}\frac{2\pi}{\sqrt{\Phi''(x_3)}}. \tag{8.6.16}$$

Employing (8.6.15) in (8.6.16), we find that

$$\lambda_1 \sim -\frac{\sqrt{\Phi''(x_3)|\Phi''(x_2)|}}{2\pi}e^{-[\Phi(x_2)-\Phi(x_3)]/\varepsilon}. \tag{8.6.17}$$

We observe that $-1/\lambda_1$ is the expected exit time of $x_\varepsilon(t)$ from the potential well at x_3.

In the case of $\Phi(x_3)=\Phi(x_1)$ (i.e., of two equally deep potential wells), similar calculations lead to the result that

$$C_1 = -\sqrt{\frac{\Phi''(x_1)}{\Phi''(x_3)}}, \qquad C_3 = 1 \tag{8.6.18}$$

in (8.6.8), so that

$$\lambda_1 \sim -\frac{\sqrt{\Phi''(x_3)}+\sqrt{\Phi''(x_1)}}{2\pi}\sqrt{|\Phi''(x_2)|}\,e^{-[\Phi(x_3)-\Phi(x_1)]/\varepsilon}. \tag{8.6.19}$$

Next we consider the case of three stable equilibrium points. We assume for definiteness that

$$\Phi(x_1) < \Phi(x_3) < \Phi(x_2),$$

where $x_i (i=1,2,3)$ are the local minima of Φ, and that $\Phi(y_1) > \Phi(y_2)$, $x_1 < y_1 < x_2 < y_2 < x_3$, where $y_i (i=1,2,3)$ are the local maxima of Φ (see Fig.

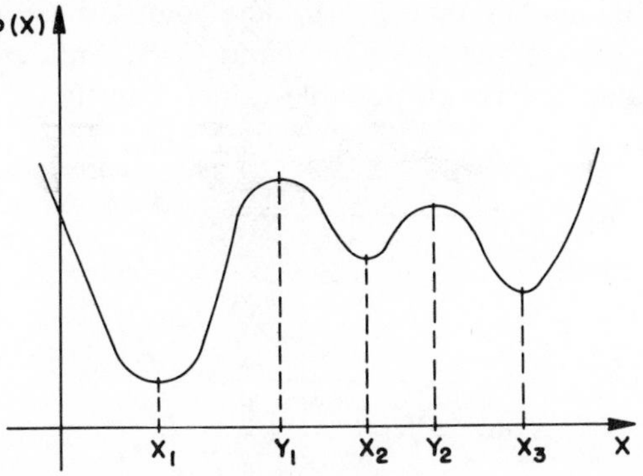

Figure 8.6.3

8.6.3). The uniform expansion of v is given by

$$v \approx \frac{C_2 - C_1}{2B_1} \int_0^{(x-y_1)/\sqrt{\varepsilon}} \exp\left[\frac{\Phi''(y_1)s^2}{2}\right] ds$$
$$+ \frac{C_3 - C_2}{2B_2} \int_0^{(x-y_2)/\sqrt{\varepsilon}} \exp\left[\frac{\Phi''(y_2)s^2}{2}\right] ds + \frac{C_1 + C_3}{2}$$

where

$$B_i = \sqrt{\frac{\pi}{2|\Phi''(y_i)|}} \qquad (i=1,2) \qquad \text{and} \qquad v(x) \to \begin{cases} C_1 & \text{if } x < y_1 \\ C_2 & \text{if } y_1 < x < y_2 \\ C_3 & \text{if } x > y_2. \end{cases}$$

Integrating as before, from $-\infty$ to $+\infty$, from y_1 to ∞, and from y_2 to ∞, respectively, we obtain the following set of equations:

$$(8.6.20) \qquad \sum_{i=1}^{3} \frac{C_i e^{-\Phi(x_i)/\varepsilon}}{\sqrt{\Phi''(x_i)}} = 0,$$

$$(8.6.21) \qquad -(C_2 - C_1)e^{-\Phi(y_1)/\varepsilon}\sqrt{|\Phi''(y_1)|} = 2\pi\lambda_1 \sum_{i=2}^{3} \frac{e^{-\Phi(x_i)/\varepsilon}C_i}{\sqrt{\Phi''(x_i)}},$$

and

$$(8.6.22) \qquad -(C_3 - C_2)e^{-\Phi(y_2)/\varepsilon}\sqrt{|\Phi''(x_2)|} = 2\pi\lambda_1 \frac{C_3 e^{-\Phi(x_3)}}{\sqrt{\Phi''(x_3)}}$$

Equation (8.6.20) implies that $C_1=0$. The solutions $C_{2,3}$ of (8.6.20) and (8.6.21) with $C_1=0$ are given by $C_3=0$ and $C_2=1$ without loss of generality. The eigenvalue λ_1 has two possible values, namely

$$\lambda_{1,1} \sim -\frac{\sqrt{\Phi''(x_3)|\Phi''(y_1)|}}{2\pi} e^{-(Q_2+Q_3-P_2)/\varepsilon}$$

and

$$\lambda_{1,2} \sim -\frac{\sqrt{\Phi''(x_2)|\Phi''(y_2)|}}{2\pi} e^{-P_2/\varepsilon}$$

where

$$P_i = \Phi(y_i) - \Phi(x_i)$$

and

$$Q_i = \Phi(y_i) - \Phi(x_{i+1}).$$

Clearly, $|\lambda_{1,1}| \ll |\lambda_{1,2}|$. The quantity $Q=Q_3+Q_2-P_2$ is the height $\Phi(y_1)-\Phi(x_3)$ of the highest potential barrier the Brownian particle has to overcome in order to reach the most stable equilibrium at x_1. Consider now the case of n wells, with $x_1<y_1<\cdots<x_{n-1}<y_{n-1}<x_n$, where x_i are the local minima and y_i are the local maxima. Then the uniform expansion of v is given by

$$v_\varepsilon \sim \sum_{i=1}^{n-1} \frac{(C_{i+1}-C_i)}{2B_i} \int_0^{(x-y_i)/\sqrt{\varepsilon}} e^{\Phi''(y_i)s^2/2}\, ds + \frac{C_1+C_n}{2},$$

since, by assumption $v_\varepsilon(x) \sim C_i$ if $y_{i-1}<x<y_i (i=1,2,\dots n)$ as $\varepsilon \to 0$. We set $y_0=-\infty$, $y_n=+\infty$. To determine λ_1 and the constants C_i, we integrate as above over the intervals (y_{i-1}, y_i). We obtain the system

(8.6.23)

$$e^{-\Phi(y_i)/\varepsilon}(C_{i+1}-C_i)\sqrt{|\Phi''(y_i)|} - e^{-\Phi(y_{i-1})/\varepsilon}(C_i-C_{i-1})\sqrt{|\Phi''(y_{i-1})|}$$
$$= \frac{2\pi\lambda_i C_i e^{-\Phi(x_i)/\varepsilon}}{\sqrt{\Phi''(x_i)}} \qquad (i=1,,\dots,n).$$

Here we set $\Phi''(y_0)=\Phi''(y_n)=0$. Setting $\Phi(y_i)-\Phi(x_i)=\varepsilon P_i$, $\Phi(y_{i-1})-\Phi(x_i)=\varepsilon Q_i$, and $\sqrt{\Phi''(x_i)|\Phi''(y_j)|}=k_{i,j}$, we see that the system (8.6.23) can be written in the form

$$(A-\lambda_1 I)C=0$$

where

$$C=(C_1,\ldots,C_n)^T \qquad \text{and} \qquad A=\{A_{ij}\},$$

where

$$A_{i,i}=-\left(k_{i,i}e^{-P_i}+k_{i,i-1}e^{-Q_i}\right)$$

$$A_{i,i-1}=k_{i,i-1}e^{-Q_i}$$

$$A_{i,i+1}=k_{ii}e^{-P_i}$$

$$A_{i,j}=0 \qquad \text{if} \quad |i-j|>1.$$

Hence $-\lambda_1$ is the smallest nonzero eigenvalue of $-A$. Evaluating the characteristic polynomial $S(\lambda)$ asymptotically, we find that

$$S(\lambda)\sim\lambda M+N,$$

where N is the modulus of the sum of the principal minors of order $n-1$ of A, and M is the modulus of the sum of the principal minors of order $n-2$ of A. Thus

$$(8.6.24)\quad N=\sum_{j=0}^{n-1}\left(\prod_{i=1}^{j}k_{ii}\right)\left(\prod_{i=j+2}^{n}k_{i,i-1}\right)e^{-(P_1+\cdots+P_j+Q_{j+2}+\cdots+Q_n)}$$

and

$$(8.6.25)\quad M=\sum\left(\prod_j k_{i_j,i_j}\right)\left(\prod_j k_{i_j,i_j-1}\right)e^{-(P_{i_1}+\cdots+P_{i_k}+Q_{i_{k+1}}+\cdots+Q_{i_{n-2}})},$$

where the summation extends over all indices

$$i_1<i_2\cdots<i_k$$

$$i_{k+1}<\cdots<i_{n-2}$$

such that $i_j \neq i_h$ if $j \neq h$ and $i_j + 1 \neq i_{k+h}$ for all $j \leqslant k$. Evaluating N and M asymptotically, we see that

$$\lambda = -\frac{N_1}{M_1}$$

where N_1 and M_1 are the maximal terms in the respective sums N and M. Hence

$$\lambda_1 = \frac{\left(\prod_i k_{ii}\right)\left(\prod_i k_{i,i-1}\right)}{\left(\prod_j k_{jj}\right)\left(\prod_j k_{j,j-1}\right)} e^{-Q/\varepsilon},$$

where Q is the height of the highest barrier a particle must overcome to reach the deepest well (i.e., Q is the activation energy of the reaction). The products extend over the respective indices corresponding to the minimal exponents in (8.6.24) and (8.6.25).

Next we compute λ_1 in two dimensions. Let $\mathbf{x} = (x^1, x^2), \mathbf{w} = (w^1, w^2) =$ Brownian motion in the plane. We consider next the stabilization problem for the process $\mathbf{x}_\varepsilon(t)$, defined by the system of stochastic differential equations

$$\dot{\mathbf{x}} = -\nabla\Phi(\mathbf{x}) + \sqrt{2\varepsilon}\,\mathbf{dw}. \tag{8.6.26}$$

We assume that $\Phi(\mathbf{x}) \to \infty$ as $|\mathbf{x}| \to \infty$ and that

$$\int e^{-\Phi/\varepsilon}\mathbf{dx} < \infty$$

for all $\varepsilon > 0$. Furthermore, we assume that $\Phi(\mathbf{x})$ has only two points of minimum $\mathbf{x}_1$ and $\mathbf{x}_2$, $\Phi(\mathbf{x}_1) < \Phi(\mathbf{x}_2)$, and we denote by Ω_i the domain of attraction of $\mathbf{x}_i (i = 1, 2)$ for the dynamical system

$$\dot{\mathbf{x}} = -\nabla\Phi(\mathbf{x}) \tag{8.6.27}$$

and denote by Γ the curve $\partial\Omega_1 \cap \partial\Omega_2$ as in Fig. 8.6.4. The rate of convergence of the solution $u_\varepsilon(\mathbf{x}, \mathbf{x}_0, t)$ of the Fokker–Planck equation for (8.6.26),

$$\varepsilon\Delta u + \nabla\cdot(\nabla\Phi u) = u_t$$

$$u(\mathbf{x}, \mathbf{x}_0, 0) = \delta(\mathbf{x} - \mathbf{x}_0)$$

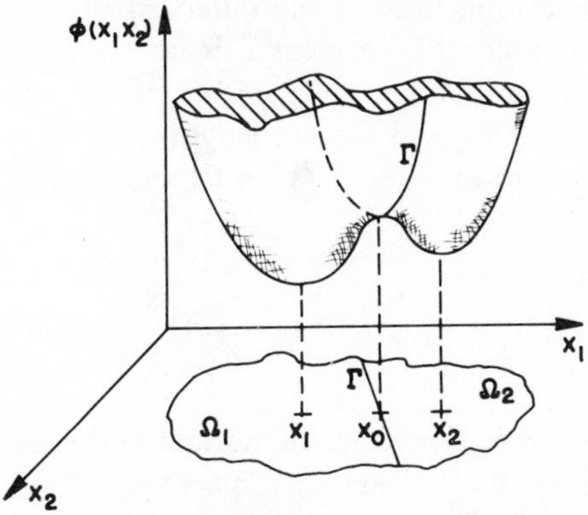

Figure 8.6.4

to the stationary solution

$$u^0(\mathbf{x}) = \frac{e^{-\Phi/\varepsilon}}{\int e^{-\Phi/\varepsilon}\,d\mathbf{x}}$$

is determined as above by the first nontrivial eigenvalue λ_1 of the problem

$$\varepsilon\,\Delta u + \nabla\cdot(\nabla\Phi u) = \lambda u$$

or equivalently, setting $u = e^{-\Phi/\varepsilon}v$, of the problem

$$\varepsilon\,\Delta v - \nabla\Phi\cdot\nabla v = \lambda v. \tag{8.6.28}$$

It can be shown (cf. Exercise 8.6.1*) that $\lambda_1 = O(e^{-K/\varepsilon})$, so that the reduced equation for (8.6.28) is

$$\nabla\Phi\cdot\nabla u = 0. \tag{8.6.29}$$

Equation (8.6.29) can be written in the form

$$\frac{du(\mathbf{x}(t))}{dt} = 0,$$

where $\mathbf{x}(t)$ is any solution of the system (8.6.27). Since all trajectories of (8.6.27) that begin in Ω_i converge to $\mathbf{x}_i$ at $t\to\infty$, we must have

$$u_0(\mathbf{x}) = \begin{cases} C_1 & \text{if } \mathbf{x}\in\Omega_1 \\ C_2 & \text{if } \mathbf{x}\in\Omega_2, \end{cases}$$

where $u_0(\mathbf{x})$ is the leading term in the outer expansion of the first eigenfunction $u_1(\mathbf{x})$. The internal layer near Γ is constructed as in Section 8.4. We introduce local coordinates $\mathbf{x}=(s,y)$ near Γ, where $y=-\text{dist}(\mathbf{x},\Gamma)$, $\mathbf{x}\in\Omega_1$, $y=\text{dist}(\mathbf{x},\Gamma)$, $\mathbf{x}\in\Omega_2$, and s is arc length in Γ. We then introduce the stretching transformation $\eta=-\varepsilon^{-1/2}y$, in terms of which (8.6.28) takes the form

$$\frac{\partial^2 v}{\partial\eta^2}+\eta b_1(s,\sqrt{\varepsilon}\,\eta)\frac{\partial v}{\partial\eta}+b_2(s,\sqrt{\varepsilon}\,\eta)\frac{\partial v}{\partial s}+o(1)=O(e^{-K/\varepsilon}). \tag{8.6.30}$$

Here $\eta b_1(s,\sqrt{\varepsilon}\,\eta)$ and b_2 represent the normal and tangential components of the vector $b(\mathbf{x})=-\nabla\Phi(\mathbf{x})$ near Γ, respectively. The leading term v^0 in the internal layer expansion near Γ, as $\varepsilon\to 0$, is the solution of

$$\frac{\partial^2 v^0}{\partial\eta^2}+\eta B(s)\frac{\partial v^0}{\partial\eta}+C(s)\frac{\partial v^0}{\partial s}=0, \tag{8.6.31}$$

$B(s)=b_1(s,0)>0, C(s)=b_2(s,0)$. Since the function $C(s)$ vanishes at any saddle point of $\Phi(\mathbf{x})$ on Γ, (8.6.31) is a degenerate parabolic equation that must be solved for all s. There exists a unique solution to the boundary value problem for (8.6.31) determined from the matching conditions (Schuss 1972). The matching conditions are

$$v^0(\eta,s)\to\begin{cases} C_1 & \text{if } \eta\to-\infty \\ C_2 & \text{if } \eta\to\infty \end{cases} \tag{8.6.32}$$

The solution $v^0(\eta,s)$ is given by

$$v^0(\eta,s)=\sqrt{\frac{2}{\pi}}\,(C_2-C_1)\int_0^{\eta\gamma(s)}e^{-s^2/2}\,ds+\frac{C_1+C_2}{2} \tag{8.6.33}$$

where $\gamma(s)$ is the solution of the Bernoulli equation

$$\frac{d\gamma}{ds}+\frac{B(s)}{C(s)}\gamma-\frac{1}{C(s)}\gamma^3=0$$

with $\gamma(0)=\sqrt{B(0)}$, so that $v^0(\eta,0)$ is the solution of (8.6.32) at $s=0$. We choose the point $s=0$ to be the minimum point $\mathbf{x}_0$ of Φ on Γ. Hence

$$\gamma(s)=\left[2\int_0^s C(\sigma)^{-1}\exp\left(-2\int_\sigma^s\frac{B(t)}{C(t)}\,dt\right)d\sigma\right]^{-1/2}.$$

Therefore

$$\frac{\partial v^0}{\partial \eta}(0,s)=\frac{C_2-C_1}{2\sqrt{\pi}}\gamma(s). \tag{8.6.34}$$

Next we determine the relationship between C_1 and C_2. Multiplying (8.6.28) by $e^{-\Phi/\varepsilon}$ and integrating we obtain

$$\int e^{-\Phi/\varepsilon} v\, d\mathbf{x}=0.$$

Using the expansion (8.6.33), we again obtain $C_1=0$ and we choose $C_2=1$. Next we multiply (8.6.28) by $e^{-\Phi/\varepsilon}$ and integrate over Ω_2 to obtain

$$\varepsilon \int_\Gamma e^{-\Phi/\varepsilon}\frac{\partial v^0}{\partial \boldsymbol{\nu}}\, ds=\lambda_1 \int_{\Omega_2} e^{-\Phi/\varepsilon} v^0\, d\mathbf{x}.$$

Using (8.6.33) and (8.6.34), we obtain

$$\lambda_1 \cong \frac{e^{-\Phi(\mathbf{x}_0)/\varepsilon}\gamma(0)/\sqrt{2\Phi''(\mathbf{x}_0)}}{2\pi\left[e^{-\Phi(\mathbf{x}_2)/\varepsilon}/H^{1/2}(\mathbf{x}_2)\right]} \tag{8.6.35}$$

$$=\frac{\left[-\partial^2\Phi(\mathbf{x}_0)/\partial \boldsymbol{\nu}^2\right]^{1/2} e^{[-\Phi(\mathbf{x}_0)-\Phi(\mathbf{x}_2)]/\varepsilon} H^{1/2}(\mathbf{x}_2)}{2\pi\sqrt{\Phi''(\mathbf{x}_0)}}$$

where $\boldsymbol{\nu}$ is the normal to Γ at $\mathbf{x}_0$,

$$H(\mathbf{x}_2)=\det\left(\frac{\partial^2\Phi}{\partial x^i \partial x^j}\right)_{i,j=1,2}(\mathbf{x}_2)$$

and

$$\Phi''(\mathbf{x}_0)=\left.\frac{d^2\Phi}{ds^2}\right|_{s=0}=\sum_{i,j=1}^{2}\frac{\partial^2\Phi(\mathbf{x}_0)}{\partial x^i\partial x^j}\dot{x}_0^i\dot{x}_0^j+\Phi_{x^1}(\mathbf{x}_0)\ddot{x}_0^1$$

$$+\Phi_{x^2}(\mathbf{x}_0)\ddot{x}_0^2, \qquad \dot{\mathbf{x}}_0=\left(\frac{dx^1}{ds},\frac{dx^2}{ds}\right)$$

at $s=0$ on Γ.

In the case of n wells, a system similar to (8.6.23) is obtained. In this case the factors $H^{1/2}(\mathbf{x}_i)$ and $\Phi''(\mathbf{x}_i)$ appear in an equation analogous to

(8.6.35). They are the products of the principal frequencies of vibration at the bottoms of the wells and at the saddle points respectively. The second normal derivatives at the saddle points also appear as in (8.6.1). The exponent in the expression for λ_1 is the activation energy in this case as well.

The n-dimensional case can be treated along the same lines (see Sect. 8.4).

EXERCISE 8.6.1*

Use the Rayleigh quotient (Courant and Hilbert 1962) to show that $\lambda_1 = O(e^{-K/\varepsilon})$ as $\varepsilon \to O$ in (8.6.28). Use the asymptotic solution as a minimizing function.

8.7 THE DIFFUSION TENSOR FOR ATOMIC MIGRATION IN CRYSTALS, THE NERNST–EINSTEIN FORMULA, AND HOMOGENIZATION OF THE SMOLUCHOWSKI EQUATION

We now compute the diffusion matrix for atomic migration in crystals, by employing the results (8.5.15) for the expected exit time and (8.4.22) for the probability distribution of the exit points, in the expression (8.2.5) for D_{ij}. Thus let p_i be the probability of exit through the isolated transition point $\mathbf{z}_i$. Then, by (8.4.22),

$$p_i = \frac{\left[\partial^2\phi(\mathbf{z}_i)/\partial\boldsymbol{\nu}^2\right]^{1/2} H^{-1/2}(\mathbf{z}_i)}{\sum_j \left[\partial^2\phi(\mathbf{z}_j)/\partial\boldsymbol{\nu}^2\right]^{1/2} H^{-1/2}(\mathbf{z}_j)}. \tag{8.7.0}$$

Combining (8.7.0) and (8.5.15) with (8.2.5), we thus obtain the result.

$$D_{ij} = \frac{1}{\pi\beta m}\mathcal{H}^{1/2}(\mathbf{0})\sum_l \left[\frac{\partial^2\Psi}{\partial\boldsymbol{\nu}^2}(z_l)\right]^{1/2} \times H^{-1/2}(\mathbf{z}_l) z_l^i z_l^j \cdot e^{-\hat{\Psi}(\mathbf{z}_l)/KT}, \tag{8.7.0'}$$

where $\mathbf{z}_l^j$ denotes the jth component of the vector $\mathbf{z}_l$. The terms $\mathcal{H}(\mathbf{0})$, $H(\mathbf{z}_l)$, and $\hat{\Psi}$ are interpreted as in the previous discussion. A single term of the form $\mathcal{H}^{1/2}(\mathbf{0})H^{-1/2}(\mathbf{z})$ is present in the formulas of Vineyard (1957) and Glyde (1967), who considered a different model for vacancy diffusion.

We note that in many problems the diffusion coefficients in different directions differ in their exponential rates. For example, in the hexagonal zinc lattice, the diffusion is isotropic in the hexagonal plane, but its rate is different by an exponential factor from the rate from one plane to another. The resulting diffusion equation is of the form (8.2.3). To account for this, it is necessary to consider, respectively, the two-dimensional diffusion in the plane and the one-dimensional diffusion from one plane to another. Thus we compute a different exit time corresponding to each case.

The term β represents the slowing down of particles injected into the lattice. The slowing down is due to the particle's giving up kinetic energy, by interaction with the lattice, thus raising the temperature of the crystal. The "viscous" effect of β therefore is to lower the diffusion coefficients. Experiments indicate that, in fact, the diffusion coefficients are lower than those predicted by the calculations based solely on Ψ and m, as given in the physical literature. The viscous factor β is proposed here as a possible explanation of this effect.

Next we analyze equations (6.1.24) and (6.1.25), which have been shown to be equivalent to (6.1.28) and (6.1.30), respectively. Equation (6.1.25) has the equilibrium solution $a=$constant. Thus, after a finite relaxation time, we expect a to be nearly constant (i.e., a_σ and $\nabla_{\mathbf{r}}$ to be small). We therefore take

(8.7.1)
$$a \sim \sum_{n=0}^{\infty} \delta^n f_n(\mathbf{r}, \mathbf{r}', \sigma'')$$

with

(8.7.2)
$$\mathbf{r}' = \delta \mathbf{r},$$
$$\sigma'' = \delta^2 \sigma,$$

and

$$\delta \ll 1.$$

The variables $\mathbf{r}'$ and σ'' describe slow changes of a in space and time; δ is an artificial small parameter that does not appear in the final results. Since (6.1.25) for a contains $\phi(\mathbf{r})$, which is periodic across a cell, we allow each f_n to depend on $\mathbf{r}$ periodically.

Introducing (8.7.1) and (8.7.2) into (6.1.25) and equating the coefficients of different powers of δ, we obtain the following sequence of equations:

$$\begin{aligned}0=&\lambda\Delta_{\mathbf{r}}f_n-\nabla_{\mathbf{r}}\phi\cdot\nabla_{\mathbf{r}}f_n\\&+2\lambda\nabla_{\mathbf{r}'}\cdot\nabla_{\mathbf{r}}f_{n-1}-\nabla_{\mathbf{r}}\phi\cdot\nabla_{\mathbf{r}'}f_{n-1}\\&+\lambda\Delta_{\mathbf{r}'}f_{n-2}-\frac{\partial}{\partial\sigma''}f_{n-2}.\end{aligned}\tag{8.7.3}$$

Setting $n=0$ in (8.7.3) yields

$$0=Lf_0=\lambda\Delta_{\mathbf{r}}f_0-\nabla_{\mathbf{r}}\phi\cdot\nabla_{\mathbf{r}}f_0,\tag{8.7.4}$$

which has the "constant" solution

$$f_0=F(\mathbf{r}',\sigma'').\tag{8.7.5}$$

The operator L acts on a space of functions $f(\mathbf{r})$, each of which is periodic in $\mathbf{r}$ across a cell C. To be more precise, we shall assume that ∂C consists of $2n$ planar faces, and that

$$f(\mathbf{r})=f(\mathbf{r}+\mathbf{l}_i),\qquad i=1,\ldots,n,$$

where $\mathbf{l}_i$ are vectors that connect a point on the ith face to its related point on the opposite face [see Fig. 8.7.1]. Boundary conditions for the functions f can be formulated. If $\mathbf{r}$ is a point on the ith face of ∂C, then $\tilde{\mathbf{r}}=\mathbf{r}+\mathbf{l}_i$ is the related point on the opposite face, and

$$f(\mathbf{r})=f(\tilde{\mathbf{r}}),\qquad \mathbf{r}\in\partial C.\tag{8.7.6}$$

It can be shown that L^*, the adjoint of L, is

$$L^*f(\mathbf{r})=\lambda\Delta_{\mathbf{r}}f+\nabla_{\mathbf{r}}\cdot f\nabla_{\mathbf{r}}\phi,\tag{8.7.7}$$

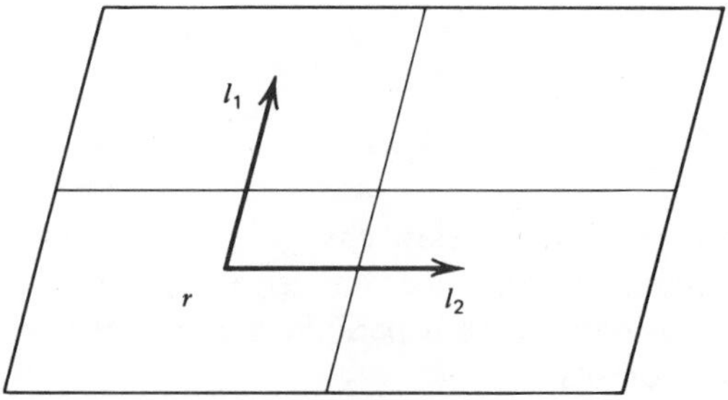

Figure 8.7.1

and that

$$L^*\{e^{-\phi(\mathbf{r})/\lambda}\}=0. \tag{8.7.8}$$

Next we set $n=1$ in (8.7.3) and get

$$0=Lf_1-\nabla_{\mathbf{r}}\phi(\mathbf{r})\cdot\nabla_{\mathbf{r}'}F(\mathbf{r}',\sigma''). \tag{8.7.9}$$

For a solution f_1 to exist, a solvability condition must be met (see Courant and Hilbert 1962). This condition is obtained by multiplying (8.7.9) by $e^{-\phi/\lambda}$ and integrating over a cell. By (8.7.8), the resulting condition is

$$\begin{aligned}0&=-\int_C e^{-\phi/\lambda}\nabla_{\mathbf{r}}\phi\cdot\nabla_{\mathbf{r}'}F\,d^3\mathbf{r}\\&=\lambda\int_C\nabla_{\mathbf{r}'}F(\mathbf{r}',\sigma'')\cdot\nabla_{\mathbf{r}}e^{-\phi(\mathbf{r})/\lambda}\,d^3\mathbf{r}\\&=\lambda\nabla_{\mathbf{r}'}F(\mathbf{r}',\sigma'')\cdot\int_{\partial C}\mathbf{n}(\mathbf{r})e^{-\phi/\lambda}\,d^2\mathbf{r}.\end{aligned}$$

But the surface integral vanishes, so this solvability condition is automatically satisfied.

Thus, (8.7.9) has the solution

$$f_1=\left[L^{-1}\nabla_{\mathbf{r}}\phi(\mathbf{r})\right]\cdot\nabla_{\mathbf{r}'}F(\mathbf{r}',\sigma''), \tag{8.7.10}$$

where the scalar operation L^{-1} is the pseudo-inverse of L, defined uniquely as follows for a periodic function $f(\mathbf{r})$ satisfying

$$0=\int_C e^{-\phi/\lambda}f\,d^3\mathbf{r},$$

$L^{-1}f$ is the unique solution of the equation

$$L(L^{-1}f)=f,$$

which satisfies

$$0=\int_C e^{-\phi/\lambda}L^{-1}f\,d^3\mathbf{r}.$$

Next, we set $n=2$ in (8.7.3) and apply the solvability condition, obtaining

$$(8.7.11)\quad \frac{\partial F}{\partial\sigma''}\left\{\int_C e^{-\phi/\lambda}\,d^3\mathbf{r}\right\}$$
$$=\nabla_{\mathbf{r}}'\cdot\left\{\int_C e^{-\phi/\lambda}\left[\lambda\mathbf{I}-(\nabla_{\mathbf{r}}\phi)L^{-1}(\nabla_{\mathbf{r}}\phi)+2\lambda\nabla_{\mathbf{r}}L^{-1}(\nabla_{\mathbf{r}}\phi)\right]d^3\mathbf{r}\right\}$$
$$\cdot\nabla_{\mathbf{r}}'F.$$

However, Green's theorem gives

$$\int_C e^{-\phi/\lambda}2\lambda\nabla_{\mathbf{r}}L^{-1}(\nabla_{\mathbf{r}}\phi)\,d^3\mathbf{r}$$
$$=\int_C e^{-\phi/\lambda}2(\nabla_{\mathbf{r}}\phi)L^{-1}(\nabla_{\mathbf{r}}\phi)\,d^3\mathbf{r},$$

so that (8.7.11) reduces to

$$(8.7.12)\quad \frac{\partial F}{\partial\sigma''}=\nabla_{\mathbf{r}'}\cdot\mathbf{M}\cdot\nabla_{\mathbf{r}'}F,$$

where the constant diffusion tensor $\mathbf{M}$ is defined by

$$(8.7.13)\quad \mathbf{M}=\frac{\int_C e^{-\phi/\lambda}\left[\lambda\mathbf{I}+(\nabla_{\mathbf{r}}\phi)L^{-1}(\nabla_{\mathbf{r}}\phi)\right]d^3\mathbf{r}}{\int_C e^{-\phi/\lambda}\,d^3\mathbf{r}}.$$

To leading order in δ, (8.7.1), (8.7.3), and (8.7.5) imply that

$$a(\mathbf{r},\sigma)=P(\delta\mathbf{r},\delta^2\sigma),$$

and then (8.7.12) yields

$$(8.7.14)\quad \frac{\partial a}{\partial\sigma}=\nabla_{\mathbf{r}}\cdot\mathbf{M}\cdot\nabla_{\mathbf{r}}a.$$

Next, (6.1.27) can be written as

$$(8.7.15)\quad p(\mathbf{x},t)=e^{-(m/kT)\Phi(\mathbf{x})}u(\mathbf{x},t),$$

with

$$u(\mathbf{x},t)=a\left[\frac{1}{l}\mathbf{x},\frac{v_0^2}{\beta_l^2}t\right]l^3v_0^3, \tag{8.7.16}$$

and by (8.7.14) the equation governing u is

$$\beta\frac{\partial u}{\partial t}=\nabla_{\mathbf{r}}\cdot\mathbf{D}\cdot\nabla_{\mathbf{x}}u, \tag{8.7.17}$$

where

$$\mathbf{D}=v_0^2\mathbf{M}. \tag{8.7.18}$$

Equation (8.7.15) shows that the solution $p(\mathbf{x},\xi,t)$ of Smoluchowski's equation can be approximated by a solution of the purely diffusion equation (8.7.17), with the initial condition

$$u(\mathbf{x},\xi,t)\to\delta(\mathbf{x}-\xi) \quad \text{as} \quad t\downarrow 0. \tag{8.7.19}$$

The process of approximating the solution of a diffusion equation in a periodic potential field (i.e., the Smoluchowski equation), by a solution of a diffusion equation without any force field is called "homogenization" (Benssousan et al. 1978).

As a check, we see from (8.7.13) that for $\Phi=0$, $\mathbf{M}=\lambda\mathbf{I}$, so that $\mathbf{D}=v_0^2\lambda\mathbf{I}=kT/m\mathbf{I}$. Thus, for $\Phi=0$, the homogenized Smoluchowski–Kramers equation (8.7.17) reduces to the original Smoluchowski–Kramers equation (6.1.28), as it should. Of course, the dependent variables in these equations, u and p, are also identical by virtue of (6.1.30).

Finally, we note that we have not derived the initial condition (8.7.19) from (6.1.29), but, as before, we remark that it can be derived by means of an initial layer analysis of (6.1.28), (6.1.29)

Next we analyze the results (8.7.13), (8.7.17), and (8.7.18) for large activation energy [i.e., we assume that $\Phi=O(1/\varepsilon)$]. Thus, as in (6.1.12), we set

$$\phi=\frac{1}{\varepsilon}\phi_0, \tag{8.7.20}$$

and then (8.7.13) and (8.7.4) become

$$\mathbf{M}=\frac{\int_C e^{-\phi/\varepsilon\lambda}\left[\lambda\mathbf{I}+(1/\varepsilon)(\nabla_r\phi)\Psi\right]d^3\mathbf{r}}{\int_C e^{-\phi/\varepsilon\lambda}\,d^3\mathbf{r}}, \tag{8.7.21}$$

where $\boldsymbol{\Psi}(\mathbf{x})$ is periodic across C and is defined by the equation

$$\lambda\Delta_{\mathbf{r}}\boldsymbol{\Psi}-(\nabla_{\mathbf{r}}\phi\cdot\nabla_{\mathbf{r}})\boldsymbol{\Psi}=\nabla_{\mathbf{r}}\phi. \tag{8.7.22}$$

We shall evaluate $\boldsymbol{\Psi}$ asymptotically for $\varepsilon \ll 1$, and then, using (8.7.21), evaluate $\mathbf{M}$ asymptotically. First, we solve (8.7.22) with periodic boundary conditions across a cell. The method of matched asymptotics is used as follows. The outer solution (i.e., the solution of the reduced equation)

$$-(\nabla_{\mathbf{r}}\phi\cdot\nabla_{\mathbf{r}})\boldsymbol{\Psi}=\nabla_{\mathbf{r}}\phi$$

is given by

$$\Psi_1=-\mathbf{x}+\text{constant}$$

To satisfy the periodicity condition, we add a boundary layer $\boldsymbol{\Psi}_2$, so that

$$\boldsymbol{\Psi}\sim\boldsymbol{\Psi}_1+\boldsymbol{\Psi}_2.$$

We introduce local coordinates near the boundary ∂C by setting $y=$ distance to the boundary measured along normals to ∂C, $\mathbf{x}'=$tangential coordinates in ∂C. In the system of coordinates $(\mathbf{x}', y)$, we have

$$\frac{\partial\boldsymbol{\Psi}_2}{\partial y}=O(\lambda^{-1/2}),\qquad \frac{\partial\boldsymbol{\Psi}_2}{\partial\mathbf{x}'}=O(1),$$

near and on ∂C.

In local coordinates near ∂C, the boundary layer $\boldsymbol{\Psi}_2$ satisfies the homogeneous equation (8.7.22).

$$\lambda\frac{\partial^2\boldsymbol{\Psi}_2}{\partial y^2}-\frac{\partial\phi}{\partial\boldsymbol{\nu}}\frac{\partial\boldsymbol{\Psi}_2}{\partial y}+\lambda A(\mathbf{x}', y)\boldsymbol{\Psi}_2+\mathbf{b}_2(\mathbf{x}', y)\cdot\nabla_{\mathbf{x}'}\boldsymbol{\Psi}_2=\mathbf{0}, \tag{8.7.23}$$

where $A(\mathbf{x}', y)$ is a second-order operator not containing the term $\partial^2/\partial y^2$. The term $-\partial\phi/\partial\boldsymbol{\nu}$ vanishes on the boundary $y=0$ and becomes positive for positive y (in the cell).

Writing $-\partial\phi/\partial\boldsymbol{\nu}=yb_1(\mathbf{x}')+O(y^2)$, $b_1(\mathbf{x}')>0$, we obtain

$$b_1(\mathbf{x}')=\frac{1}{2}\frac{\partial^2\phi}{\partial\boldsymbol{\nu}^2},$$

so that (8.7.23) takes the form

$$\text{(8.7.24)} \qquad \lambda\frac{\partial^2\Psi_2}{\partial y^2}+yb_1(\mathbf{x}')\frac{\partial\Psi_2}{\partial y}$$
$$+\lambda A(\mathbf{x}',y)\Psi_2+\mathbf{b}_2(\mathbf{x}',y)\cdot\nabla_{\mathbf{x}'}\Psi_2=\mathbf{0}.$$

We introduce the stretching transformation $y=\lambda^{1/2}\eta$ in the normal direction. Then (8.7.24) becomes

$$\frac{\partial^2\Psi_2}{\partial\eta^2}+\eta b_1(\mathbf{x}')\frac{\partial\Psi_2}{\partial\eta}+\sqrt{\lambda}\,A\big(\mathbf{x}',\sqrt{\lambda}\,\eta\big)\Psi_2+\mathbf{b}_2\big(\mathbf{x}',\sqrt{\lambda}\,\eta\big)\cdot\nabla_{\mathbf{x}'}\Psi_2=0.$$

We may neglect the term $\sqrt{\lambda}\,A(\mathbf{x}',\sqrt{\lambda}\,\eta)\Psi_2$ for small λ, and replace $\mathbf{b}_2(\mathbf{x}',\lambda\eta)$ by $\mathbf{b}_2(\mathbf{x}')=\mathbf{b}_2(\mathbf{x}',0)$. The expression $\mathbf{b}_2(\mathbf{x}')$ denotes differentiation in ∂C in the characteristic directions determined by

$$\frac{d\mathbf{x}}{dt}=-\nabla\phi(\mathbf{x})$$

in ∂C. The leading term Ψ_2^0 in the expansion of Ψ_2 as $\lambda\to 0$ is the solution of the boundary layer equation

$$\text{(8.7.25)} \qquad \frac{\partial^2\Psi_2^0}{\partial\eta^2}+\eta b_1(\mathbf{x}')\frac{\partial\Psi_2^0}{\partial\eta}+\mathbf{b}_2(\mathbf{x}')\cdot\nabla_{\mathbf{x}'}\Psi_2^0=\mathbf{0}.$$

The matching conditions imply that $\Psi_2^0(\mathbf{x}',\eta)\to 0$ as $\eta\to\infty$ and that $-\mathbf{x}+\Psi_2^0(\mathbf{x}',0)$ is periodic across the cell. Thus $\Psi\sim-\mathbf{x}+\Psi_2^0(\mathbf{x}',y/\sqrt{\lambda}\,)\zeta(y)$, where $\zeta(y)=1$ for $0\leqslant y\leqslant\gamma$ and $\zeta(y)=0$ for $y>2\gamma$ for some small $\gamma>0$. The term $-\mathbf{x}\cdot\nabla\phi/\lambda$ cancels with the term $\lambda\mathbf{I}$ after an integration by parts. The main contribution to the remaining integral comes from a neighborhood of ∂C, in particular from the neighborhood of the saddle points of ϕ. At such saddle points, $\mathbf{x}_i'$ say, the term $\mathbf{b}_2(\mathbf{x}_i')=\mathbf{0}$ in (8.7.24). It follows that it suffices to determine $\Psi_2^0(\mathbf{x}',\eta)$ at the saddle points $\mathbf{x}_i$ only. For such points (8.7.24) takes the form

$$\frac{\partial^2\Psi_2^0}{\partial\eta^2}+\eta b_1(\mathbf{x}_i')\frac{\partial\Psi_2^0}{\partial\eta}=\mathbf{0}.$$

The solution satisfying boundary conditions is given by

$$\Psi_2^0(\mathbf{x}', y) \sim \frac{\mathbf{x}_i' b_1^{1/2}(\mathbf{x}_i')}{\sqrt{\pi/2}} \int_{y/\sqrt{\lambda}}^{\infty} e^{-s^2 b_1(\mathbf{x}_i')/2}\, ds.$$

Expanding the integral in the numerator of (8.7.21) about the saddle points $\mathbf{z}_1, \ldots, \mathbf{z}_n$, and the integral in the denominator about the minimum of ϕ [assume that $\min_C \phi = \phi(\mathbf{0}) = 0$], we obtain

$$(8.7.26)\quad \mathbf{M}_{kl} = \frac{1}{\pi} \mathcal{H}^{1/2}(\mathbf{0}) \sum_i \left[\frac{\partial^2 \phi}{\partial \boldsymbol{\nu}^2}(\mathbf{z}_i) \right]^{1/2} H^{-1/2}(\mathbf{z}_i) z_i^k z_i^l e^{-m\phi(\mathbf{z}_i)/kT}$$

where

$$H(\mathbf{z}) = \det \frac{\partial^2 \phi}{\partial x_i' \partial x_j'}(\mathbf{z}) \qquad (i, j = 1, 2)$$

is the Hessian of ϕ in ∂C, and

$$\mathcal{H}(\mathbf{0}) = \det \frac{\partial^2 \phi}{\partial x^i \partial x_j}(\mathbf{0}) \qquad (i, j = 1, 2, 3)$$

is the Hessian of ϕ at $\mathbf{x} = \mathbf{0}$. In terms of the potential $\Phi = m\phi$, we obtain an additional factor of m^{-1} in (8.7.26). The diffusion tensor is given by $\mathbf{D} = (1/\beta)\mathbf{M}$; hence

$$(8.7.27)\quad \mathbf{D}_{kl} = \frac{1}{\pi m \mu} \mathcal{H}^{1/2}(\mathbf{0}) \sum_i \left[\frac{\partial^2}{\partial \boldsymbol{\nu}^2}(\mathbf{z}_i) \right]^{1/2} H^{-1/2}(\mathbf{z}_i) z_i^k z_i^l e^{-\Phi(\mathbf{z}_i)/kT},$$

where μ is the friction or viscosity coefficient in the crystal (i.e., $\beta = m\mu$). Formula (8.7.27) is identical to formula (8.7.0′).

EXERCISE 8.7.1

Repeat the calculations of Section 8.7 in the one-dimensional case. Solve the differential equation explicitly and compare the results with (8.7.27) (Larsen and Schuss 1978).

Next we derive the Nernst–Einstein formula (see Section 8.2). In ionic crystals electrical conduction is due to the motion of ions through the crystal, which is subjected to a uniform electrostatic field E, say. Thus we may assume that the ion of charge q is moving in a potential field. In a simplified one dimensional model, the potential Φ, which appears in the Smoluchowski equation is given by

(8.7.28) $$m\Phi(x) = -qEx + qK\sin\omega x, A < x < T$$

and

$$\Phi \to \infty \text{ as } |x| \to \infty,$$

(see Fig. 8.7.2). Here the periodic term represents the internal potential of the crystal, and the linear term represents the potential of the external electrostatic field.

In the absence of diffusion, a particle trapped at R cannot move to S. Therefore the mere presence of an external field E will not cause conductance, unless it is sufficiently strong. Nevertheless, ionic conductivity is observed even for weak electrostatic fields. Thus we attribute ionic conductivity to diffusion. We observe that the potential barrier to be overcome for a motion from R to S to occur is $Q = \Phi(P) - \Phi(R)$, which is lower than the potential barrier $Q^* = \Phi(P) - \Phi(S)$, which must be overcome for a motion from S to R to occur. Therefore the net flow will be from R to S. To compute the conductivity of the crystal, we compute the net current $i = i_{RS} - i_{SR}$ flowing from R to S, and employ it in

(8.7.29) $$\sigma = \left| \frac{\partial i}{\partial V} \right|,$$

where $V = \Phi(S) - \Phi(R)$, is the potential difference between S and R. We note that V is given by

(8.7.30) $$V = Q - Q^* = -\frac{2\pi qE}{\omega}.$$

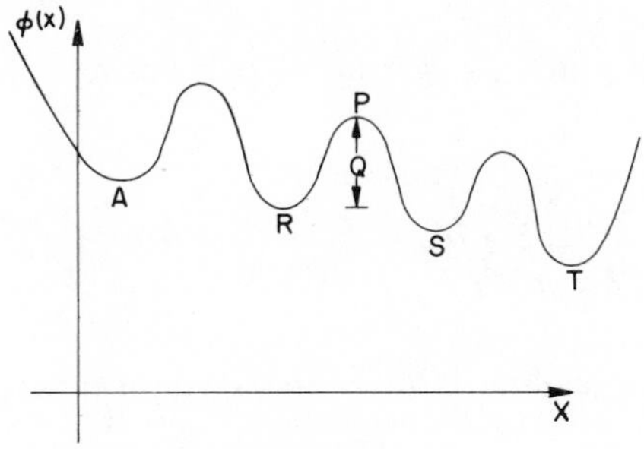

Figure 8.7.2

Now, the current flowing to the right is

$$(8.7.31) \qquad i_{RS}=\frac{Cq}{\bar{\tau}_R},$$

while the current flowing to the left is

$$(8.7.32) \qquad i_{SR}=\frac{Cq}{\bar{\tau}_L}$$

where $\bar{\tau}_R$ and $\bar{\tau}_L$ are the average times required to overcome the potential barriers Q and Q^* respectively, and C denotes the concentration of ions. We observe that $\bar{\tau}_R=\lambda_1^{-1}$, where λ_1 is the second eigenvalue corresponding to Φ. Therefore, employing (8.6.17), we have

$$(8.7.33) \qquad \bar{\tau}_R=\frac{2\pi}{\sqrt{\Phi''(R)|\Phi''(P)|}}e^{mQ/kT}.$$

Similarly,

$$(8.7.34) \qquad \bar{\tau}_L=\frac{2\pi}{\sqrt{\Phi''(S)|\Phi''(P)|}}e^{mQ^*/kT}.$$

Therefore, employing (8.7.30)–(8.7.33) in (8.7.29) we obtain

$$(8.7.35) \qquad \sigma=\frac{\partial}{\partial V}\frac{Cq}{2\pi m}\sqrt{\Phi''(R)|\Phi''(P)|}\,e^{-mQ/kT}(1-e^{-V/kT}).$$

Here we have used the fact that $\Phi''(S)=\Phi''(R)$, which follows from (8.7.28). Now (8.7.28) implies that

$$(8.7.36) \qquad \sqrt{\Phi''(R)|\Phi''(P)|}=\frac{\omega^2Kq}{m}\sqrt{1-\left(\frac{E}{\omega K}\right)^2}.$$

Then, employing (8.7.30) and (8.7.36) in (8.7.35), we obtain

$$(8.7.37) \qquad \sigma=\frac{\partial}{\partial E}\frac{Cq\omega^2K}{4\pi^2m}\sqrt{1-\left(\frac{E}{\omega K}\right)^2}\,e^{-mQ/kT}\left(1-e^{\frac{-2\pi qE}{\omega kT}}\right).$$

If $\frac{E}{\omega K}\ll 1$, then the leading term in the asymptotic expansion of (8.7.37) with respect to the parameter $\frac{E}{\omega K}$, is given by

$$\sigma_0 = \frac{Cq^2\omega K}{2\pi kTm} e^{-H/kT} \tag{8.7.38}$$

where $H = 2Kq$ is the value of mQ when $E = 0$. We recall from Exercise 8.7.1 that the quantity $\frac{\omega K}{2\pi m} e^{-H/kT}$ is the one dimensional diffusion coefficient for atomic migration in crystals. Therefore (8.7.38) may be written as

$$\sigma_0 = \frac{Cq^2 D}{kT}. \tag{8.7.39}$$

This formula for ionic conductance was derived by Nernst and Einstein (Girifalco 1964). Thus formula (8.7.37) is a generalization of the Nernst-Einstein formula, since (8.7.37) reduces to (8.7.39) for small $\frac{E}{\omega K}$. In nonisotropic crystals, $D = \{D_{ij}\}$ is a diffusion tensor so that $\sigma = \{\sigma_{ij}\}$ is a conductance tensor. (see Matkowsky and Schuss, to appear)

EXERCISE 8.7.2

Set $\varepsilon = \frac{kT}{2qK}$, $x = \frac{E}{\omega K}$. Show that

$$\sigma = \frac{\sigma_0}{\pi}\left\{\pi\sqrt{1-x^2}\, e^{-\pi x/\varepsilon} + (1 - e^{-\pi x/\varepsilon})\left(\sqrt{1-x^2}\cos^{-1}x - \frac{\varepsilon}{\omega K\sqrt{1-x^2}}\right)\right\}\exp\left(\frac{1 + x\cos^{-1}x - \sqrt{1-x^2}}{\varepsilon}\right).$$

EXERCISE 8.7.3

Find the generalization of the Nernst–Einstein formula for the conductivity tensor σ_{ij} in three dimensions.

CHAPTER 9

Filtering Theory

9.1 INTRODUCTION

In many problems of engineering the following situation arises. A deterministic or random signal is sent through a noisy channel, and it is to be estimated, or filtered, from the noisy measurements. For example, a constant, but unknown frequency is transmitted by radio, and the received signal usually contains additional noise, such as atmospheric static, internal noise of the transmitter and the receiver, and so on. This is the case, for example, in the transmission of speech: the voice of the speaker is sampled 5000 times per second and a constant signal is transmitted for $1/5000$ of a second. The filtering problem in this case is to construct a realizable estimation mechanism to determine the transmitted frequency. If the transmitted signal is speech or music, say, it cannot be considered deterministic, so that a stochastic model for the signal has to be constructed. The signals usually undergo some modulation [e.g., amplitude modulation (AM) or frequency modulation (FM)] before they are transmitted. A simple model for the filtering problem is shown in Fig. 9.1.1.

The possibly random signal $x(t)$ (e.g., the voltage produced by a microphone) undergoes a memoryless nonlinear transformation *called modulation*, so that a modulated signal $g(x(t),t)$ is transmitted. The signal $g(x(t),t)$ acquires additional noise $n(t)$, so that the received signal $r(t)$ consists of the modulated signal $g(x(t),t)$ and the additional noise $n(t)$. Thus

$$r(t)=g(x(t),t)+n(t).$$

An example of modulation is that of FM transmission. The relatively slowly varying signal $x(t)$, in the audio range, 16 to 3000 Hz, say, is integrated, amplified, and a high-frequency carrier is added to it, so that it

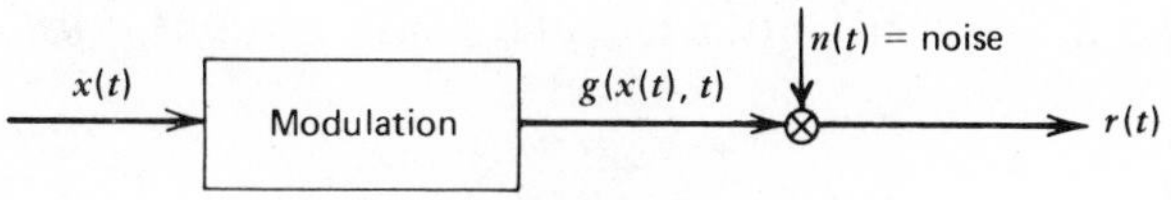

Figure 9.1.1

is transformed by

$$x(t) \to \omega_0 t + d_f \int_0^t x(s)\, ds.$$

Next the transformation $C\sin(\cdot)$ is applied to it, so that

$$g(x(t), t) = C\sin\left(\omega_0 t + d_f \int_0^t x(s)\, ds\right).$$

Here C is the strength of the transmission, d_f is called *frequency deviation*, and ω_0 is the high frequency of the carrier (in the MHz range). If, for example, $x(t) = x_0 = \text{constant}$, we see that

$$g(x(t), t) = C\sin(\omega_0 + d_f x_0)t,$$

so that $d_f x(t)$ indeed modulates the frequency ω_0.

Some basic texts on this subject are Jazwinski (1970), Van Trees (1970), Lindsey (1972), and Snyder (1969). The theory of FM and the phase-locked loop can be found in Viterbi (1968), Lindsey (1972), Snyder (1969), Tausworth (1972), Bobrovsky and Schuss (to appear), and Bobrovsky et al. (to appear).

9.2 A MODEL OF THE SIGNAL

In communication theory the signal (e.g., speech or music) is often modeled as a random process $x(t)$. More precisely, the voltage entering the modulator is a random process $x(t)$ and we denote by

$$P(x(t_1) < b_1, \ldots, x(t_n) < b_n)$$

the probability that the voltage entering the modulator at times $t_1 < t_2 < \cdots < t_n$ satisfies the inequalities $x(t_i) < b_i (i = 1, 2, \ldots, n)$. It is common practice to assume that $x(t)$ is a stationary Gaussian process. Furthermore, it is often assumed to have a power spectrum function $S_x(\omega)$ given by

(9.2.1) $$S_x(\omega) = \begin{cases} c & \text{if } |\omega| < k \\ 0 & \text{otherwise,} \end{cases}$$

where c and k are constants. Here the *power spectral density function* is defined by

$$S_x(\omega)=\int_{-\infty}^{\infty} e^{-is\omega}Ex(t+s)x(t)\,ds.$$

It represents the power output of the signal $x(t)$ at frequency ω. We shall assume that the signal $x(t)$ is a stationary process [i.e., the *autocorrelation function* $Ex(t+s)x(t)$ is independent of t]. The constant k represents the *bandwidth* of the message. The assumption (9.2.1) means that all frequencies in the band $|\omega|<k$ appear in $x(t)$ with the same power output. Thus *white noise* is assumed to have all frequencies with the same power output; that is, for the white noise $\dot{w}(t)$, we have

$$S_{\dot{w}}(\omega)=1 \qquad \text{for} \quad -\infty<\omega<\infty$$

as

$$E\dot{w}(t)\dot{w}(t+s)=\delta(s).$$

This implies that the total power output of a white noise is infinite, which makes no sense from the physical point of view. Since the realization of a process $x(t)$ such that (9.2.1) holds, as a solution of a stochastic differential equation is unknown, we shall proceed as follows. Following Van Trees (1970), we consider the Butterworth family of spectra

$$S_n(\omega)=\frac{2n\sin(\pi/2n)}{k\left[(\omega/k)^{2n}+1\right]} \tag{9.2.2}$$

Obviously, $S_n(\omega)\to S_x(\omega)$ as $n\to\infty$. A process $x_n(t)$ with power spectral density function $S_n(\omega)$, given by (9.2.2), can be realized as a solution of Itô-type equation, and subsequently by a computer or an analog circuit as follows. Set

$$H_n(i\omega)=\frac{\sqrt{2n[\sin(\pi/2n)]k^{2n-1}}}{\prod_{j=1}^{n}(\omega-\zeta_j k)},$$

where ζ_j are the $2n$th roots of -1 in the half-plane $\operatorname{Re}\omega<0$. Then the solution of the differential equation

$$L_n x_n(t)=\dot{w}(t), \tag{9.2.3}$$

where

$$\int_0^\infty L_n x_n(t) e^{-st}\, dt = H_n(s)\hat{x}_n(s)$$

$$\hat{x}_n(s) = \int_0^\infty e^{-st} x_n(t)\, dt,$$

has the spectral density (9.2.2). Equation (9.2.3) is equivalent to the system of Itô stochastic differential equations given by

(9.2.4) $$d\mathbf{x}_n(t) = \mathbf{A}_n \mathbf{x}_n(t)\, dt + \mathbf{B}\, dw,$$

where

$$\mathbf{x}_n(t) = \left[x_n(t), \dot{x}_n(t), \ldots, x_n^{(n-1)}(t)\right]^T,$$

$\mathbf{B}$ is a constant vector, and $\mathbf{A}_n$ is a constant matrix (Zadeh and Desoer 1963).

EXERCISE 9.2.1

Write down (9.2.3) and (9.2.4) for $n=1$ and $n=2$.

In general, a stationary scalar Gaussian process $x(t)$ with power spectral density $S(\omega)$, which is a rational function of ω, such that $S(\omega) \to 0$ as $|\omega| \to \infty$, can be represented as a solution of the mth-order stochastic differential equation

(9.2.5)

$$x^m(t) + \alpha_1 x^{(m-1)}(t) + \cdots + \alpha_m x(t) = \gamma_1 w^{(m-1)}(t) + \gamma_2 w^{(m-2)}(t) + \cdots + \gamma_m w(t),$$

where α_j and γ_j are constants. The solution can be realized as the steady-state response of the filter shown in Fig. 9.2.1. An Itô representa-

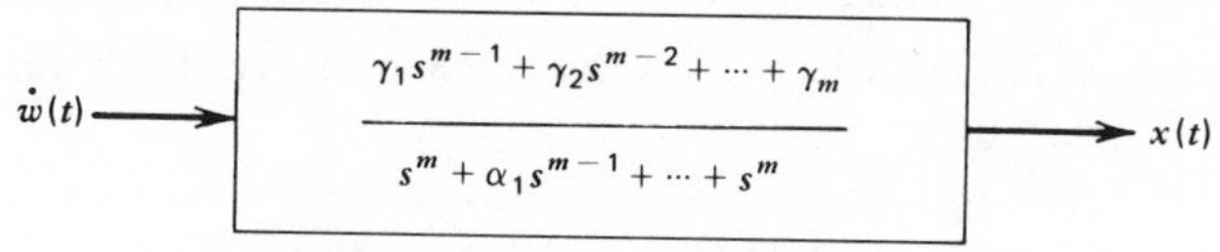

Figure 9.2.1 (From Snyder 1969)

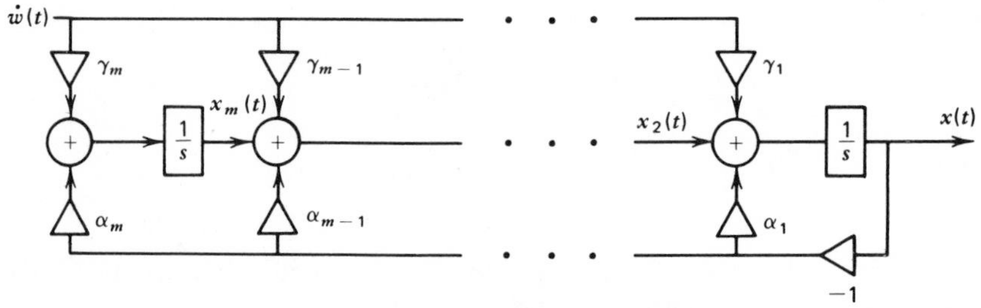

Figure 9.2.2 (From Snyder 1969)

tion is given by $dx_j = [-\alpha_j x_1(t) + x_{j+1}(t)]\,dt + \gamma_j\,dw$, $1 \leqslant j \leqslant m$, where $x_{m+1}(t) \equiv 0$. Here $x(t) = x_1(t)$ (Zadeh and Desoer 1963; Snyder 1969). An alternative realization of $x(t)$ is given in Fig. 9.2.2. The matrix $\mathbf{A}_m$ of (9.2.4) is given by

$$\mathbf{A}_m = \begin{bmatrix} -\alpha_1 & 1 & & \\ -\alpha_2 & & 1 & \\ \vdots & & & \ddots \\ -\alpha_m & & & & 1 \end{bmatrix}$$

and the vector $\mathbf{B}_m$ is given by $\mathbf{B}_m = [\gamma_1, \ldots, \gamma_m]^T$.

We shall assume for simplicity that $x(t)$ has a Butterworth spectral density of order 1; that is,

$$dx(t) = -kx + \sqrt{2k}\,dw.$$

9.3 MODULATION AND MEASUREMENT OF THE SIGNAL

The signal $x(t)$ undergoes modulation before being sent out. The most common types of modulation are *amplitude modulation* (AM), where the signal is subject to the linear transformation

$$g(x(t), t) = x(t)\sqrt{2}\sin\omega_0 t, \tag{9.3.1}$$

where ω_0 is the high frequency of the carrier. The measured signal contains additive noise $\sqrt{2N_0}\,\dot{w}$, where N_0 is the noise intensity. Thus the measured signal $y(t)$ is given by

$$dy(t) = \left[x(t)\sqrt{2}\sin\omega_0 t\right]dt + \sqrt{2N_0}\,dw. \tag{9.3.2}$$

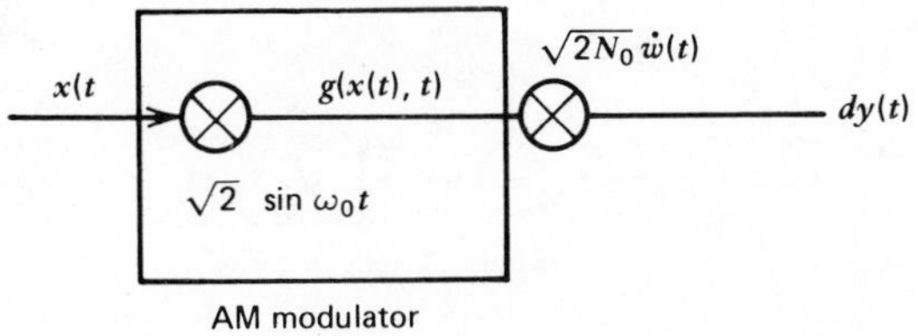

Figure 9.3.1

The factor $\sqrt{2}$ is chosen to make the signal strength equal 1. A schematic description of AM transmission is shown in Fig. 9.3.1. Another type of modulation is *phase modulation* (PM), where $x(t)$ undergoes the nonlinear transformation

$$g(x(t),t)=\sqrt{2}\ \sin[\omega_0 t+\beta x(t)]. \tag{9.3.3}$$

Assuming that $\operatorname{Var} x(t)=1$, we call β the *modulation* index. The noisy measurements are given by

$$dy=\sqrt{2}\ \sin[\omega_0 t+\beta x(t)]\,dt+\sqrt{2N_0}\,dw \tag{9.3.4}$$

(see Fig. 9.3.2).

Finally, *frequency modulation* is given by the transformation

$$g(x(t),t)=\sqrt{2}\ \sin\left[\omega_0 t+d_f\int_0^t x(s)\,ds\right] \tag{9.3.5}$$

and

$$dy=\sqrt{2}\ \sin\left[\omega_0 t+d_f\int_0^t x(s)\,ds\right]dt+\sqrt{2N_0}\,dw \tag{9.3.5}$$

(see Fig. 9.3.3).

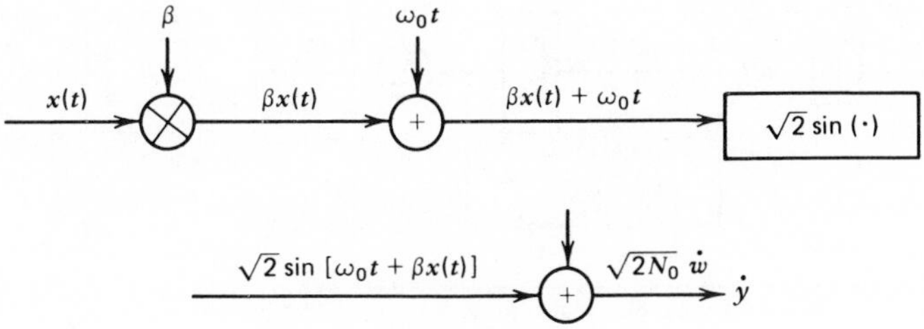

Figure 9.3.2

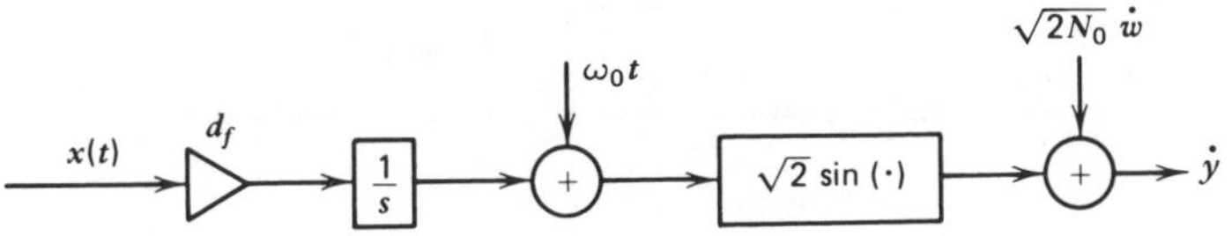

Figure 9.3.3

Setting

$$x(t) = x_1(t)$$

$$x_2(t) = d_f \int_0^t x_1(s)\, ds$$

we have the following system of equations

$$dx_1 = -kx_1\, dt + \sqrt{2k}\, dw_1$$

$$dx_2 = d_f x_1\, dt$$

$$dy = \sqrt{2}\, \sin(\omega_0 t + x_2)\, dt + \sqrt{2N_0}\, dw_2,$$

where w_1 and w_2 are two independent standard Brownian motions.

In general, we have the following model for a communication system:

$$d\mathbf{x} = \mathbf{A}\mathbf{x}\, dt + \mathbf{B}\, dw_1$$

$$dy = g(\mathbf{x}(t), t)\, dt + \sqrt{2N_0}\, dw_2.$$

A block diagram of the communication model is shown in Fig. 9.3.4.

Note that if $\mathbf{x}(t)$ is *known*, then the function $y(t) - y_0 = \int_0^t g(\mathbf{x}(s), s)\, ds$ is a Gaussian variable; that is,

$$p(dy\,|\mathbf{x}(t)) = \frac{1}{(2\pi N_0\, dt)^{n/2}} \exp\left[\frac{|dy - g(\mathbf{x}(t), t)\, dt|^2}{2N_0\, dt} \right].$$

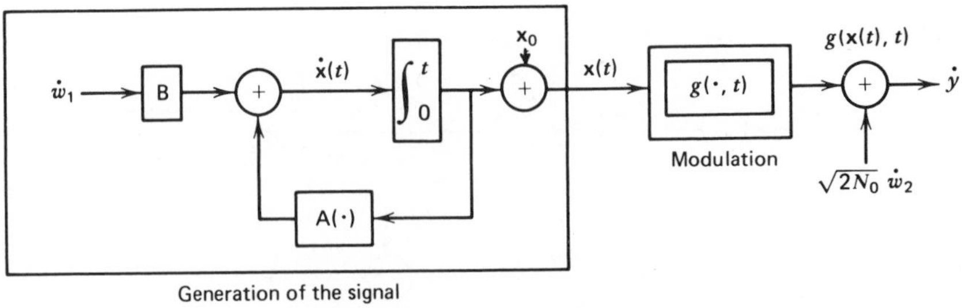

Figure 9.3.4 (From Snyder 1969)

The filtering problem is to construct an estimator $\hat{x}(t)$ such that the mean square error

$$E|x(t)-\hat{x}(t)|^2$$

is minimal, where the expectation is taken with respect to the a posteriori distribution, given the observations $y(s), 0 \leqslant s \leqslant t$.

9.4 THE OPTIMAL ESTIMATOR AND KUSHNER'S EQUATION

The optimal estimator $\hat{x}(t)$ is given by

$$\begin{aligned}\hat{x}(t) &= \int_{-\infty}^{\infty} xp(x(t)=x|y(s), 0 \leqslant s \leqslant t)\,dx \\ &= E(x(t)|y(s), 0 \leqslant s \leqslant t).\end{aligned}$$

Indeed, let $\tilde{x}(t)$ be any other estimator than

$$\begin{aligned}\widehat{|x(t)-\tilde{x}(t)|^2} &= \widehat{|x(t)-\hat{x}(t)|^2} + \widehat{[x(t)-\hat{x}(t)][\hat{x}(t)-\tilde{x}(t)]} \\ &\quad + \widehat{|\tilde{x}(t)-\hat{x}(t)|^2}.\end{aligned}$$

Since $\tilde{x}(t)$ and $\hat{x}(t)$ are constant with respect to expectation conditioned on $y(s), 0 \leqslant s \leqslant t$, we have

$$\widehat{[x(t)-\hat{x}(t)][\hat{x}(t)-\tilde{x}(t)]} = [\hat{x}(t)-\tilde{x}(t)]\widehat{[x(t)-\hat{x}(t)]} = 0.$$

Hence

$$\widehat{|x(t)-\tilde{x}(t)|^2} \geqslant \widehat{[x(t)-\hat{x}(t)]}^2.$$

The a posteriori density

$$p(x,t,y,t) \equiv p(x(t)=x|y(s), 0 \leqslant s \leqslant t)$$

satisfies Kushner's equation (Kushner 1964)

$$(9.4.1) \quad dp = Mp\,dt + [g(x(t),t) - E_y g(x(t),t)][dy - E_y g(x(t),t)\,dt]\frac{p}{N_0},$$

where M is the operator in the Fokker–Planck equation and E_y indicates expectation with respect to $p(x,t,y,t)$. Equation (9.4.1) is a stochastic integro-partial differential equation. We shall use Kushner's derivation, which also yields another derivation of the Fokker–Planck equation. We shall assume for simplicity that $x(t)$ satisfies a one-dimensional stochastic differential equation. We have, by Bayes' rule (see Exercise 1.3.2),

(9.4.2)

$$p(x,t|y,t+\Delta t)=p(x,t|y,t,\Delta y)\frac{-p(\Delta y|x,t,y,t)p(x,t|y,t)}{\int_{-\infty}^{\infty}p(\Delta y|x,t,y,t)p(x,t|y,t)\,dx}$$

$$=\frac{p(x,t|y,t)\exp\{-|\Delta y-g(x,t)\Delta t|^2/4N_0\Delta t\}}{\int_{-\infty}^{\infty}p(x,t|y,t)\exp\{-|\Delta y-g(x,t)\Delta t|^2/4N_0\Delta t\}(1+O(\Delta t))\,dx}$$

Thus, after some cancellations, we get

(9.4.3)
$$z(\Delta y,\Delta t)\equiv\frac{p(x,t|y,t+\Delta t)}{p(x,t|y,t)}$$

$$=\frac{\exp\left\{\dfrac{g(x,t)\,\Delta y}{2N_0}-\dfrac{g^2(x,t)\,\Delta t}{4N_0}\right\}}{\int_{-\infty}^{\infty}p(x,t|y,t)\exp\left\{\dfrac{g(x,t)\,\Delta y}{2N_0}-\dfrac{g^2(x,t)\,\Delta t}{4N_0}\right\}dx}$$

Expanding z about $\Delta y=0$, $\Delta t=0$, and noting that $E\,\Delta y^2=O(\Delta t)$, we get

(9.4.4)
$$z(\Delta y,\Delta t)=z(0,0)+z_{\Delta t}(0,0)\,\Delta t+z_{\Delta y}(0,0)\,\Delta y$$
$$+\tfrac{1}{2}z_{\Delta y\,\Delta y}(0,0)\,\Delta y^2+O(\Delta t).$$

From (9.4.3) we see that

(9.4.5)
$$z(0,0)=1$$

$$z_{\Delta t}(0,0)=\frac{E_y g^2(x(t),t)-g^2(x,t)}{4N_0}.$$

$$z_{\Delta y}(0,0)=\frac{g(x,t)-E_y g(x(t),t)}{2N_0},$$

and

$$4N_0 z_{\Delta y\,\Delta y}(0,0)=g^2(x,t)-2g(x,t)E_y g(x(t),t)$$
$$+2\big[E_y g(x(t),t)\big]^2-E_y g^2(x(t),t).$$

Substituting (9.4.5) in (9.4.4) and replacing Δy^2 by its mean value $2N_0\,\Delta t$, we obtain

$$z(\Delta y,\Delta t)=1+\frac{\Delta y g(x,t)}{2N_0}-\frac{\Delta y E_y g(x(t),t)}{2N_0}-\frac{g(x,t)E_y g(x(t),t)}{2N_0}$$
$$+\frac{\big[E_y g(x(t),t)\big]^2}{2N_0}+O(\Delta t)$$
$$=\frac{1+\big[\Delta y-E_y g(x(t),t)\big]\big[g(x,)-E_y g(x(t)-t)\big]}{2N_0}+O(\Delta t).$$

Hence, by (9.4.3),

$$p(x,t|y,t+\Delta t)-p(x,t|y,t)$$
$$=\frac{p(x,t|y,t)\big[g(x,t)-E_y g(x(t),t)\big]\big[\Delta y-E_y g(x(t),t)\big]}{2N_0}+O(\Delta t)$$
$$\equiv\Delta q+O(\Delta t).$$

Now, by the Chapman–Komogorov equation (4.3.2) and by the Markov property (4.3.1), we have

$$p(x,t+\Delta t|y,t+\Delta t)=\int_{-\infty}^{\infty}p(x,t+\Delta t|u,t,y,t,\Delta y)p(u,t|y,t+\Delta t)\,du$$
$$=\int_{-\infty}^{\infty}p(x,t+\Delta t|u,t)p(u,t|y,t+\Delta t)\,du,$$

as shown in Chapter 2. For any test function φ, we have

$$\int_{-\infty}^{\infty}\varphi(x)p(x,t+\Delta t|y,t+\Delta t)\,dx$$
$$=\int_{-\infty}^{\infty}\varphi(x)\big[p(x,t|y,t)+Mp\big]\,dx+O(\Delta t)$$

where M is the Fokker–Planck operator; hence

(9.4.6) $p(x,t+\Delta t|y,t+\Delta t)-p(x,t|y,t)=\Delta q(x,t)+Mp+O(\Delta t).$

Dividing by Δt and taking the limit as $\Delta t\to 0$, we obtain Kushner's equation.

EXERCISE 9.4.1

Derive Kushner's equation for the system (9.3.7) of stochastic differential equations. Let

(9.4.7) $$d\mathbf{x}(t)=\mathbf{A}(\mathbf{x}(t),t)\,dt+\mathbf{B}(\mathbf{x}(t),t)\,d\mathbf{w}_1,$$

where $\mathbf{x}(t)$ and $\mathbf{A}$ are n-vectors, $\mathbf{B}$ is an $n\times r$ matrix, and $\mathbf{w}_1$ is an r-vector of Brownian motions such that

$$Ew_{1i}(t)w_{1j}(t)=\int_0^t Q_{ij}(s)\,ds,\qquad \mathbf{Q}(t)=\{Q_{ij}(t)\}$$

and let

(9.4.8) $$d\mathbf{y}(t)=\mathbf{g}(\mathbf{x}(t),t)\,dt+d\mathbf{w}_2,$$

where $\mathbf{y}$ and $\mathbf{g}$ are m-vectors and $\mathbf{w}_2$ is an m-vector of Brownian motions such that

$$Ew_{2i}(t)w_{2j}(t)=\int_0^t R_{ij}(s)\,ds,\qquad \mathbf{R}(t)=\{R_{ij}(t)\}.$$

Show that Kushner's equation takes the form

(9.4.9)

$$dp=Mp\,dt+\left[\mathbf{g}(\mathbf{x},t)-E_{\mathbf{y}}\mathbf{g}(\mathbf{x}(t),t)\right]^T\mathbf{R}^{-1}(t)\left[d\mathbf{y}-E_{\mathbf{y}}\mathbf{g}(\mathbf{x}(t),t)\right]p$$

EXERCISE 9.4.2

Give another derivation of Kushner's equation based on the following calculations: set

$$d\zeta(t)=\frac{\frac{1}{2}g^2(x(t),t)\,dt+g(x(t),t)\,dw_2(t)}{N_0}$$

and show that

$$p(x,t\,|\,y,t)=\frac{\left(E_{x,t}e^{\zeta(t)}\right)p(x,t)}{E_{x,t}e^{\zeta(t)}}\equiv\frac{Q}{p}.$$

Show that

$$dQ=Q_t\,dt+Q^{\zeta}\,d\zeta+\frac{\frac{1}{2}g^2(x(t),t)Q_{\zeta\zeta}\,dt}{N_0}$$

and

$$Q_t=\left(E_{x,t}e^{\zeta(t)}\right)\frac{\partial p(x,t)}{\partial t}=MQ$$

$$Q_{\zeta}=Q_{\zeta\zeta}=Q.$$

Hence show that

$$dQ=M(Q)\,dt+\frac{g(x(t),t)Q\,dy(t)}{N_0}.$$

Using $P=\int Q\,dx$, obtain

$$dP=\int(dQ)\,dx=P\left[M\left(\int QP^{-1}\,dx\right)dt+\frac{1}{N_0}\left(\int QP^{-1}g(x,t)\,dx\right)dy(t)\right]$$

$$=\frac{PE_yg(x(t)t)\,dy}{N_0}.$$

Next show that

$$d(QP^{-1})=M(QP^{-1})\,dt$$

$$+QP^{-1}\left[g(x,t)-E_yg(x(t),t)\right]\frac{\left[dy-E_yg(x(t),t)\,dt\right]}{N_0}.$$

Hence obtain (9.4.1) (Jazwinski 1970).

9.5 THE ESTIMATION EQUATIONS, LINEAR THEORY, AND AM TRANSMISSION

We can easily obtain a stochastic differential equation for the optimal estimate $\hat{\mathbf{x}}(t)$ from Kushner's equation. Furthermore, an analog of Itô's formula for the conditional expectation of functions of $\mathbf{x}(t)$ can be derived from Kushner's equation. Indeed, let $\varphi(\mathbf{x})$ be a sufficiently smooth function of $\mathbf{x}$; then

(9.5.1)

$$dE_{\mathbf{y}}\varphi(\mathbf{x}(t)) = \int \varphi(\mathbf{x})(dp)\,d\mathbf{x} = \left[\int \varphi(\mathbf{x})M(p)\,d\mathbf{x}\right]dt$$

$$+\int \varphi(\mathbf{x})\left[\mathbf{g}(\mathbf{x},t) - E_{\mathbf{y}}\mathbf{g}(\mathbf{x}(t),t)\right]\left[d\mathbf{y} - E_{\mathbf{y}}\mathbf{g}(\mathbf{x}|t),t)\,dt\right]p\,d\mathbf{x}.$$

Writing $E_{\mathbf{y}}\varphi(\mathbf{x}(t)) = \hat{\varphi}(\mathbf{x}(t))$, we obtain from (9.4.7)–(9.4.9) and (9.5.1),

(9.5.2)

$$d\hat{\varphi} = \widehat{\nabla\varphi^T\mathbf{A}\,dt} + \frac{1}{2}\sum_{i,j=1}^{n} \sigma_{ij}\widehat{\frac{\partial^2\varphi}{\partial x_i\partial x_j}}dt + \left[\widehat{\varphi\mathbf{g}} - \hat{\varphi}\hat{\mathbf{g}}\right]^T\mathbf{R}^{-1}(t)\left[d\mathbf{y}(t) - \hat{\mathbf{g}}\,dt\right],$$

where $\boldsymbol{\sigma} = \mathbf{BQB}^T$ (Jazwinski 1970). Setting $\varphi(\mathbf{x}) = x_i$, we obtain a stochastic differential equation for $\hat{\mathbf{x}}(t)$:

$$d\hat{x}_i(t) = \hat{A}_i(\mathbf{x}(t),t)\,dt + \left[\widehat{x_i\mathbf{g}} - \hat{x}_i\hat{\mathbf{g}}\right]^T\mathbf{R}^{-1}(t)\left[d\mathbf{y} - \hat{\mathbf{g}}\,dt\right]. \tag{9.5.3}$$

Setting

$$\mathbf{V}(t) = \widehat{(\mathbf{x}(t) - \hat{\mathbf{x}}(t))(\mathbf{x}(t) - \hat{\mathbf{x}}(t))^T} \equiv \widehat{\mathbf{x}(t)\mathbf{x}^T(t)} - \hat{\mathbf{x}}(t)\hat{\mathbf{x}}^T(t),$$

we have

$$d\mathbf{V} = d\widehat{\mathbf{x}(t)\mathbf{x}^T(t)} - d\hat{\mathbf{x}}(t)\hat{\mathbf{x}}^T(t). \tag{9.5.4}$$

Now, using (9.5.2) with $\varphi(\mathbf{x}) = x_ix_j$, we obtain

$$d(\widehat{x_ix_j}) = \left[\widehat{x_iA_j} + \widehat{A_ix_j} + \hat{\sigma}_{ij}\right]dt \tag{9.5.5}$$

$$+\left[\widehat{x_ix_j\mathbf{g}} - \widehat{x_ix_j}\hat{\mathbf{g}}\right]^T\mathbf{R}^{-1}(t)\left[d\mathbf{y}_j - \hat{\mathbf{g}}\hat{x}_j\right]dt.$$

Since $\hat{\mathbf{x}}(t)$ is a solution of the stochastic differential equation (9.5.3), we obtain, from Itô's formula,

$$(9.5.6)\quad d(\hat{x}_i\hat{x}_j)=\hat{x}_i\,d\hat{x}_j+\hat{x}_j\,d\hat{x}_i+\left[\widehat{x_i\mathbf{g}}-\hat{x}_i\hat{\mathbf{g}}\right]^T\mathbf{R}^{-1}(t)\left[\widehat{x_j\mathbf{g}}-\hat{x}_j\hat{\mathbf{g}}\right]dt.$$

Hence, by (9.5.4)–(9.5.6), we have

$$(9.5.7)\qquad \begin{aligned} dV_{ij}=&\Big\{\widehat{x_iA_j}-\hat{x}_i\hat{A}_j+\widehat{x_jA_i}-\hat{x}_j\hat{A}_i\\ &+\hat{\sigma}_{ij}-\left[\widehat{x_i\mathbf{g}}-\hat{x}_i\hat{\mathbf{g}}\right]^T\mathbf{R}^{-1}(t)\left[\widehat{x_j\mathbf{g}}-\hat{x}_j\hat{\mathbf{g}}\right]\Big\}\,dt\\ &+\Big[\widehat{x_ix_j\mathbf{g}}-\widehat{x_ix_j}\hat{\mathbf{g}}-\hat{x}_i\widehat{x_j\mathbf{g}}-\hat{x}_j\widehat{x_i\mathbf{g}}\\ &+2\hat{x}_i\hat{x}_j\hat{\mathbf{g}}\Big]^T\mathbf{R}^{-1}(t)\left[d\mathbf{y}(t)-\hat{\mathbf{g}}\,dt\right].\end{aligned}$$

Equation (9.5.3) is called the *processor equation* and (9.5.7) is called the *variance equation.*

In the linear case the variance equation becomes an ordinary differential equation, since setting

$$d\mathbf{y}(t)=\mathbf{M}(t)\mathbf{x}(t)\,dt+d\mathbf{w}_2$$

in (9.4.7), equation (9.5.7) yields

(9.5.8)

$$\frac{d\mathbf{V}}{dt}=\mathbf{A}(t)\mathbf{V}(t)+\mathbf{V}\mathbf{A}^T(t)+\mathbf{B}(t)\mathbf{Q}(t)\mathbf{B}^T(t)-\mathbf{V}(t)\mathbf{M}^T(t)\mathbf{R}^{-1}(t)\mathbf{M}(t)\mathbf{V}(t),$$

where $\mathbf{A}(\mathbf{x},t)=\mathbf{A}(t)\mathbf{x}$ and $\mathbf{B}(\mathbf{x},t)=\mathbf{B}(t)\mathbf{x}$. The processor equation becomes

$$(9.5.9)\quad d\hat{\mathbf{x}}(t)=\mathbf{A}(t)\hat{\mathbf{x}}(t)\,dt+\mathbf{V}(t)\mathbf{M}^T(t)\mathbf{R}^{-1}(t)\left[d\mathbf{y}(t)-\mathbf{M}(t)\hat{\mathbf{x}}(t)\,dt\right].$$

The matrix $\mathbf{V}(t)$ determines the gains in the filter described by (9.5.8) and (9.5.9) (see Fig. 9.5.1). This filter was discovered by Kalman and Bucy and it is called a *Kalman–Bucy filter*. The matrix

$$\mathbf{K}(t)=\mathbf{V}(t)\mathbf{M}^T(t)\mathbf{R}^{-1}(t)$$

is called the Kalman gain and can be precomputed, since it is independent of the observations $\mathbf{y}(t)$.

The blocks in Fig. 9.5.1 are analogous to those in Fig. 9.3.4.

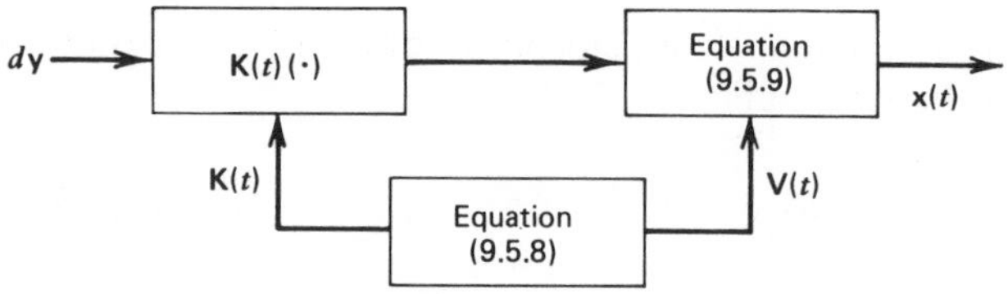

Figure 9.5.1 Kalman–Bucy filter.

EXERCISE 9.5.1

Show that the nonlinear Riccati equation (9.5.8) is equivalent to the linear system

$$\frac{d\mathbf{X}}{dt} = -\mathbf{A}^T(t)\mathbf{X} + \mathbf{M}^T(t)\mathbf{R}^{-1}(t)\mathbf{M}(t)\mathbf{Y}$$

$$\frac{d\mathbf{Y}}{dt} = \mathbf{B}(t)\mathbf{Q}(t)\mathbf{B}^T(t)\mathbf{X} + \mathbf{A}(t)\mathbf{Y}$$

$$\mathbf{X}(t_0) = \mathbf{I}, \qquad \mathbf{Y}(t_0) = \mathbf{V}(t_0)$$

$$\mathbf{V}(t) = \mathbf{Y}(t)\mathbf{X}^{-1}(t).$$

EXERCISE 9.5.2

Consider the case of no modulation and additive white noise channel

$$dx = -kx\,dt + \sqrt{2k}\,dw_1$$

$$dy = x\,dt + \sqrt{2N_0}\,dw_2.$$

Write down the processor and variance equations and solve them. Construct a block diagram for the message, observation, and the Kalman filter. Use the steady-state solution of the variance equation to determine the Kalman gain.

The linear theory can be used to construct a Kalman filter for AM transmission. For an AM signal we have, by (9.3.2),

(9.5.10) $$dx = -kx\,dt + \sqrt{2k}\,dw_1$$

(9.5.11) $$dy = x(t)\sqrt{2}\sin\omega_0 t\,dt + \sqrt{2N_0}\,dw_2.$$

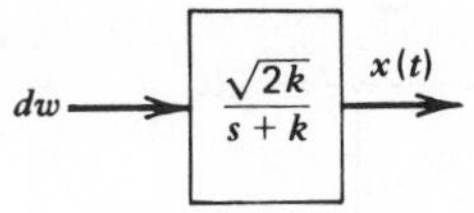

Figure 9.5.2 First-order Butterworth signal.

The realization of $x(t)$ is shown in Fig. 9.5.2.
The variance equation is given by

$$\dot{V}=\frac{-2k(V-1)-V^2\sin^2\omega_0 t}{N_0}. \tag{9.5.12}$$

Since

$$\sin^2\omega_0 t=\tfrac{1}{2}(1-\cos 2\omega_0 t),$$

the high-frequency term $\cos^2\omega_0 t$ will not propagate through a low-pass filter to be included in the demodulator; thereby it can be neglected. The simplified variance equation is now given by

$$\dot{V}\approx -2k(V-1)-\frac{V^2}{2N_0}. \tag{9.5.13}$$

A realization of (9.5.12) is given in Fig. 9.5.3.
The processor equation is given by

$$d\hat{x}=-k\hat{x}\,dt+\frac{V(t)\sqrt{2}\sin\omega_0 t\left[dy-\hat{x}(t)\sqrt{2}\sin\omega_0 t\,dt\right]}{2N_0}. \tag{9.5.14}$$

Using the steady-state solution of (9.5.13), we can replace $V(t)$ by

$$V=\frac{-1+\sqrt{1+\Lambda}}{\Lambda/2}=\frac{2}{1+\sqrt{1+\Lambda}},$$

where $\Lambda=1/kN_0$ is the *signal-to-noise ratio* (SNR) measured in bandwidth.

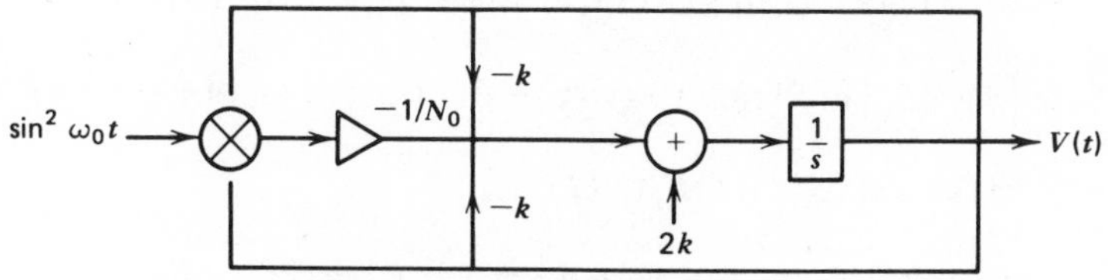

Figure 9.5.3 Realization of (9.5.12).

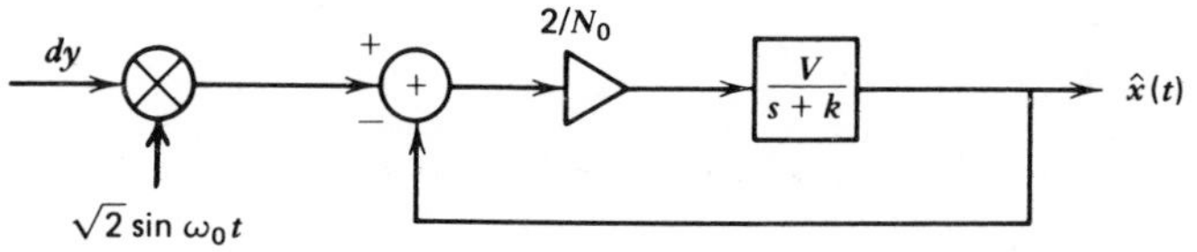

Figure 9.5.4

We have, after neglecting the term $\hat{x}(t)\cos 2\omega_0 t$ in (9.5.14) (resulting from $\sin^2\omega_0 t$), a simplified processor equation

(9.5.15) $$d\hat{x} = -k\hat{x}(t)\,dt + \frac{V\left[\sqrt{2}\,\sin\omega_0 t\,dy - \hat{x}(t)\,dt\right]}{2N_0}.$$

A realization of 9.5.15 is shown in Fig. 9.5.4.

EXERCISE 9.5.3

Construct a block diagram for the full realization of the optimal AM receiver.

9.6 NONLINEAR FILTERING, PM TRANSMISSION

In the nonlinear case, both the processor and the variance equations are nonlinear stochastic differential equations in general and are usually hard to integrate. Approximate methods are therefore called for. Thus, assuming that the error $x(t)-\hat{x}(t)$ is small, we expand $g(x(t),t)$ in (9.5.3) in Taylor's series about $\hat{x}(t)$. We shall deal with the one-dimensional case for simplicity. We obtain

(9.6.1)

$$\begin{aligned} d\hat{x}(t) = & -\left[k\hat{x}(t) - k\left[x(t)-\hat{x}(t)\right]\right]dt \\ & + E_y\{x(t)-\hat{x}(t))\left[g(\hat{x}(t),t)+(x(t)-\hat{x}(t))g_x(\hat{x}(t),t)+\cdots\right]\} \\ & \times\left[dy - E_y\{g(\hat{x}(t),t)+(x(t)-\hat{x}(t))g_x(\hat{x}(t),t)+\cdots\}\right](2N_0)^{-1} \\ = & -k\hat{x}(t)+g_x(\hat{x}(t),t)E_y(x(t) \\ & -\hat{x}(t))^2\left[dy - g(\hat{x}(t),t)\,dt\right](2N_0)^{-1}+\cdots. \end{aligned}$$

since $g(\hat{x}(t),t)$ is constant with respect to E_y and $\widehat{x(t)-\hat{x}(t)}=0$. Setting $V(t)=\widehat{[x(t)-\hat{x}(t)]^2}$ and using the same procedure in the variance equation, we obtain

(9.6.2)

$$dV=-2k(V-1)\,dt+V^2g_{xx}(\hat{x}(t),t)[dy-g(\hat{x}(t),t)\,dt](2N_0)^{-1}+\cdots.$$

Let $x^*(t)$ and V^* be the solutions of the system (9.6.1), (9.6.2) where the higher-order terms are deleted, and assume that the initial conditions

$$x^*(0)=\hat{x}_0=\int_{-\infty}^{\infty}xp(x,0|y,0)\,dx,$$

where $p(x,0|y,0)$ is the a priori density of x_0 (e.g., take x_0 to be a zero mean Gaussian variable with $Ex_0^2=1$); we also take $V^*(0)=V_0=Ex_0^2=1$. The value of $x^*(t)$ is the approximate minimum mean-square-error estimate of $x(t)$ that we shall use. Equation (9.6.1) without the higher-order terms will be called the approximate processor equation, and (9.6.2) without the higher order terms will be called the *approximate variance equation.* Note that if both $x(t)$ and $y(t)$ satisfy linear equations; then (9.6.1) and (9.6.2) become exact.

We consider next the case of phase modulation (PM). The state equations are given by

(9.6.3) $$dx=-kx\,dt+\sqrt{2k}\,dw_1$$

(9.6.4) $$dy=\sqrt{2}\,\sin(\omega_0t+\beta x(t))dt+\sqrt{2N_0}\,dw_2.$$

Setting $\omega_0t+\beta x(t)=\theta(t)$ and $\omega_0t+\beta x^*(t)=\theta^*(t)$, we can write the approximate variance equation in the form

(9.6.5)

$$\dot{V}^*=-2k(V^*-1)-\frac{\beta^2V^2}{N_0}\sin\theta^*\sin\theta-\frac{-\beta^2V^{*2}}{\sqrt{N_0}}\dot{w}_2\sin\theta-\frac{\beta^2V^{*2}}{N_0}\sin^2\theta^*.$$

Now

$$\sin\theta^*\sin\theta=\tfrac{1}{2}[\cos(\theta-\theta^*)+\cos(\theta+\theta^*)]\qquad\text{and}$$

$$\theta+\theta^*=2\omega_0t+\beta[x(t)+x^*(t)],$$

so the low-pass filter will eliminate the high-frequency terms $\cos^2\theta^*$ and $\cos(\theta+\theta^*)$, thus simplifying (9.6.5) into

$$(9.6.6)\quad \dot{V}^* = -2k(V^*-1) - \frac{\beta^2 V^{*2}}{2N_0}\left[\cos(\theta-\theta^*)+1+\sqrt{N_0}\,\dot{w}_2 \sin\theta\right].$$

The process of suppressing the high-frequency terms by a low-pass filter is called *heterodyning*. In the approximation we use we can replace $\cos(\theta-\theta^*)$ by 1, and since

$$P\left(\left|\sqrt{N_0}\,\dot{w}_2\right| > \varepsilon\right) \to 0 \qquad \text{as} \quad N_0 \to 0$$

for any $\varepsilon > 0$, we may neglect the last term in (9.6.6), thus obtaining the simplified equation

$$\dot{V}^* = -2k(V^*-1) - \frac{\beta^2 V^{*2}}{2N_0}.$$

In the steady state $\dot{V}^* = 0$, so that

$$V^* = \frac{2}{1+\sqrt{1+2\beta^2\Lambda}}.$$

The processor equation is given by

$$d\hat{x} = -k\hat{x}\,dt + \frac{\beta^2 V^* \cos(\omega_0 t + \beta x^*(t))\,dy}{N_0}.$$

EXERCISE 9.6.1

Construct a block diagram for the message, modulation, and demodulation of PM transmission.

9.7 FM TRANSMISSION AND THE PHENOMENON OF CYCLE SLIPPING IN PLLs

In FM transmission the measured signal is given by

$$dy = \sqrt{2}\,\sin\left[\omega_0 t + d_f \int_0^t x(s)\,ds\right] dt + \sqrt{2N_0}\,dw_2 \equiv g(u(t),t) + \sqrt{2N_0}\,dw_2,$$

where

$$u(t)=\int_0^t x(s)\,ds.$$

We assume that $x(t)$ satisfies the equation

$$dx(t)=-kx(t)\,dt+\sqrt{2k}\,dw_1.$$

Thus we have the following system:

(9.7.1) $$dx=-kx\,dt+\sqrt{2k}\,dw_1$$

(9.7.2) $$du=x\,dt$$

(9.7.3) $$dy=\sqrt{2}\,\sin\left[\omega_0 t+d_f u(t)\right]dt+\sqrt{2N_0}\,dw_2.$$

Setting $u(t)=x_0(t)$, $x_1(t)=x(t)$,

$$\mathbf{x}(t)=\begin{pmatrix}x_0\\x_1\end{pmatrix};\qquad \mathbf{A}=\begin{bmatrix}0&1\\0&-k\end{bmatrix};\qquad \mathbf{b}(t)=\begin{pmatrix}0\\\sqrt{2k}\end{pmatrix},$$

we have in (9.7.1) and (9.7.2),

(9.7.4) $$d\mathbf{x}=\mathbf{A}\mathbf{x}\,dt+\mathbf{b}\,dw_1.$$

There are several FM demodulators known in practice, some of which are based on the so-called *phase-locked loop* (PLL), which has been extensively studied in the literature (Lindsey 1972; Viterbi 1966).

The object of our study is an FM receiver that contains a demodulator based on a PLL, shown schematically in Fig. 9.7.1.

The PLL is an important device, widely used in communications, sonar, radar, and so on. We shall study next the workings of the PLL, the (quasi) optimal structure of the filter F, and the errors in reception.

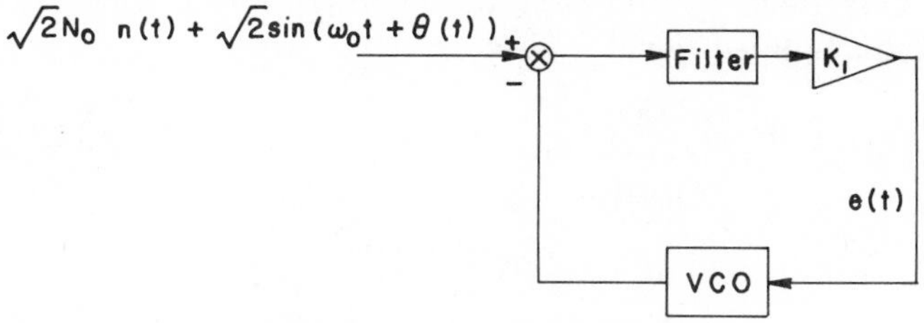

Figure 9.7.1 Phase-locked loop.

The signal $g+\sqrt{2N_0}\,n(t)$ entering the loop contains additional noise $\sqrt{2N_0}\,n(t)$ (e.g., atmospheric disturbances, internal noise in the transmitter, etc.).

It is known (see Exercise 3.4.9) that $n(t)$ can be represented by

$$n(t)=\sqrt{2}\,(n_1(t)\sin\omega_0 t+n_2(t)\cos\omega_0 t)$$

where $n_1(t)$ and $n_2(t)$ are independent white noise processes. The parameter N_0 measures the noise intensity. The term $1/N_0$ is called the signal-to-noise ratio (SNR). The output of the voltage-controlled oscillator (VCO) produces a cosine wave whose frequency is controlled by the input voltage $e(t)$; that is,

$$H(t)=\sqrt{2}\,K_3\cos[\omega_0 t+\theta_2(t)],$$

where $\dot{\theta}_2(t)=K_2 e(t)$. The constants $K_i(i=1,2,3)$ represent gains. The device $\otimes$ represents multiplication of the received signal $g+\sqrt{2N_0}\,n(t)$ by $H(t)$, with the result

$$\begin{aligned}\left[g+\sqrt{2N_0}\,n(t)\right]H(t)=2K_3\Big\{&\sin[\theta_1(t)-\theta_2(t)]-\sqrt{2N_0}\,n_1(t)\sin\theta_2(t)\\&+\sqrt{2N_0}\,n_2(t)\cos\theta_2(t)+\sin[2\omega_0 t\\&\qquad+\theta_1(t)+\theta_2(t)]\\&+\sqrt{2N_0}\,n_1(t)\sin[2\omega_0 t+\theta_2(t)]\\&+\sqrt{2N_0}\,n_2(t)\cos[2\omega_0 t+\theta_2(t)]\Big\}.\end{aligned}$$

The low-pass filter F suppresses the high-frequency terms, so that the filtered and amplified signal is given by

$$\begin{aligned}e(t)=K_1K_3F(s)\Big\{&\sin[\theta_1(t)-\theta_2(t)]-\sqrt{2N_0}\,n_1(t)\sin\theta_2(t)\\&+\sqrt{2N_0}\,n_2(t)\cos\theta_2(t)\Big\},\end{aligned}$$

where $F(s)$ is a linear operator that represents the effect of the linear filter

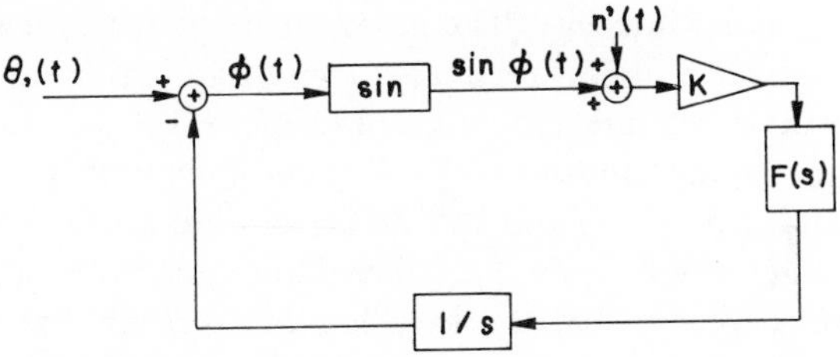

Figure 9.7.2 Block diagram of the phase-locked loop.

F (see Fig. 9.7.2). The term

$$n'(t) = -n_1(t)\sin\theta_2(t) + n_2(t)\cos\theta_2(t)$$

is also a white noise. The block diagram model is given by Fig. 9.7.2. Here $K = K_1K_2K_3$. The filter $F(s)$ can be chosen in various ways. Consider, for example, the case $F(s) = 1$ and $x(t) = \omega - \omega_0 =$ constant. Then, setting $\varphi(t) = \theta_1(t) - \theta_2(t)$ and noting that $\theta_1(t) = \omega t$ and $\dot{\theta}_2 = K_2 e(t)$, we obtain

$$\dot{\varphi} = \dot{\theta}_1(t) - K_2 e(t) = \dot{\theta}_1(t) - K\sin\varphi - Kn'(t)$$

or

$$\dot{\varphi} = \omega - \omega_0 - K_2\sin\varphi - Kn'(t).$$

If $\omega = \omega_0$ and there is no noise in the loop, we obtain for the phase error φ the ordinary differential equation

$$\dot{\varphi} = -K\sin\varphi.$$

The critical points of this equation are

$$\varphi = n\pi \qquad (n = 0, \pm 1, \pm 2, \ldots).$$

The points $\varphi = (2n+1)\pi$ are unstable equilibria, while the points $\varphi = 2n\pi$ are stable equilibria. Thus the loop is locked on the correct phase (mod $2n\pi$). The frequency error is now given by $e_x(t) = \dot{\varphi} = 0$ as $\varphi =$ constant.

The presence of the noise in the loop leads in this case to the equation

$$\dot{\varphi} = -K\sin\varphi - Kn'(t)$$

or

$$d\varphi = -K\sin\varphi\, dt - \sqrt{2N_0}\, dw, \tag{9.7.5}$$

where $2N_0$ is the noise intensity. The noise is sure to drive the solution $\varphi(t)$ away from a given equilibrium state, $\varphi=0$, say, in finite time. More specifically, if the noise intensity is small (SNR$\gg$1), as is usually the case in communications, the solution $\varphi(t)$ will spend long time intervals near an equilibrium state (e.g., $\varphi=0$) and will be driven into a neighboring state, $\varphi=2\pi$, say, where it will spend another long time interval. When this transition occurs, we say that the loop has *lost lock* or that the loop has *slipped a cycle*. The result of slipping a cycle is an audible "click," for the slip causes a sharp change in the estimate of the frequency, which produces the click. The phenomenon of cycle slipping may be quite annoying and may cause serious errors in measurements of distances by radar, for example. It is important therefore to know the dependence of the frequency of cycle slips on the parameters of the loop, especially for loop designing purposes. The problem now is identical with the exit problem, or the problem of atomic migration in crystals in one dimension. Its solution is given in the following exercises.

EXERCISE 9.7.1

Identify the problem of cycle slipping with that of the motion of a Brownian particle in a potential field, where the potential is given by $\Psi = K\cos\varphi$. Write down and solve Dynkin's equation for the expected time of a cycle slip (Viterbi 1966).

We return now to the problem of choosing the filter $F(s)$ in a possibly optimal way. We use the approximate processor and variance equations for the system (9.7.1)–(9.7.3). We see that in this case we have to consider a system of processor equations and therefore also a system of covariance equations.

EXERCISE 9.7.2

Show that if $\mathbf{x}(t)$ satisfies the linear system

$$d\mathbf{x}(t)=\mathbf{A}(t)\mathbf{x}(t)\,dt+\mathbf{B}(t)\,d\mathbf{w}_1$$

and $\mathbf{y}(t)$ satisfies the nonlinear system

$$d\mathbf{y}(t)=\mathbf{g}(\mathbf{x}(t),t)\,dt+\sqrt{\mathbf{N}_0}\,d\mathbf{w}_2(t),$$

where $\mathbf{N}_0$ is the covariance matrix for $\mathbf{w}_2(t)$, then the approximate processor and covariance equations are given by

$$d\mathbf{x}^*(t)=\mathbf{A}(t)\mathbf{x}^*(t)\,dt+\mathbf{V}^*(t)\mathbf{g}_{\mathbf{x}}(\mathbf{x}^*(t),t)\mathbf{N}_0^{-1}\left[d\mathbf{y}(t)-\mathbf{g}(\mathbf{x}^*(t),t)\,dt\right]$$

and

$$d\mathbf{V}^*(t)=\left[\mathbf{A}(t)\mathbf{V}^*(t)+\mathbf{V}^*(t)\mathbf{A}^T(t)+\mathbf{B}(t)\mathbf{B}^T(t)\right]dt.$$

An approximately optimal FM demodulator can now be found from the system (9.7.3), (9.7.4). The approximate processor and variance equations for (9.7.3) and (9.7.4) are given by

$$(9.7.6)\qquad d\mathbf{x}^*=\mathbf{A}\mathbf{x}^*dt+\begin{bmatrix} v_{00}^* \\ v^*_{01}\end{bmatrix}\frac{d_f\sqrt{2}}{2N_0}\cos\left[\omega_0 t+d_f u^*(t)\right]dy$$

and

$$(9.7.7)\quad d\mathbf{V}^*(t)=\left[\mathbf{A}\mathbf{V}^*(t)+\mathbf{V}^*(t)\mathbf{A}^T+\begin{pmatrix}0 & 0\\ 0 & 2k\end{pmatrix}\right]dt$$

$$-\frac{d_f^2\sqrt{2}}{2N_0}\begin{bmatrix} v_{00}^{*2} & v_{00}^*v_{01}^* \\ v_{01}^*v_{00}^* & v_{01}^{*2}\end{bmatrix}\sin\left[\omega_0 t+d_f u^*(t)\right]dy$$

respectively, where

$$\mathbf{V}^*=\begin{bmatrix} v_{00}^* & v_{01}^* \\ v_{01}^* & v_{11}^*\end{bmatrix}$$

(Snyder 1969).

Replacing dy by $g(u(t),t)\,dt+\sqrt{2N_0}\;dw_2$ and suppressing the high-frequency terms (heterodyning), as in the case of PM, the variance equation takes the form

$$\dot{\mathbf{V}}^*=\mathbf{A}\mathbf{V}^*+\mathbf{V}^*\mathbf{A}^T-\begin{pmatrix}0 & 0\\ 0 & 2k\end{pmatrix}-\frac{d_f^2}{2N_0}\begin{bmatrix} v_{00}^{*2} & v_{00}^*v_{01}^* \\ v_{01}^*v_{00}^* & v_{01}^*\end{bmatrix}.$$

The solution of the steady-state equation is given by

$$d_f^2 v_{00}^* = \frac{4\beta\Lambda^{-1/2}}{1+\sqrt{1+2\beta\Lambda^{1/2}}} \tag{9.7.8}$$

$$d_f v_{01}^* = \frac{4\beta}{\left(1+\sqrt{1+2\beta\Lambda^{1/2}}\right)^2}$$

$$v_{11}^* = \frac{1-4\beta^2\Lambda}{\left(1+\sqrt{1+2\beta\Lambda^{1/2}}\right)^4}$$

where $\beta = d_f/k$ and $\Lambda = 1/kN_0$.

For high SNR we have the approximate formulas

$$d_f^2 v_{00}^* = \frac{4\beta\Lambda^{-1/2}}{1+\sqrt{1+2\beta\Lambda^{1/2}}} \approx 2\sqrt{2\beta}\ \Lambda^{-3/4} \tag{9.7.9}$$

$$d_f v_{01}^* = \frac{4\beta}{\left(1+\sqrt{1+2\beta\Lambda^{1/2}}\right)^2} \cong 2\Lambda^{-1/2}, \tag{9.7.10}$$

since in most practical cases $\beta\Lambda^{1/2} \gg 1$.

The block diagram of the quasi-optimal FM demodulator corresponding to (9.7.6) and (9.7.7) is shown in Fig. 9.7.3. We use $x^*(t)$ as the loop

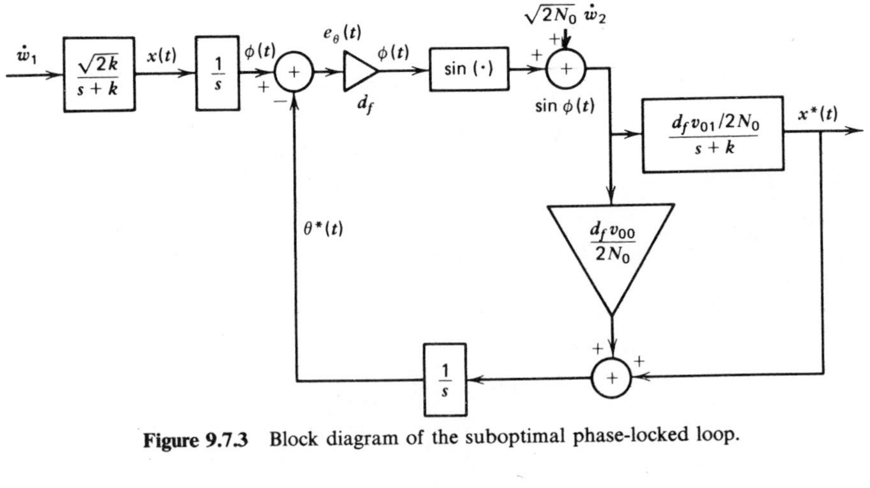

Figure 9.7.3 Block diagram of the suboptimal phase-locked loop.

estimate of $x(t)$, so that $e_x(t)=x(t)-x^*(t)$ is the error of the estimation. The process $\theta(t)$ represents the phase and $\theta^*(t)$ is the loop estimate of $\theta(t)$. The total phase error is given by $\varphi(t)=d_f(\theta(t)-\theta^*(t))$. The system of equations describing the PLL is therefore given by

$$(9.7.11)\qquad de_x = -\left[ke_x + \frac{v_{01}^* d_f}{2N_0}\sin\varphi\right]dt + \left[\sqrt{2k}\,dw_1 - \frac{v_{01}^* d}{\sqrt{2N_0}}\,dw_2\right]$$

$$(9.7.12)\qquad d\varphi = \left[d_f e_x - \frac{v_{00}^* d_f^2}{\sqrt{2N_0}}\sin\varphi\right]dt - \frac{v_{00}^* d_f^2}{\sqrt{2N_0}}\,dw_2$$

(see Snyder (1969). Using (9.7.9), (9.7.10), setting

$$\varepsilon = 2\sqrt{2\beta}\,\Lambda^{-3/4}, \qquad \xi = \sqrt{2\beta}\,e_x\Lambda^{-1/4},$$

and scaling the time by

$$t' = \frac{t}{\gamma}, \qquad \gamma = \frac{1}{\sqrt{2\beta}\,k\Lambda^{1/4}},$$

we get from (9.7.11) and (9.7.12) the scaled system

$$(9.7.13)\qquad d\xi(t') = -(\delta\xi + \sin\varphi)\,dt' + (dw_1' - dw_2')\sqrt{\varepsilon}$$

$$(9.7.14)\qquad d\varphi(t') = \left(\tfrac{1}{2}\xi - \sin\varphi\right)dt' - dw_2'\sqrt{\varepsilon}$$

where $\delta = (\varepsilon/8\beta^2)^{1/3}$ and $dw_i'(t') = (1/\sqrt{\gamma}\,)\,dw_i(t)(i=1,2)$, according to the Brownian scaling law (2.1.16 ii). The coefficients in (9.7.13) and (9.7.14) are periodic in φ; therefore, a "skipping" of 2π in φ will leave the system unchanged. Since the frequency $x(t)$ is proportional to $\dot{\theta}$, an error of 2π in the estimate θ^* of θ will cause a sharp change in x^* which is heard as a "click." Obviously, frequent "clicks" of this type tend to obscure the transmitted message. A natural measure of click frequency is the mean time $E\tau$ between clicks, whose computation is the object of the next section. From a mathematical point of view, the system (9.7.13), (9.7.14) represents a small stochastic perturbation of the dynamical system.

$$\dot{\xi} = -\sin\varphi - \delta\xi \tag{9.7.15}$$

$$\dot{\varphi} = \tfrac{1}{2}\xi - \sin\varphi, \tag{9.7.16}$$

which has stable equilibria at $\xi=0, \varphi=2\pi j (j=0, \pm 1, \ldots)$. All solutions of (9.7.15) and (9.7.16) that remain bounded as $t \to \infty$ converge to equilibrium points, so that any trajectory that begins in the domain of attraction of a given stable equilibrium point will not cross into the domain of attraction of another one. However, even the slightest stochastic perturbation is sure to cause such a crossing in finite time. Thus the phenomenon of slipping cycles by the PLL can be described mathematically as an instability caused by stochastic forcing of a stable system. More precisely, the solution $(\xi(t), \varphi(t))$ spends long time intervals near an attractive point, $\xi=0, \varphi=0$, say, and we shall say therefore that a cycle slipping has occurred whenever a trajectory crosses to the domain of attraction of $\xi=0, \varphi=\pm 2\pi$, say. If D is the domain of attraction of $\xi=0, \varphi=0$, then the slip time is defined by

$$\tau = \inf\{t \,|\, (\xi(t), \varphi(t)) \in \partial D\},$$

where ∂D is the boundary of D (see Bobrovsky and Schuss, (to appear)).

9.8 COMPUTATION OF THE MEAN CYCLE SLIP TIME IN A PLL

We start by describing the domain of attraction D of the point $\xi=0, \varphi=0$, and its boundary ∂D. The attractive points are located on the φ-axis and are separated there by saddle points at $\varphi=(2n+1)\pi (n=0, \pm 1, \ldots)$. The four trajectories that converge to the points $\xi=0, \varphi=\pm\pi$ form the boundary ∂D of the domain of attraction D of the origin (cf. Fig. 9.8.1).

Linearizing (9.7.15) and (9.7.16) about the point $\xi=0, \varphi=\pi$, we see that δ can be neglected if ε is the small and we see that the separating curves are the solutions of (9.7.15), (9.7.16) with

$$\frac{d\varphi}{d\xi} = \frac{1-\sqrt{3}}{2} \quad \text{at} \quad \xi=0, \quad \varphi=\pm\pi.$$

The expected time between cycle slips is the expected time of first exit from D for the system (9.7.13), (9.7.14), where φ is taken $\mathrm{mod}(2\pi)$. Thus the expected slip time

$$v(\xi, \varphi) = E\{\tau \,|\, \xi(0)=\xi, \varphi(0)=\varphi\}$$

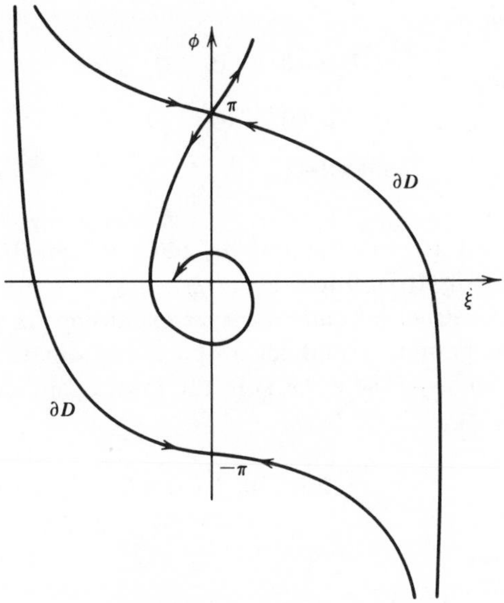

Figure 9.8.1 The domain D.

is the solution of Dynkin's equation (5.4.5)

$$L_\varepsilon v=\varepsilon\left(v_{\xi\xi}+b_{\xi\varphi}+\tfrac{1}{2}v_{\varphi\varphi}\right)-(\sin\varphi+\delta\xi)v_\xi \tag{9.8.1}$$

$$+\left(\tfrac{1}{2}\xi-\sin\varphi\right)v_\varphi=-1 \quad \text{in} \quad D$$

$$v=0 \quad \text{in} \quad \partial D.$$

Since δ is small, we neglect the term $\delta\xi v_\xi$. The construction of an asymptotic solution as $\varepsilon\to 0$ will follow the method of Chapter 8. We note first that the maximum principle for (9.8.1) gives the bound

$$v\leqslant C(\varepsilon)e^{H/\varepsilon},$$

where H is a constant and

$$C(\varepsilon)=o(e^{H/\varepsilon}) \quad \text{as} \quad \varepsilon\to 0. \tag{9.8.2}$$

Therefore, we scale v by setting

$$v(\xi,\varphi)=C(\varepsilon)e^{H/\varepsilon}u(\xi,\varphi), \tag{9.8.3}$$

where H is a constant to be chosen, $C(\varepsilon)$ satisfies (9.8.2), and $\max u(\xi,\varphi)=1$. This is our ansatz. The equation for $u(\xi,\varphi)$, up to transcendentally small terms is now given by

$$L_\varepsilon u \sim 0 \quad \text{in } D$$
$$u = 0 \quad \text{on } \partial D$$
$$\max_D u = 1.$$

Since $v \to \infty$ as $\varepsilon \to 0$ at each point in D, but $v=0$ on ∂D, the function $u(\xi, \varphi)$ changes rapidly from 0 to 1 near ∂D.

We now shall construct a boundary layer expansion for u. Let y be the distance from the boundary and let x' be a coordinate in a direction tangent to the boundary. Then (9.8.1) in the local coordinates (x, y) near ∂D is given by

(9.8.4) $$\varepsilon u_{yy} + y b_0(x') u_y + L_1 u = 0,$$

where

$$\begin{pmatrix} -\sin\varphi \\ \frac{1}{2}\xi - \sin\varphi \end{pmatrix} \cdot \boldsymbol{\nu} = y b_0(x') + 0(y^2) \quad \text{as } y \to 0$$

$\boldsymbol{\nu}$ = outer normal to ∂D, and $b_0(x')$ is the coefficient of the first term in Taylor's expansion of the function

$$b(x', y) = \begin{pmatrix} -\sin\varphi \\ \frac{1}{2}\xi - \sin\varphi \end{pmatrix} \cdot \boldsymbol{\nu}$$

near $y=0$. We note that $b(x', 0) = 0$, since the vector

$$\begin{pmatrix} -\sin\varphi \\ \frac{1}{2}\xi - \sin\varphi \end{pmatrix}$$

is tangent to ∂D, by (9.7.15) and (9.7.16) (with $\delta = 0$). The expression $L_1 u$ contains tangential and mixed derivatives of u. Setting $\eta = y/\sqrt{\varepsilon}$, we obtain the equation

(9.8.5) $$u_{\eta\eta} + \eta b_0(x') u_\eta + L_1 u = 0.$$

The second order terms in $L_1 u$ are $O(\sqrt{\varepsilon})$ as $\varepsilon \to 0$. The boundary conditions are

$$u(x', \eta) \to 0 \quad \text{as } \eta \to 0$$
$$u(x', \eta) \to 1 \quad \text{as } \eta \to \infty.$$

To determine $C(\varepsilon)$ and H in (9.8.3), we construct a solution $w(\xi,\varphi)$ to the adjoint equation

$$\varepsilon\left(w_{\xi\xi}+w_{\xi\varphi}+\tfrac{1}{2}w_{\varphi\varphi}\right)+\sin\varphi w_{\xi}-\left[\left(\tfrac{1}{2}\xi-\sin\varphi\right)w\right]_{\varphi}=0$$

such that $w(0,0)=1$. Then, multiplying (9.8.1) by $w(\xi,\varphi)$ and integrating by parts, we obtain

$$\text{(9.8.6)}\qquad \int_D\int wLv\,d\xi\,d\varphi=\varepsilon\oint_{\partial D}\left[v_{\xi}\left(\nu_1+\tfrac{1}{2}\nu_2\right)\right.$$

$$\left.+\tfrac{1}{2}v_{\varphi}(\nu_1+\nu_2)\right]w\,ds=-\int_D\int w(\xi,\varphi)\,d\xi\,d\varphi,$$

where $\boldsymbol{\nu}=(\nu_1,\nu_2)^T$ is the outer normal at ∂D. We have used here the facts that $v=0$ on ∂D [see (9.8.1)] and $L^*w=0$, where L^* is the adjoint of L. Inserting (9.8.3) into (9.8.6), we obtain

$$\text{(9.8.7)}\qquad C(\varepsilon)e^{H/\varepsilon}\oint_{\partial D}\left[u_{\xi}\left(\nu_1+\tfrac{1}{2}\nu_2\right)+\tfrac{1}{2}u_{\varphi}(\nu_1+\nu_2)\right]w\,ds$$

$$=-\int_D\int w\,d\xi\,d\varphi.$$

We shall construct $w(\xi,\varphi)$ using the "ray method" (Cohen and Lewis 1967), and evaluate the integral asymptotically using the Laplace method (Olver 1976), and thus obtain H and $C(\varepsilon)$. We assume that $w(\xi,\varphi)$ has the form

$$\text{(9.8.8)}\qquad w(\xi,\varphi)=e^{-\Psi(\xi,\varphi)/\varepsilon}g(\xi,\varphi,\varepsilon),$$

where

$$\text{(9.8.9)}\qquad g(\xi,\varphi,\varepsilon)\sim\sum_{j=0}^{\infty}g_j(\xi,\varphi)\varepsilon^j$$

with $\Psi(0,0)=0$, $\Psi(\xi,\varphi)>0$ for $\xi^2+\varphi^2>0$ and $g_0(0,0)=1$. Inserting (9.8.8) and (9.8.9) into (9.8.1) and equating the coefficients of each power of ε separately to zero, we obtain equations for Ψ and g_j.

In particular, Ψ satisfies the nonlinear equation

$$\text{(9.8.10)}\qquad \Psi_{\xi}^2+\Psi_{\xi}\Psi_{\varphi}+\tfrac{1}{2}\Psi_{\varphi}^2-\sin\varphi\Psi_{\xi}+\left(\tfrac{1}{2}\xi-\sin\varphi\right)\Psi_{\varphi}=0,$$

while the leading term g_0 in the expansion (9.8.9) satisfies

$$(9.8.11)\quad (2\Psi_\xi+\Psi_\varphi-\sin\varphi)\frac{\partial g_0}{\partial \xi}+\left(\Psi_\xi+\Psi_\varphi+\tfrac{1}{2}\xi-\sin\varphi\right)\frac{\partial g_0}{\partial \varphi}$$

$$=-\left(\Psi_{\xi\xi}+\Psi_{\xi\varphi}+\tfrac{1}{2}\Psi_{\varphi\varphi}-\cos\varphi\right)g_0.$$

Equations (9.8.10) and (9.8.11) are equivalent to the following system of six differential equations (Sect. 10.2):

$$(9.8.12)\quad \begin{aligned} \dot{\xi}&=2p+q-\sin\varphi\\ \dot{\varphi}&=p+q+\tfrac{1}{2}\xi-\sin\varphi\\ \dot{p}&=-\tfrac{1}{2}q\\ \dot{p}&=(p+q)\cos\varphi\\ \dot{\Psi}&=p^2+pq+\tfrac{1}{2}q^2\\ \dot{g}_0&=\chi g_0, \end{aligned}$$

where $\chi=-(\Psi_{\xi\xi}+\Psi_{\xi\varphi}+\frac{1}{2}\Psi_{\varphi\varphi}-\cos\varphi)$.

Inserting (9.8.8) into (9.8.7) and comparing terms of same orders of magnitude in ε, we obtain

$$(9.8.13)\quad C(\varepsilon)e^{H/\varepsilon}\oint_{\partial D}e^{-\Psi/\varepsilon}g_0\left[u_\xi\left(\nu_1+\tfrac{1}{2}\nu_2\right)+\tfrac{1}{2}u_\varphi(\nu_1+\nu_2)\right]ds$$

$$\cong-\int_D\!\!\int e^{-\Psi/\varepsilon}g_0\,d\xi\,d\varphi.$$

Evaluating the Laplace-type integrals asymptotically, we see that the main contribution to the double integral in (9.8.13) comes from the origin; hence

$$(9.8.14)\quad \int_D\!\!\int e^{-\Psi/\varepsilon}g_0\,d\xi\,d\varphi\approx\frac{2\pi\varepsilon}{J}$$

where

$$J=\left|\det\begin{bmatrix}\Psi_{\xi\xi} & \Psi_{\xi\varphi}\\ \Psi_{\xi\varphi} & \Psi_{\varphi\varphi}\end{bmatrix}\right|_{\xi=0,\,\varphi=0}.$$

To evaluate J we use Taylor's expansion of $\Psi(\xi,\varphi)$ about the origin and find that

$$\Psi(\xi,\varphi)=\tfrac{1}{2}\xi^2-\xi\varphi+\varphi^2+O(\xi^2+\varphi^2); \tag{9.8.15}$$

hence $J=1$. The main contribution to the line integral in (9.8.13) comes from the points of minimum of Ψ on ∂D, at $(\xi,\varphi)=\pm(\xi_0,\varphi_0)$, say. The values of ξ_0 and φ_0 can be found by numerical integration of (9.8.12), as will be explained below. The asymptotic expansion of the boundary integral about (ξ_0,φ_0) in (9.8.13) is given by

$$\oint_{\partial D} e^{-\Psi/\varepsilon}g_0\left[u_\xi\left(\nu_1+\tfrac{1}{2}\nu_2\right)+\tfrac{1}{2}u_\varphi(\nu_1+\nu_2)\right]ds \tag{9.8.16}$$

$$\simeq 2\frac{\sqrt{2\pi\varepsilon}\,e^{-\Psi(\xi_0,\varphi_0)/\varepsilon}}{\sqrt{|\Psi''(\xi_0,\varphi_0)|}}g_0(\xi_0,\varphi_0)\cdot\left[u_\xi\left(\nu_1+\tfrac{1}{2}\nu_2+\tfrac{1}{2}u_\varphi(\nu_1+\nu_2\right)\right]\Big|_{(\xi_0,\varphi_0)},$$

where the factor 2 comes from the invariance of the integrand and the boundary under reflection in the origin. The term $\Psi''(\xi_0,\varphi_0)$ is the second derivative of Ψ with respect to arc length in ∂D at (ξ_0,φ_0). The asymptotic form of $u(\xi,\varphi)$ near the point (ξ_0,φ_0) is given by (9.8.6),

$$u^0(\xi,\varphi)\simeq b_0^{1/2}(x')\int_0^{y/\sqrt{\varepsilon}}\frac{e^{-s^2b_0(x')/2}\,ds}{\sqrt{\pi/2}};$$

hence

$$\frac{\partial u^0}{\partial\xi}(\xi_0,\varphi_0)\simeq\frac{b_0^{1/2}(0)\partial y/\partial\xi}{\sqrt{\pi\varepsilon/2}}\Bigg|_{(\xi_0,\varphi_0)},$$

where the point $x'=0$ corresponds to the point (ξ_0,φ_0), and

$$\frac{\partial u^0}{\partial\xi}(\xi_0,\varphi_0)\simeq\frac{-b_0^{1/2}(0)\partial y/\partial\varphi}{\sqrt{\pi\varepsilon/2}}\Bigg|_{(\xi_0,\varphi_0)}.$$

We have

$$\mathbf{b}(x',y)\cdot\boldsymbol{\nu}=y\left[-\tfrac{1}{2}\sin\varphi_0\nu_1(\xi_0,\varphi_0)+\left(\tfrac{1}{2}\xi_0\cos\varphi_0+\tfrac{1}{2}\sin\frac{\varphi_0}{2}\right)\nu_2\right];$$

hence

$$b_0(0) = -\tfrac{1}{2}\sin\varphi_0 \nu_1(\xi_0, \varphi_0) + \left(\tfrac{1}{2}\xi_0 \cos\varphi_0 + \tfrac{1}{2}\sin\frac{\varphi_0}{2}\right)\nu_2(\xi_0, \varphi_0).$$

Here $\nu_1(\xi_0, \varphi_0) = \frac{1}{2}\xi_0 - \sin\varphi_0$ and $\nu_2(\xi_0, \varphi_0) = \sin\varphi_0$. Thus the quantity

$$K \equiv u_\xi^0\left(\nu_1 + \tfrac{1}{2}\nu_2\right) + \tfrac{1}{2}u_\varphi^0(\nu_1 + \nu_2)|_{\xi_0, \varphi_0}$$

can be found, provided that (ξ_0, φ_0) is known. To compute $\Psi''(\xi_0, \varphi_0)$, we differentiate with respect to arc length s

$$\frac{d^2\Psi}{ds^2} = \frac{d^2\xi}{ds^2}p + \frac{d^2\varphi}{ds^2}q + \left(\frac{d\xi}{ds}\right)^2 p_\xi + \frac{d\xi}{ds}\frac{d\varphi}{ds}(p_\xi + q_\varphi) + \left(\frac{d\varphi}{ds}\right)^2 q_\varphi,$$

where $p = \partial\Psi/\partial\xi$ and $q = \partial\Psi/\partial\varphi$,

$$\frac{d\xi}{ds} = \frac{\dot{\xi}}{\sqrt{\dot{\xi}^2 + \dot{\varphi}^2}} = \frac{-\sin\varphi}{\sqrt{\sin^2\varphi + \left(\frac{1}{2}\xi - \sin\varphi\right)^2}},$$

$$\frac{d^2\xi}{ds^2} = -\frac{d}{dt}\left[\frac{\sin\varphi\sqrt{\sin^2\varphi + \left(\frac{1}{2}\xi - \sin\varphi\right)^2}}{\sqrt{\sin^2\varphi + \left(\frac{1}{2}\xi - \sin\varphi\right)^2}}\right].$$

Similar expressions for the derivatives of φ are obtained. Thus $\Psi''(\xi_0, \varphi_0)$ can be evaluated if the values of p, q, p_ξ, p_φ, q_ξ, and q_φ are known at (ξ_0, φ_0). The values of Ψ, p, q, q_ξ, p_φ, q_ξ, q_φ and g_0 at (ξ_0, φ_0), as well as the point (ξ_0, φ_0), can be found numerically as follows. The expansion (9.8.15) can be used to evaluate Ψ, and its partial derivatives up to second order on a small circle C about the origin, thus initial conditions for (9.8.12) are given on C. For g_0 one obtains a similar expansion on C, taking $g_0(0,0) = 1$. Differentiating the first four equations of (9.8.12) with respect to the initial condition $\xi = \xi^0$ and setting $\xi_1 = \partial\xi/\partial\xi^0$, $\varphi_1 = \partial\varphi/\partial\xi^0$, $p_1 = \partial p/\partial\xi^0$, and $q_1 = \partial q/\partial\xi^0$, we obtain four additional equations,

$$\dot{\xi}_1 = 2p_1 + q_1 - \varphi_1\cos\varphi$$

$$\dot{\varphi}_1 = p_1 + q_1 + \tfrac{1}{2}\xi_1 - \varphi_1\cos\varphi$$

$$\dot{p}_1 = -\tfrac{1}{2}q_1$$

$$\dot{q}_1 = (p_1 + q_1)\cos\varphi - (p + q)\varphi_1\sin\varphi.$$

The corresponding initial conditions on C are given by $\xi_1=1, \varphi_1=0, =p_1=1, q_1=-1$. Now, since

(9.8.17) $$p_1=\frac{\partial p}{\partial \xi^0}=\left(\frac{\partial p}{\partial \xi}\right)\left(\frac{\partial \xi}{\partial \xi^0}\right)+\left(\frac{\partial p}{\partial \varphi}\right)\left(\frac{\partial \varphi}{\partial \xi^0}\right)=\Psi_{\xi\xi}\xi_1+\Psi_{\xi\varphi}\varphi_1$$

(9.8.18) $$q_1=\Psi_{\xi\varphi}\xi_1+\Psi_{\varphi\varphi}\varphi_1$$

(9.8.19) $$\dot{p}=\Psi_{\xi\xi}\dot{\xi}+\Psi_{\xi\varphi}\dot{\varphi},$$

we have a system of three linear algebraic equations for the three unknown second partials of Ψ, (9.8.17)–(9.8.19), on each characteristic curve. Drawing level curves of Ψ, we find that the minimum of Ψ on D is achieved at the point $p=(\xi_0, \varphi_0)\simeq(2.1;2.4)$. Now, using (9.8.16) in (9.8.14), we obtain from (9.8.13)

$$C(\varepsilon)Ke^{H/\varepsilon}g_0(\xi_0,\varphi_0)e^{-\Psi(\xi_0,\varphi_0)/\varepsilon}=\frac{2\pi}{J}=2\pi;$$

hence $H=\Psi(\xi_0,\varphi_0)$ and $C(\varepsilon)=2\pi/Kg_0(\xi_0,\varphi_0)$, or

(9.8.20)

$$kE[\tau|\xi(0)=0,\varphi(0)=0]=\frac{2\pi\Lambda^{-1/4}}{Kg_0(\xi_0,\varphi_0)\sqrt{2\beta}}\exp\left[\frac{\Psi(\xi_0,\varphi_0)}{2\sqrt{2\beta}}\Lambda\right]^{3/4}$$

(see Bobrovsky and Schuss, to appear).

9.9 MULTIPLE CYCLE SLIPS AND THE THRESHOLD PHENOMENON IN A PLL

It is known (Snyder 1969) that the mean-square frequency estimation error $\overline{e_x^2}$ varies linearly with SNR (signal-to-noise ratio) Λ if Λ is large. As Λ decreases (i.e., as noise increases), there is a sharp increase in $\overline{e_x^2}$ (see Fig. 9.9.1). For $\Lambda<\Lambda_0$, where Λ^0 is the critical value of Λ, the error $\bar{e}_x^2$ is so large that the signal cannot be distinguished from the noise, and we say that Λ_0 (or the corresponding value of ε_0 of ε) is the threshold, beyond which the PLL is no longer an efficient demodulator. We shall explain this phenomenon by describing the influence of cycle slips on the variance of e_x. We begin with the description of the behavior of trajectories during and after cycle slips. For small values of ε, the cycle slips are infrequent

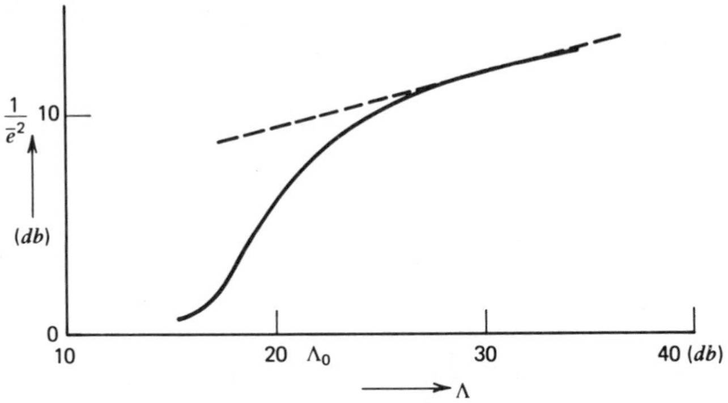

Figure 9.9.1 Threshold in FM reception.

and appear as single slips, separated by long time intervals when the loop is locked about $\xi=0, \varphi=2\pi n$. As ε increases, the slips become more frequent and some of the trajectories, having crossed ∂D at the neighborhood of P (see Fig. 9.9.2), go on to cross the boundary of the next domain before relocking takes place. This happens when $|\xi(t)|$ increases, which in turn causes $|\varphi(t)|$ to increase by a multiple of 2π within a short time period. Next, we compute the probability of a second passage. After crossing D near P the trajectory fluctuates about the deterministic path, which is nearly parallel to D for a period of time of length t_0. During this time period the average distance between the point $(\xi(t), \varphi(t))$ and ∂D is ρ. The mean time $\bar{\tau}$ of passage to the boundary from a point whose distance to the boundary is ρ is given by

$$\bar{\tau} \simeq \bar{\tau}_0 \operatorname{erfc}\left[\frac{\rho\sqrt{b_0(x')}}{2\varepsilon}\right],$$

where $\bar{\tau}_0$ is the average first exit time from the origin, given by (9.8.20), and x' is the tangential coordinate of the point under consideration (the normal coordinate is ρ). Therefore, the probability of passage to ∂D during this period, of length t_0, is given by

$$p = 1 - e^{-t_0/\bar{\tau}}.$$

Hence the mean frequency f of second passages is given by

$$f = \frac{1 - e^{-t_0/\bar{\tau}}}{\bar{\tau}_0} = \frac{p}{\bar{\tau}_0}.$$

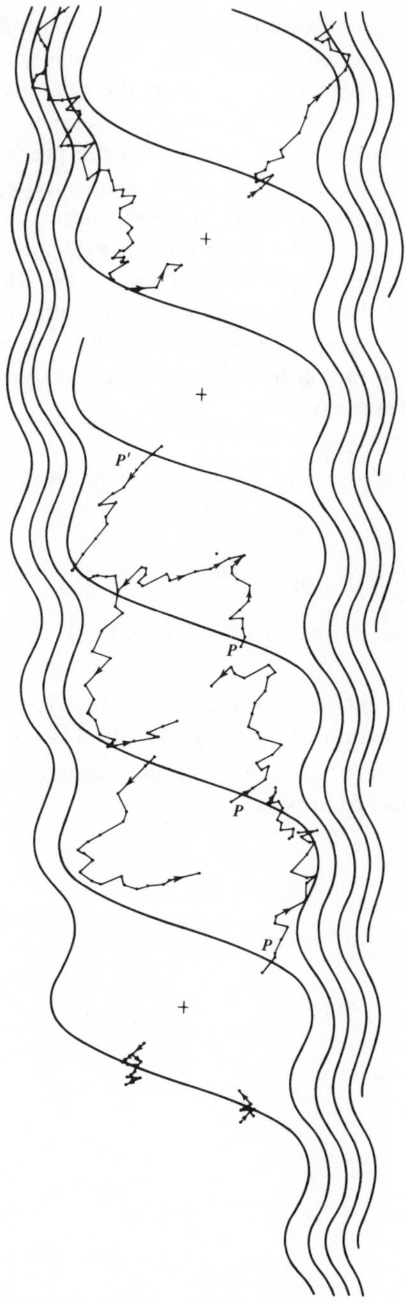

Figure 9.9.2 Single and multiple cycle slips.

Finally, we present a rough estimate of the contribution of the excursions (of the cycle slips) to the variance of $\xi(t)$. The following assumptions are made: (i) the average distance S between the consecutive boundaries is constant for large ξ, and (ii) the frequency of passages from a boundary to the next one is given by $e^{-d/\varepsilon}/K$, where K is the preexponential factor in (9.8.20) and d is the increment in Ψ from one boundary to the next. The value of $\xi(t)$ in the nth strip is $\xi(t)\simeq x_0+nS$, where x_0 is the half-width of D. The speed of motion of the deterministic system (9.7.15)–(9.7.16) for large ξ is $\dot{\varphi}\simeq\frac{1}{2}\xi$, since the contribution of $\sin\varphi$ in (9.7.16) can be neglected. The distance traveled by the trajectory is then equal to $2\pi n$, so that the time to absorption in the nth domain is approximately $2\pi n/\frac{1}{2}\xi$. Since $\xi(t)\simeq x_0+nS$, we have the following expression for the contribution of the excursions to the variance of $\xi(t)$:

$$\overline{\xi^2}(t)\simeq 2\left[4\pi x_0+4\pi(x_0+S)p+4\pi\sum_{n=2}^{\infty}(x_0+nS)npe^{-nd/\varepsilon}/K\right]+2\varepsilon$$

$$\simeq\frac{8\pi e^{-Q/\varepsilon}}{K}\left\{\begin{array}{l} x_0+(x_0+S)p+e^{-d/\varepsilon}(2-e^{-d/\varepsilon})x_0p/K(1-e^{-d/\varepsilon})^2\\ +pSe^{-d/\varepsilon}\left[(1+e^{-d/\varepsilon})/(1-e^{-d/\varepsilon})^2-1\right]\end{array}\right\}+2\varepsilon,$$

where Q is the exponent in (9.8.20). The term 2 is the variance of $\xi(t)$ obtained from linearization of (9.7.13) and (9.7.14) about the origin. The graph of $1/\overline{\xi^2}(t)$ vs. $1/\varepsilon$ on a logarithmic scale is given in Fig. 9.9.1 (Bobrovsky and Schuss to appear).

CHAPTER 10

Topics in Classical Mechanics and in Differential Equations

10.1 HAMILTON'S EQUATIONS OF MOTION, LIOUVILLE'S EQUATION, POINCARÉ'S THEOREM, THE LOSCHMIDT AND ZERMELO PARADOXES, AND SMOLUCHOWSKI'S THEORY

Let $\mathbf{q}=(q_1,\ldots,q_n)^T$ be the generalized coordinates of a system of particles and let $\mathbf{p}=(p_1,\ldots,p_n)^T$ be the momentum vector of this system. Assume that the potential energy of the system is $V(\mathbf{q})$. Then the function

$$H(\mathbf{p},\mathbf{q})=\mathbf{p}\cdot\mathbf{q}+V(\mathbf{q}) \tag{10.1.1}$$

is the total energy of the system and H is called the Hamiltonian of the system. The equations of motion of the system (Newton's second law of motion) were given by Hamilton:

$$\dot{\mathbf{p}}=-\frac{\partial H}{\partial \mathbf{q}} \quad \text{and} \quad \dot{\mathbf{q}}=\frac{\partial H}{\partial \mathbf{p}} \tag{10.1.2}$$

$$\mathbf{p}(0)=\mathbf{p}_0, \qquad \mathbf{q}(0)=\mathbf{q}_0.$$

Let $D(\mathbf{p},\mathbf{q})$ be the density of particles in phase space [i.e., in the space $(\mathbf{p},\mathbf{q})$]. From (10.1.2), we have

$$\frac{dD}{dt}=\frac{\partial D}{\partial \mathbf{p}}\dot{\mathbf{p}}+\frac{\partial D}{\partial \mathbf{q}}\dot{\mathbf{q}}+\frac{\partial D}{\partial t}=\frac{\partial H}{\partial \mathbf{p}}\frac{\partial D}{\partial \mathbf{q}}-\frac{\partial H}{\partial \mathbf{q}}\frac{\partial D}{\partial \mathbf{p}}+\frac{\partial D}{\partial t}\equiv\frac{\partial D}{\partial t}-\{H,D\}.$$

The expression $\{H, D\}$ is called Poisson's brackets. Liouville's theorem asserts that $dD/dt=0$; hence Liouville's equation for the density of particles in phase space follows:

$$\frac{\partial D}{\partial t} = \{H, D\}. \tag{10.1.3}$$

To prove the theorem, we note that by observing the velocity vector $(\dot{\mathbf{p}}, \dot{\mathbf{q}})$ at a given point $(\mathbf{p}, \mathbf{q})$ is phase space, we may use "hydrodynamic" considerations by viewing $(\dot{\mathbf{p}}, \dot{\mathbf{q}})$ as a function of $(\mathbf{p}, \mathbf{q}, t)$. We shall compute the divergence of $(\dot{\mathbf{p}}, \dot{\mathbf{q}})$:

$$\operatorname{div}(\dot{\mathbf{p}}, \dot{\mathbf{q}}) = \sum_i \left(\frac{\partial \dot{p}_i}{\partial p_i} + \frac{\partial \dot{q}_i}{\partial q_i} \right) = \sum_i \left(\frac{\partial^2 H}{\partial q_i \, \partial p_i} - \frac{\partial^2 H}{\partial p_i q_i} \right) = 0$$

by (10.1.2). Now, for any element of volume Ω in phase space, we have, obviously,

$$\int_\Omega \operatorname{div}(\dot{\mathbf{p}}, \dot{\mathbf{q}}) \, d\mathbf{p} \, d\mathbf{q} = -\frac{d}{dt} \int_\Omega D \, d\mathbf{p} \, d\mathbf{q};$$

hence Liouville's theorem follows.

One obvious consequence of Liouville's theorem is as follows. Let $T(t)$ be the family of transformations defined by (10.1.2); that is,

$$(\mathbf{p}(t), \mathbf{q}(t)) = T(t)(\mathbf{p}_0, \mathbf{q}_0).$$

Obviously $T(t)$: "phase space"→"phase space" and by the classical uniqueness theorem for ordinary differential equations $T(t)$ is a one-to-one mapping. Liouville's theorem implies that $T(t)$ is volume preserving in phase space. It can also be shown that $T(t)$ preserves surface area in any dimension lower than $2n$ (Landau and Lifshitz 1954). Consider the surface Γ:

$$H(\mathbf{p}, \mathbf{q}) = E;$$

that is, consider all points in phase space that correspond to constant energy E. The measure [the $(2n-1)$-dimensional surface area] of any subset A of Γ is then given by

$$\mu(A) = \int_A |\nabla H|^{-1} \, dS,$$

where

$$|\nabla H|^2 = \sum \left[\left(\frac{\partial H}{\partial p_i} \right)^2 + \left(\frac{\partial H}{\partial q_i} \right)^2 \right]$$

and dS is a surface element on Γ.

By Liouville's theorem we have

$$\mu(\mathrm{T}(t)A) = \mu(A).$$

We assume that $\mu(\Gamma) = 1$ without loss of generality. Poincare's recurrence theorem states that if $A \subset \Gamma$ and $\mu(A) > 0$, then for almost all points $\gamma \in A$, there exists an arbitrarily large t such that $T(t)\gamma \in A$. Before proving this theorem, let us relate it to the kinetic theory of gases and classical thermodynamics. The second law of thermodynamics states that "heat cannot by itself be transferred from a colder to a hotter body," or that "the entropy of a closed system must never decrease." Consider, for example, a hermetically closed box in which there is no gas, and assume that at a given moment a small balloon filled with gas is blown up at a given corner of the box. According to the second law of thermodynamics, expansion of the gas is an irreversible process, for the entropy of the gas in the corner is lower than that of the expanded gas. Poincaré's theorem leads to the following paradox, discovered by Zermelo. Since the motion of the gas is governed by the system (10.1.2), which is a conservative system in this case, Poincaré's theorem implies that the system will be arbitrarily close to its initial state at finite time intervals. This is an apparent contradiction to the second law of thermodynamics. Boltzmann estimated the length of the time the system takes to return to the neighborhood of the initial state so that the positions agree to within 10% of the average distance between the molecules and the velocities to agree within 0.2%. He considered a cubic centimeter of air containing 10^{18} molecules moving with the velocity 500 m/sec. The average distance between the molecules is of the order $(10^{18})^{-1/3} = 10^{-6}$ cm. Under normal conditions each molecule will suffer about 4×10^9 collisions per second, so that the total number of collisions will be

$$C = \frac{4 \times 10^9 \times 10^{18}}{2} = 2 \times 10^{27} \text{ collisions/sec.}$$

The difference in position should be of the order of 10^{-7} cm at the end of the Poincaré cycle, while the difference in velocity should not exceed 1 m/sec. The recurrence time needs not be less than the time necessary for all molecules to take on all possible values of velocity, with the understanding that we distinguish between two velocities only if at least one of the components differs by at least 1 m/sec. The first molecule has all velocities between zero and $v=500\times10^9$ m/sec as there are 10^{18} molecules. The second molecule can have velocities in the range zero to $(v^2-v_1^2)^{1/2}$, where v_1 is the velocity of the first molecule. The third one can have velocities in the range 0 to $(v^2-v_1^2-v_2^2)^{1/2}$, where v_2 is the velocity of the second, and so on. The number of all possible combinations is, therefore,

$$N=(4\pi)^{n-1}\int^{v} v_1^2\,dv_1\int^{(v^2-v_1^2)^{1/2}} v_2^2\,dv_2\int^{(v^2-v_1^2-v_2^2)^{1/2}} v_3^2\,dv_3$$

$$\times\cdots\int^{(v^2-v_1^2\ldots-v_{n-1}^2)^{1/2}} v_{n-1}^2\,dv_{n-1}$$

$$=\left\{\frac{2(2\pi)^{(3n-4)/2}}{[3(n-1)]!!}\right\}v^{3(n-1)}$$

if n is odd, where $v=500\times10^9$, $n=10^{10}$, $[3(n-1)]!!=3\cdot5\cdot7\cdot\ldots\cdot3(n-1)$. Since each of the N combinations can be taken after a collision, the total time required for the velocities to run through all possibilities is N/C. The time for the positions to recur within the specified distance is of the same order of magnitude. When confronted with this paradox, Boltzmann exclaimed, "you should wait so long" (Chandrasekhar 1954; Kac 1959).

This calculation does not reflect the possibility that the distances recur within the specified neighborhood while the velocities take on arbitrary values. Such a situation may occur many times during a single Poincaré cycle, thus causing the expected time of approximate recurrence of the initial distances to be much shorter than estimated by Boltzmann. Loschmidt pointed at a similar paradox by stating that in view of the symmetry of the laws of mechanics under time reversal, all molecular processes should be reversible from the point of view of statistical mechanics. This apparently contradicts the second law of thermodynamics. The resolution of the Loschmidt and Zermelo paradoxes based on the concept of the Brownian motion was given by Smoluchowski (see Exercises 10.1.4–10.1.12).

To prove Poincaré's theorem, let us consider the sets

$$A_n=A\cap\bigcap_{k=1}^{n-1}(T^{-k}A)^c\cap T^{-n}A,$$

where $T=T(1)$, and obviously $T^n=T(n)$. We have

$$\chi_{A_n}(\gamma)=\chi_A(\gamma)\chi_A(T^n\gamma)\prod_{k=1}^{n-1}\left[1-\chi_A(T^k\gamma)\right];$$

hence

$$\mu(A_n)=\int_\Gamma \chi_A(\gamma)\chi_A(T^n\gamma)\prod_{k=1}^{n-1}\left[1-\chi_A(T^k\gamma)\right]d\mu(\gamma).$$

Setting

$$\nu_n=\int_\Gamma \prod_{k=0}^{n}\left[1-\chi_A(T^k\gamma)\right]d\mu(\gamma),$$

we see that $\{\nu_n\}$ is a nonincreasing sequence and we have by induction

$$\text{(10.1.4)} \qquad \mu(A_n)=\nu_{n-1}-2\nu_n+\nu_{n+1} \qquad (\nu_0=1).$$

In the proof of (10.1.3), we use the fact that T is measure-preserving in the identity

$$\int_\Gamma \prod_{k=1}^{n}\left[1-\chi_A(T^k\gamma)\right]d\mu(\gamma)=\int_\Gamma \prod_{k=0}^{n-1}\left[1-\chi_A(T^k\gamma)\right]d\mu(\gamma).$$

Hence

$$\sum_{k=1}^{n}\mu(A_k)=1-\nu_1-(\nu_n-\nu_{n+1}).$$

Since $\{\nu_n\}$ is convergent, monotone, and bounded, we have $\nu_n-\nu_{n+1}\to 0$; hence

$$\sum_{n=1}^{\infty}\mu(A_n)=1-\nu_1=\mu(A).$$

Let B be the set of all points γ is A such that only a finite number of the points $T^n\gamma\in A$. Then

$$B=\bigcup_{k=1}^{\infty}B_k,$$

where $B_k=\{\gamma\in A|\mathrm{T}^n\gamma\notin A$ for all $n\geqslant k\}$. Setting $T_1=T$, we have, by the previous argument, $\mu(B_k)=0$; hence $\mu(B)=0$. Thus for almost all $\gamma\notin A$, there is an infinite sequence of $\{\mathrm{T}^n\gamma\}$ in A. The paradoxes of Loschmidt and Zermelo were resolved by Smoluchowski, who proposed a statistical theory of density fluctuations in fluids. The following exercises describe this theory (Chandrasekhar 1954; Kac 1959).

EXERCISE 10.1.1

Consider an element of volume v in a very much larger volume containing a large number of particles in equilibrium. Suppose that we observe the number of particles contained in v systematically at constant intervals of time τ apart. It is well known (Feller 1971) that the frequency with which different numbers of particles will be observed in v will follow the Poisson distribution.

(i) Consider the number of particles n and m contained in v at an interval of time τ apart. Following Smoluchowski, make the following two assumptions: (1) the motions of the individual particles are not mutually influenced and are independent of each other and (2) all positions in v have equal a priori probability. Define the *probability after effect* P as the probability that a particle somewhere inside v will have emerged from it during the time τ. Let $X(t)$ be the number of particles in v at time t. Show that $X(t)$ is the sum of a binomial variable $B(n,p)$ and a Poisson variable $S(\nu P)$; that is, $P(X(t+\tau)=m|X(t)=n)=P(B(n,p)+S(\nu p)=m-n)$ (the probability of observing m particles in v at time $t+\tau$ if n particles were observed at time t). Set $\Delta=X(t+\tau)-X(t)$. Show that $E_n\Delta=E(\Delta|x(t)=n)=(\nu-n)P$ and $E_n\Delta^2=nP(1-P)+(\nu-n)^2P^2$. Average over n with respect to $S(\nu)$ to obtain $EE_n\Delta=(\nu-ES(\nu))P=0$ and

$$(10.1.5)\quad EE_n\Delta^2=P^2\left[E(\nu-S(\nu))^2-ES^2(\nu)\right]=E(S(\nu)+\nu)P=2\nu P.$$

(ii) Interpret (i) as a direct method for the experimental determination of the probability after effect P from the simple evaluation of $EE_n\Delta^2$ from long sequences of observations of $X(t)$. Show that if there is no correlation between the numbers that will be observed on two occasions at an interval τ apart, then $P=1$. Hence $EE_n\Delta^2=E(S_1(\nu)-S_2(\nu))^2=2\nu$, where $S_1(\nu)$ and $S_2(\nu)$ are independent Poisson variables.

EXERCISE 10.1.2

Make systematic counts of the number of pedestrians in a block every 5 sec. Define a probability after-effect factor $P = v\tau/a$, where v is the average speed of a pedestrian, τ the chosen interval of time, and a the length of the block. Apply Smoluchowski's theory (Exercise 10.1.1) to estimate v and compare to statistical data (Fürth 1918).

EXERCISE 10.1.3

Show that in Smoluchowski's theory, the probability of the pair (n, m) is given by

$$P(S(\nu)=n)P(X(t+\tau)-X(t)=m|X(t)=n)$$

(see Exercise 10.1.1).

EXERCISE 10.1.4

Use Smoluchowski's theory to determine the *mean life time* T_n *of a fluctuation*, that is,

$$\mathrm{T}_n = \sum_{k=1}^{\infty} k\tau P[X(t+j\tau)=n, j=1,2,\ldots,k-1, \text{and } X(t+k\tau)\neq n].$$

Determine the average time of recurrence Θ_n for a given state, that is,

$$\Theta_n = \sum_{k=1}^{\infty} k\tau P[(t+j\tau)\neq n, j=1,\ldots,k-1 \text{ and } X(t+k\tau)=n|X(t)\neq n].$$

Derive the formulas

$$\mathrm{T}_n = \frac{\tau}{1-P(X(t+\tau)=n|X(t)=n)}$$

and

$$\Theta_n = \mathrm{T}_n \frac{1-P(S(\nu)=n)}{P(S(\nu)=n)}. \tag{10.1.6}$$

Show that $\Theta_{17} \approx 10^{13}$ in case $\nu = 1.428$, $P = 0.374$. Use Table 10.1.1 of the observed frequencies for the pair (n, m) in this case for comparison with (10.1.6).

Table 10.1.1

n \ m	0	1	2	3	4	5	6
0	210	126	35	7	0	1	0
1	134	281	117	29	1	1	0
2	27	138	108	63	16	3	0
3	10	20	76	38	24	6	0
4	2	2	14	22	13	11	3
5	0	0	2	10	10	1	3

EXERCISE 10.1.5

Use the result of Exercise 10.1.4 to resolve Loschmidt's reversibility paradox.

EXERCISE 10.1.6

Smoluchowski's theory was based on Svedberg's (1911) observations. Westgren (1918) conducted experiments for the verification of Smoluchowski's theory. He made observations by means of an ultramicroscope on the numbers of particles in a well-defined element of volume in a colloidal solution. These observations were made at constant time intervals τ apart. Assume that $\tau \gg 1/\beta$; hence the motion of the colloidal particles is Brownian and the diffusion coefficient is given by

$$D = \frac{k\mathrm{T}}{m\beta} = \frac{k\mathrm{T}}{6\pi a\eta}.$$

(i) Show that the probability that such a Brownian particle somewhere inside the given element of volume v (with uniform probability) at time $t=0$ will find itself outside of it at time $t=\tau$ is given by

$$\text{(10.1.7)} \qquad \frac{1}{(4\pi D\tau)^{3/2} v} \int_{v^c}\int_{v} \left(\exp\left(-\frac{|\mathbf{r}_1-\mathbf{r}_2|^2}{4D\tau}\right)\right) d\mathbf{r}_1\, d\mathbf{r}_2,$$

where $\mathbf{r}_1 = (x_1, y_1, z_1)^T$ ranges over v and $\mathbf{r}_2 = (x_2, y_2, z_2)^T$ ranges over the complement v^c of v. Compare P from (10.1.7) to (10.1.5) as follows.

(ii) Show that if v is a long rectangular parallelpiped of width h, then

$$P=1-\frac{1}{\sqrt{\pi}}\int_0^{\alpha} e^{-x^2}\,dx+\frac{1}{\alpha\sqrt{\pi}}\left(1-e^{-\alpha^2}\right), \tag{10.1.8}$$

where $\alpha=h/2\sqrt{D\tau}$.

(iii) Show that if v is a cylinder of radius r_0, then

$$P=e^{-2\sqrt{\alpha}}\left[I_0(2\sqrt{\alpha}\,)+I_1(2\sqrt{\alpha}\,)\right], \tag{10.1.9}$$

where $\alpha=r_0/2\sqrt{D\tau}$ and I_0, I_1 are Bessel's functions with imaginary arguments ($e^{-x}I_{0,1}(x)$ are tabulated in Watson, 1922).

EXERCISE 10.1.7

Table 10.1.2 is a sample of Westgren's observations in the rectangular arrangement of Exercise 10.1.7(ii):

Table 10.1.2

21111102211123230000001100101112233453422121320
22102221232223222222133422

with the following values for the various physical parameters: $h=6.56\ \mu$, $D=3.96\times10^{-8}$, $\tau=1.39$ sec, $T=290.0$°K, $a=49.5\ \mu\mu$, and $\nu=1.428$.

(i) Use the table to see how well the Poisson distribution of $S(\nu)$ represents the observed frequencies of the different values of n. The frequencies in the experiment are given in Table 10.1.3. Construct a calculated table for $S(1.428)$ for comparison with Table 10.1.3. Calculate the mean and the variance from Table 10.1.3 and compare with $\nu=1.428$.

(ii) Compute from Table 10.1.2 $EE_n\Delta^2$ for various values of τ (e.g., $\tau=\tau_0=1.39$ sec, $\tau=2\tau$, etc.) and plot $P=P(\tau)$ from the calculations using (10.1.8) and the ones using $P=EE_n\Delta^2/\tau$.

(iii) In the cylindrical arrangement of Exercise 10.1.6(iii), the observations listed in Table 10.1.4 were made, with $r_0=10.0\ \mu$, $a=63.5\ \mu\mu$, $T=290$°K, $D=3.024\times10^{-8}$, and $\nu=1.933$. Construct a table of P_{calc} using (10.1.9) and P_{obs} using Table 10.1.4. Plot the observed and calculated $P(\tau)$.

Table 10.1.3

n	0	1	2	3	4	5	6	7
Frequency	381	568	357	175	67	28	5	2

Table 10.1.4

τ(sec)	1.50	3.00	4.50	6.00	7.50
$EE_n\Delta^2$	0.836	1.200	1.512	1.718	1.939

EXERCISE 10.1.8

Use $D = k\mathrm{T}/6\pi a\eta = RT/6Na\eta$, where R is the gas constant to determine Avogardro's number N from Exercise 10.1.6(iii).

EXERCISE 10.1.9

Use the expression

$$P(X(t+\tau)=n \mid X(t)=n) = P(B(n,P)+S(\nu P)=0)$$

$$= e^{-\nu P}\sum_{i=0}^{n}\binom{n}{i}P^i(1-P)^{n-i}\frac{(\nu P)^i}{i!} \equiv W_n$$

to show that

$$W_n = 1-(n+\nu)P(\tau)+O(P^2) \qquad \text{as} \quad \tau\to 0 \quad \text{and} \quad P\to 0.$$

Hence show that

$$(10.1.10) \quad \phi(t)\,\Delta t = P(X(t+\Delta t)\neq n \mid X(t)=n) \approx W_n^{t/\Delta t}(1-W_n)$$

$$= \left[1-(n+\nu)P(\Delta t)+O(P^2)\right]^{t/\Delta t}(n+\nu)P(\Delta t);$$

that is, we must have $P(\Delta t) = O(\Delta t)$. Show that (10.1.7) implies that $P(\tau) = O(\sqrt{\tau}\,)$ as $\tau\to 0$. Explain the contradiction.

EXERCISE 10.1.10

Use Exercise 2.1.2 to reconcile the contradiction in Exercise 10.1.9 and derive $P(\Delta t)=P_0\,\Delta t+O(\Delta t)^2$ as $\Delta t\to 0$. Hence, using (10.1.10), obtain

$$\phi_n(t)=\exp[-(n+\nu)P_0 t](n+\nu)P_0$$

and

$$\mathrm{T}_n=\int_0^\infty t\phi_n(t)\,dt=\frac{1}{(n+\nu)P_0},$$

$$\Theta_n=\frac{1}{(n+\nu)P_0}\,\frac{1-P(S(\nu)=n)}{P(S(\nu)=n)}$$

[see (10.1.5)]. Use these results to compare T_n(calc.) and Θ_n(calc.) with the observations in Table 10.1.5.

Table 10.1.5

n	0	1	2	3	4
T_n(obs.)	1.67	1.50	1.37	1.25	1.23
Θ_n(obs.)	6.08	3.13	4.11	7.85	18.6

EXERCISE 10.1.11

$P(\Delta t)$ represents the probability that a particle initially inside v and with a Maxwellian velocity distribution will emerge from v before a time Δt. This is the same as the number of molecules striking the inner surface of v in time Δt when the molecular concentration is $1/v$. According to the familiar kinetic theory of gases, the number of molecules with velocities between $|u|$ and $|u|+d|u|$ which strike the unit area of any solid surface per unit time with a solid angle $d\Omega$ at an angle θ with the normal to the surface is given by

$$N\left(\frac{m}{2\pi k\mathrm{T}}\right)^{3/2}\exp\left[\frac{-m|u|^2}{2k\mathrm{T}}\right]|u|^3\cos\theta\, d\Omega\, d|u|,$$

where N denotes the molecular concentration. Integrate this expression to obtain

$$P_0 = \frac{\sigma}{v}\left(\frac{kT}{2\pi n}\right)^{1/2},$$

where σ is the total surface area of v. Hence obtain

$$T_n = \frac{v}{\sigma(n+\nu)}\left(\frac{2\pi m}{kT}\right)^{1/2}$$

$$\Theta_n = \frac{v}{\sigma(n+\nu)}\left(\frac{2\pi m}{kT}\right)^{1/2}\frac{1-P(S(\nu)=n)}{P(S(\nu)=n)}$$

$$\sim \pi\frac{v}{\sigma}\left(\frac{m}{kT}\right)^{1/2}\exp\left[\frac{(n-\nu)^2}{2\nu}\right]$$

for large ν and $n \sim \nu$ (see Exercise 10.1.1). Compute, following Smoluchowski, the average time of recurrence of a state of fluctuation in which the molecular concentration of oxygen in a sphere of radius a will differ from the average value by 1%. Take $T=300°\text{K}$, $\nu = 3\times 10^{19}\times \frac{4}{3}\pi a^3$. Obtain the results shown in Table 10.1.6.

Table 10.1.6

a(cm)	1	5×10^{-5}	3×10^{-6}	$2\times5\times10^{-5}$	1×10^{-5}
Θ(sec)	$10^{10^{14}}$	10^{68}	10^{6}	1	1

EXERCISE 10.1.12

Explain Zermelo's recurrence paradox.

Use Smoluchowski's theory to show that for the number 17 in Westgren's sequences, the average time of recurrence is $10^{13}\tau$, with $\nu = 1.55$. For $\tau = \frac{1}{39}$ min, $\Theta \approx 500.000$ yr. Use Table 10.1.6 to show that for oxygen in a sphere of radius $a \geqslant 5\times10^{-5}$ cm at $T = 300°\text{K}$ and $\nu = 3\times10^{10}\,\text{cm}^{-3}$, the average time of recurrence of a state n such that $(n-\nu)/\nu = 0.01$ (1%) is $\Theta > 10^{68}\,\text{sec} = 3.17\times10^{61}$ yr. Interpret this result in view of the second law of thermodynamics (Chandrasekhar 1954).

10.2 PARTIAL DIFFERENTIAL EQUATIONS OF THE FIRST ORDER

Liouville's equation (10.1.2) is an example of a formulation of classical mechanics in terms of first-order partial differential equations. It is equivalent to the system (10.1.1) and it is linear. Another formulation of classical mechanics is given by the Hamilton–Jacobi nonlinear first-order partial differential equation. Let

$$S=\int_{t_1}^{t_2} L\,dt, \tag{10.2.1}$$

where L is the *Lagrangean* of the system (10.1.1); that is,

$$L=T(\mathbf{p},\mathbf{q})-V(\mathbf{q}),$$

where T is the kinetic energy and $V(\mathbf{q})$ is the potential energy of the system.

The function S is called *action*. Since

$$\frac{\partial L}{\partial \dot{q}_2}=p_i,$$

we have

$$dS=\mathbf{p}\,d\mathbf{q}; \tag{10.2.2}$$

hence considering S a function of $\mathbf{q}$, we have

$$\frac{dS}{dt}=\frac{\partial S}{\partial t}+\nabla_q S\cdot\dot{\mathbf{q}}=\frac{\partial S}{\partial t}+\mathbf{p}\cdot\dot{\mathbf{q}}.$$

In view of (10.2.1), we have

$$\frac{dS}{dt}=L;$$

hence

$$\frac{\partial S}{\partial t}=L-\mathbf{p}\cdot\dot{\mathbf{q}}=-H(\mathbf{p},\mathbf{q}).$$

In view of (10.2.2),

$$\mathbf{p}=\nabla_{\mathbf{q}}S;$$

hence we have

$$\frac{\partial S}{\partial t}+H(\nabla_{\mathbf{q}}S,\mathbf{q})=0. \tag{10.2.3}$$

The nonlinear equation (10.2.3) is called the *Hamilton–Jacobi equation* and it is equivalent to (10.1.1) (Landau and Lifshitz 1954). The general theory of first-order partial differential equations shows the equivalence of the partial differential equation to a system of ordinary differential equations. Let $\mathbf{x}=(x_1,\ldots,x_n)^T$ and set $\mathbf{p}=\nabla u(\mathbf{x})$. A first-order partial differential equation is a relation of the form

$$F(\mathbf{x},u,\mathbf{p})=0, \tag{10.2.4}$$

where $u(\mathbf{x})$ is the unknown function. The *initial value problem* for (10.2.4) is to find a solution $u(\mathbf{x})$ such that on a given $(n-1)$-dimensional surface Γ, the solution assumes a preassigned value

$$u(\mathbf{x})=\varphi(\mathbf{x}), \qquad \mathbf{x}\in\Gamma. \tag{10.2.5}$$

Suppose that $\mathbf{t}=(t_1,\ldots,t_{n-1})^T$ are independent parameters on Γ [i.e., Γ is given by $\mathbf{x}=\mathbf{x}(\mathbf{t})$]. Then we must have on Γ

$$\nabla_{\mathbf{t}}\varphi(\mathbf{x}(\mathbf{t}))=\frac{\partial \mathbf{x}}{\partial \mathbf{t}}\mathbf{p}, \tag{10.2.6}$$

where

$$\left(\frac{\partial \mathbf{x}}{\partial \mathbf{t}}\right)_{ij}=\frac{\partial x_i}{\partial t_j} \qquad (i=1,\ldots,n;\ \ j=1,\ldots,n-1).$$

From (10.2.4)–(10.2.6) we determine $\mathbf{p}=\mathbf{p}(\mathbf{t})$ on Γ. The initial problem (10.2.4), (10.2.5) is equivalent to the system of *characteristic* ordinary differential equations

$$\frac{d\mathbf{x}}{ds}=\nabla_{\mathbf{p}}F \tag{10.2.7}$$

$$\frac{d\mathbf{p}}{ds}=-\left(\frac{\partial F}{\partial u}\mathbf{p}+\nabla_{\mathbf{x}}F\right) \tag{10.2.8}$$

$$\frac{du}{ds}=\mathbf{p}\nabla_{\mathbf{p}}F \tag{10.2.9}$$

where s is a parameter (e.g., time).

Initial conditions for (10.2.7) and (10.2.8) at $s=0$ are given at any point of Γ. Let $\mathbf{x}_0=\mathbf{x}_0(\mathbf{t})$ be any point on Γ, then we set

$$\mathbf{x}|_{s=0}=\mathbf{x}_0(\mathbf{t}), \qquad \mathbf{p}|_{s=0}=\mathbf{p}(\mathbf{t}), \qquad u|_{s=0}=\varphi(\mathbf{x}_0(\mathbf{t})),$$

where $\mathbf{p}(\mathbf{t})$ is obtained by solving (10.2.4)–(10.2.6) at $\mathbf{x}_0(\mathbf{t})$. Thus $u(\mathbf{x})$ is determined along each characteristic curve of (10.2.7)–(10.2.9) (Courant and Hilbert 1962).

EXERCISE 10.2.1

Write down the characteristic equations for Liouville's equation and for the Hamilton–Jacobi equation. What is the initial manifold Γ in each case in a physical problem?

EXERCISE 10.2.2

A material point moves in a potential field of forces. Write down Newton's equations of motion, Liouville's equation, and the Hamilton–Jacobi equation describing the motion of the particle. Introduce into the equations dynamical friction (friction$=-\beta\mathbf{v}$, where $\mathbf{v}$ is the velocity and β is a friction coefficient). Introduce into the equations an external time-dependent force. Assume that the particle possesses an electrostatic charge and in addition to the potential field there is a magnetic field $\mathbf{B}(\mathbf{q}, t)$, where $\mathbf{q}$ is the coordinate of the particle (magnetic force$=\gamma\mathbf{v}\times\mathbf{B}$, where γ is a constant). Assume that the initial density of particles in Liouville's equation is given [e.g., $D(\mathbf{p},\mathbf{q},0)=\delta(\mathbf{p},-\mathbf{p}_0,\mathbf{q}-\mathbf{q}_0)$].

EXERCISE 10.2.3

Write down and solve Liouville's and Hamilton–Jacobi's equation for the damped harmonic oscillator

$$\ddot{x}+\beta\dot{x}+\omega^2x=0.$$

10.3 ELLIPTIC AND PARABOLIC PARTIAL DIFFERENTIAL EQUATIONS

Let $\mathbf{x}=(x_1,\ldots,x_n)^T$ be the variable in a bounded domain $\Omega\subset R^n$ with a smooth boundary $\partial\Omega$. We denote partial derivatives by subscripts, thus:

$$\frac{\partial u(\mathbf{x})}{\partial x_i}\equiv u_i(\mathbf{x}),$$

and we shall use the summation convention for double indices. Thus

$$a^{ij}u_{ij}=\sum_{i,j=1}^{n} a^{ij}\frac{\partial^2 u}{\partial x_i\,\partial x_j}.$$

Let $a^{ij}(\mathbf{x})$, $b^i(\mathbf{x})$, and $c(\mathbf{x})$, $i,\ j=1,2,\ldots,n$, be smooth functions in $\bar{\Omega}=\Omega\cup\partial\Omega$. The operator

$$Lu=a^{ij}(\mathbf{x})u_{ij}(\mathbf{x})+b^i(\mathbf{x})u_i(\mathbf{x})+c(\mathbf{x})u(\mathbf{x})$$

is called *elliptic* in Ω if for every vector $\boldsymbol{\xi}\in R^n$ and any $\mathbf{x}\in\Omega$,

$$E(\mathbf{x},\boldsymbol{\xi})\equiv a^{ij}(\mathbf{x})\xi_i\xi_j\neq 0.$$

If there exists a positive constant δ such that $E(\mathbf{x},\boldsymbol{\xi})\geqslant\delta|\boldsymbol{\xi}|^2$ for all $\mathbf{x}\in\Omega$ and $\boldsymbol{\xi}\in R^n$, then L is called a *uniformly elliptic* operator. It is well known that if L is uniformly elliptic and if $c(\mathbf{x})\leqslant 0$ in Ω, then the equation

$$Lu(x)=f(\mathbf{x})\qquad \text{in } \Omega \tag{10.3.1}$$

has a unique solution satisfying the boundary condition

$$\alpha(\mathbf{x})u(\mathbf{x})+\beta(\mathbf{x})\frac{\partial u(\mathbf{x})}{\partial\boldsymbol{\nu}}=\varphi(\mathbf{x}) \tag{10.3.2}$$

for $\mathbf{x}\in\partial\Omega$, where $\alpha(\mathbf{x})$, $\beta(\mathbf{x})$, and $\varphi(\mathbf{x})$ are given smooth function, $\alpha^2+\beta^2>0$. For the case $f(\mathbf{x})=0$, there exists a *Green's function* for the problem (10.3.1) and (10.3.2), that is, a function $G(\mathbf{x},\mathbf{y})$ defined for $\mathbf{x}\in\Omega$, $\mathbf{y}\in\partial\Omega$ such that the solution is given by

$$u(\mathbf{x})=\int_{\partial\Omega}G(\mathbf{x},\mathbf{y})\varphi(\mathbf{y})\,dS_{\mathbf{y}}.$$

For example, if $L=\Delta=\partial^2/\partial x^2+\partial^2/\partial y^2$ in the circle $\Omega=\{x^2+y^2\leqslant R^2\}$,

then

$$u(x,y)=\frac{1}{2\pi}\int_0^{2\pi}\frac{(R^2-\rho^2)\varphi(\theta)}{R^2-2R\rho\cos(\theta-\phi)+\rho^2}\,d\theta,$$

where $x=\rho\cos\phi$ and $y=\rho\sin\phi$; that is,

$$G(x,y,\xi,\eta)=\frac{R^2-\rho^2}{R^2-2R\rho\cos(\theta-\phi)+\rho^2},$$

where $\xi=R\cos\theta$ and $\eta=R\sin\theta$, $(\xi,\eta)\in\partial\Omega$.

EXERCISE 10.3.1

Find the Green's function for the problem

$$\Delta u=0 \quad \text{in } |\mathbf{x}|<R$$
$$u=\varphi \quad \text{on } |\mathbf{x}|=R$$

in R^n.

EXERCISE 10.3.2

Find the Green's function for the problem

$$\Delta u=0 \qquad \text{in } x_n>0$$
$$u=\varphi \qquad \text{on } x_n=0,$$

where $\mathbf{x}=(x_1,\ldots,x_n)^T$.

EXERCISE 10.3.3

Solve

$$\Delta u=0 \qquad \text{in } x^2+y^2<R^2$$
$$u=\varphi \qquad \text{on } x^2+y^2=R^2$$

by the method of *separation of variables*, that is,

$$u(\rho,\phi)=\sum_{n=0}^{\infty} R_n(\rho)F_n(\phi).$$

EXERCISE 10.3.4

Find the Green's function for

$$\begin{aligned} \Delta u&=0 && \text{in } r^2<x^2+y^2<R^2\\ u&=\varphi_1 && \text{on } x^2+y^2=r^2\\ u&=\varphi_2 && \text{on } x^2+y^2=R^2 \end{aligned}$$

EXERCISE 10.3.5

Find the Green's function for

$$\begin{aligned} \Delta u+cu&=0 && \text{in } x^2+y^2<R^2\\ u&=\varphi && \text{on } x^2+y^2=R^2 \qquad c=\text{constant}. \end{aligned}$$

Let Lu be defined by (10.3.1). A second-order linear operator L^* is called the (formal) *adjoint* to L if for every pair of smooth functions u and v in Ω such that $u=v=0$ on $\partial\Omega$,

$$\int_\Omega uLv\,d\mathbf{x}=\int_\Omega vL^*u\,d\mathbf{x}. \tag{10.3.3}$$

Using integration by parts (i.e., using the divergence theorem), we see that

$$L^*u=\left[a^{ij}(\mathbf{x})u(\mathbf{x})\right]_{,ij}-\left[b_i(\mathbf{x})u(\mathbf{x})\right]_{,i}+c(\mathbf{x})u(\mathbf{x}).$$

Indeed,

$$\begin{aligned} uLv-vL^*u&=\left(a^{ij}uv_{,i}\right)_{,j}-\left(a^{ij}u_{,i}v\right)_{,j}\\ &\quad-\left(a^{ij}_{,j}uv\right)_{,i}+\left(b_iuv\right)_{,i}=\operatorname{div}\mathbf{P}, \end{aligned}$$

where $\mathbf{P}=(P^1,\ldots,P^n)^T$ and

$$P^j=a^{ij}uv_i-a^{ij}u_iv-a_i^{ij}uv+b^juv.$$

We have used the fact $a^{ij}=a^{ji}$. Hence

(10.3.4)
$$\int[uLv-vL^*u]\,d\mathbf{x}=\int_{\partial\Omega}\left[a^{ij}\nu^j(uv_i-u_iv)\right.$$
$$\left.+\left(b_i-a_j^{ij}\right)\nu^iuv\right]dS_{\mathbf{x}}=0,$$

where $\boldsymbol{\nu}=(\nu^1,\ldots,\nu^n)^T$ is the outer normal to $\partial\Omega$. The last equality follows from the assumption that $u=v=0$ on $\partial\Omega$.

If $v=0$ on $\partial\Omega$ but $u\neq 0$ on $\partial\Omega$, then we have

(10.3.5)
$$\int_\Omega[uLv-vL^*u]\,d\mathbf{x}=\int_{\partial\Omega}ua^{ij}\nu^jv_i\,dS_{\mathbf{x}}=\int_{\partial\Omega}u\frac{\partial v}{\partial\mathbf{n}}\,dS_{\mathbf{x}},$$

where the directional derivative

$$\frac{\partial v}{\partial\mathbf{n}}=a^{ij}\nu^jv_i$$

is called the *conormal derivative* of v. In case $a^{ij}=\delta_{ij}$ (Kronecker's δ), we have

$$\frac{\partial v}{\partial\mathbf{n}}=\frac{\partial v}{\partial\boldsymbol{\nu}}.$$

EXERCISE 10.3.6

Use (2.3.4) in case $L=\Delta$ to obtain Green's identity

$$\int_\Omega[u\Delta v-v\Delta u]\,d\mathbf{x}=\int_{\partial\Omega}\left[u\frac{\partial v}{\partial\boldsymbol{\nu}}-v\frac{\partial u}{\partial\boldsymbol{\nu}}\right]dS_{\mathbf{x}}.$$

The operator L is called *self-adjoint* if $L=L^*$. The boundary value problem

$$Lu=f \quad \text{in } \Omega$$

$$Bu=\alpha u+\beta\frac{\partial u}{\partial\boldsymbol{\nu}}=0 \quad \text{on } \partial\Omega$$

is called self-adjoint if (10.3.3) holds for all functions u and v satisfying the boundary condition $Bu=Bv=0$.

The number λ is called an *eigenvalue* of the operator L if there exists a nonzero function $\varphi_\lambda(\mathbf{x})$ such that

$$L\varphi_\lambda(\mathbf{x})+\lambda\varphi_\lambda(\mathbf{x})=0 \qquad \text{in } \Omega$$

$$\varphi_\lambda(\mathbf{x})=0 \qquad \text{on } \partial\Omega.$$

The function $\varphi_\lambda(\mathbf{x})$ is called the eigenfunction corresponding to λ.

If L is a self-adjoint elliptic operator, then it has an infinite sequence $\{\lambda_n\}$ of eigenvalues, they are real, and $\lambda_n\to\infty$ as $n\to\infty$. If $c(\mathbf{x})\geqslant 0$, all eigenvalues are nonnegative. In this case

$$Lu=\left(a_{ij}(\mathbf{x})u_i\right)_j+c(\mathbf{x})u.$$

The eigenfunctions $\varphi_{\lambda_n}(\mathbf{x})$ are orthogonal; that is,

$$\int_\Omega \varphi_{\lambda_n}(\mathbf{x})\varphi_{\lambda_m}(\mathbf{x})\,d\mathbf{x}=0 \qquad \text{if} \quad m\neq n.$$

For any square-integrable function $f(\mathbf{x})$ in Ω, we have

$$\int_\Omega \Big|f(\mathbf{x})-\sum_{n=1}^{m} c_n\varphi_{\lambda_n}(\mathbf{x})\Big|^2\,d\mathbf{x}\to 0$$

as $m\to\infty$, where

$$c_n=\frac{\int_\Omega f(\mathbf{x})\varphi_{\lambda_n}(\mathbf{x})\,d\mathbf{x}}{\int_\Omega |\varphi_{\lambda_n}(\mathbf{x})|^2\,d\mathbf{x}}.$$

The solution of the problem

$$Lu=f(\mathbf{x}) \qquad \text{in } \Omega \tag{10.3.6}$$

$$u=0 \qquad \text{on } \partial\Omega$$

in this case is given by

$$u(\mathbf{x})=\sum_{n=1}^{\infty} a_n\varphi_{\lambda_n}(\mathbf{x}),$$

where $a_n = c_n/\lambda_n$ (i.e., Green's function for the problem 10.3.6) is given by

$$G(\mathbf{x},\mathbf{y}) = \frac{\sum_{n=1}^{\infty} \varphi_{\lambda_n}(\mathbf{x})\varphi_{\lambda_n}(\mathbf{y})}{\lambda_n \int_{\Omega} |\varphi_{\lambda_n}(\mathbf{z})|^2 \, d\mathbf{z}}.$$

The boundary value problem

$$Lu = 0 \qquad \text{in } \Omega$$

$$u = g(\mathbf{x}) \qquad \text{on } \partial\Omega$$

can be reduced to (10.3.6) by the substitution

$$u(\mathbf{x}) = U(\mathbf{x}) + G(\mathbf{x}),$$

where $G(\mathbf{x})$ is any smooth function in Ω such that $G(\mathbf{x}) = g(\mathbf{x})$ on $\partial\Omega$ and U satisfies (10.3.6) with $f(\mathbf{x}) = -LG(\mathbf{x})$ (Courant and Hilbert 1962). The equation

$$\frac{\partial u}{\partial t} - Lu = f(\mathbf{x}, t), \tag{10.3.7}$$

where the coefficients of L may depend on t, is called *parabolic* if L is elliptic and $E(\mathbf{x}, \xi, t) \geqslant 0$. If $E(\mathbf{x}, \xi, t) \geqslant \delta|\xi|^2$ for all $(\mathbf{x}, t)$ and $\xi \in R^n$, then (10.3.7) is called uniformly parabolic. If (10.3.7) is uniformly parabolic in $R^n \times (0, T)$, then the problem (10.3.7) with the initial condition

$$u(\mathbf{x}, 0) = g(\mathbf{x}) \tag{10.3.8}$$

has a unique solution.

The problem (10.3.7), (10.3.8) is called the *Cauchy problem.* It has a *fundamental solution* [i.e., a function $\Gamma(\mathbf{x}, s, \mathbf{y}, t)$ $(s < t)$] such that the solution of the Cauchy problem is given by

$$u(\mathbf{y}, t) = \int_{R^n} \Gamma(\mathbf{x}, 0, \mathbf{y}, t) g(\mathbf{x}) \, d\mathbf{x}$$

$$- \int_0^t \int_{R^n} \Gamma(\mathbf{x}, s, \mathbf{y}, t) f(\mathbf{x}, s) \, d\mathbf{x} \, ds.$$

The parabolic equation

$$\frac{\partial u}{\partial t} = \tfrac{1}{2} D\, \Delta u$$

is called the *diffusion* or the *heat equation*, and $\frac{1}{2}D$ is called the diffusion coefficient. The function

$$\Gamma(\mathbf{x}, s, \mathbf{y}, t) = [2\pi D(t-s)]^{-n/2} e^{-|\mathbf{x}-\mathbf{y}|^2/2D(t-s)}$$

is the fundamental solution of the heat equation. The fundamental solution satisfies the equation

$$\frac{\partial \Gamma}{\partial t} = L_{\mathbf{y}}\Gamma, \qquad \mathbf{y} \neq \mathbf{x}, \quad t > s$$

and the initial condition

$$\Gamma(\mathbf{x}, s, \mathbf{y}, t) \to \delta(\mathbf{x}-\mathbf{y}) \qquad \text{as} \quad t \downarrow s.$$

EXERCISE 10.3.7

Find the fundamental solution for the equation

$$a_{ij}u_{ij} + b_i u_i + cu = \frac{\partial u}{\partial t},$$

where a_{ij}, b_i, and c are constants.

EXERCISE 10.3.8

Find the fundamental solution of the equation

$$a_{ij}u_{ij} + b_{ij}(x_i u)_j = \frac{\partial u}{\partial t},$$

where a_{ij} and b_{ij} are constants. Under what conditions does $\Gamma(\mathbf{x}, \mathbf{y}, t) \to \Gamma(\mathbf{y})$ as $t \to \infty$, where $\Gamma(\mathbf{y})$ is a limit function? Here use the fact that

$$\Gamma(\mathbf{x}, s, \mathbf{y}, t) = \Gamma(\mathbf{x}, 0, \mathbf{y}, t-s).$$

Consider first the case $a_{ij} = \delta_{ij}$.

EXERCISE 10.3.9

Let $\Gamma(\mathbf{x}, s, \mathbf{y}, t)$ be the fundamental solution of the problem

$$[a_{ij}(\mathbf{x}, t)u(\mathbf{x}, t)]_{ij} + [b_i\mathbf{x}, t)u(\mathbf{x}, t)]_i = \frac{\partial u}{\partial t}.$$

Show that

$$\int_{R^n} \Gamma(\mathbf{x}, s, \mathbf{y}, t)\, d\mathbf{y} = 1.$$

(Assume that $\Gamma \to 0$ as $|\mathbf{y}| \to \infty$.)

Let Ω be a bounded domain in R^n with a smooth boundary $\partial\Omega$. The initial boundary value problem

$$\text{(10.3.9)} \qquad \frac{\partial u}{\partial t} - Lu = f(\mathbf{x}, t) \qquad \text{in } \Omega \times (0, T)$$

$$u(\mathbf{x}, t) = g(\mathbf{x}, t) \qquad \mathbf{x} \in \partial\Omega$$

$$u(\mathbf{x}, 0) = h(\mathbf{x}) \qquad \mathbf{x} \in \Omega$$

has a unique solution for every $T > 0$. The problem (10.3.9) can be transformed by substitution to the case $g \equiv 0$. Then it has a fundamental solution $\Gamma(\mathbf{x}, s, \mathbf{y}, t)$ such that

$$u(\mathbf{y}, t) = \int_\Omega \Gamma(\mathbf{x}, 0, \mathbf{y}, t)h(\mathbf{x})\, d\mathbf{x} - \int_0^t \int_\Omega \Gamma(\mathbf{x}, s, \mathbf{y}, t) f(\mathbf{x}, s)\, d\mathbf{x}\, ds.$$

Consider, for example, the equation

$$\text{(10.3.10)} \qquad \frac{\partial u}{\partial t} = \frac{\partial^2 u}{\partial x^2}, \qquad 0 < x < 1, t > 0$$

$$\text{(10.3.11)} \qquad u(0, t) = u(1, t) = 0, \qquad t > 0$$

$$\text{(10.3.12)} \qquad u(x, 0) = h(x), \qquad 0 < x < 1.$$

The solution can be found by the method of separation of variables:

$$u(x, t) = \sum_{n=0}^{\infty} R_n(x) T_n(t),$$

where for each n the function $R_n(x)T_n(t)$ is a solution of (10.3.10) and

(10.3.11); that is,

$$\frac{R_n''(x)}{R_n(x)} = \frac{T'(t)}{T(t)} = -\lambda_n^2.$$

We must have $R_n(x) = a_n \sin \lambda_n x$ as $R_n(0) = 0$. Since $R_n(1) = 0$, we must have $\lambda_n = n\pi$. Thus

$$u(x,t) = \sum_{n=1}^{\infty} a_n \sin n\pi x e^{-n^2\pi^2 t}.$$

Setting $t = 0$, we must have by (10.3.12),

$$h(x) = \sum_{n=1} a_n \sin n\pi x;$$

hence

$$a_n = 2\int_0^1 h(x) \sin n\pi x \, dx.$$

In general, if L is independent of t and L is a self-adjoint uniformly elliptic operator in Ω, then

$$\Gamma(\mathbf{x},\mathbf{y},t) = \frac{\sum_{n=1}^{\infty} \varphi_{\lambda_n}(\mathbf{x}) \varphi_{\lambda_n}(\mathbf{y}) e^{-\lambda_n t}}{\int_\Omega |\varphi_{\lambda_n}(\mathbf{z})|^2 \, d\mathbf{z}},$$

where λ_n and $\varphi_{\lambda_n}(\mathbf{x})$ are the eigenvalues and eigenfunctions of L, respectively.

EXERCISE 10.3.10

Let L be independent of t and assume that

$$L_1 = G(\mathbf{x})L$$

is a self-adjoint uniformly elliptic operator in Ω. Show that the fundamental solution of (10.3.9) is given by

$$\Gamma(\mathbf{x},\mathbf{y},t)=\sum_{n=1}^{\infty} G(\mathbf{x})\varphi_{\lambda_n}(\mathbf{x})\varphi_{\lambda_n}(\mathbf{y})e^{-\lambda_n t},$$

where $\varphi_{\lambda_n}(\mathbf{x})$ are suitably normalized.

EXERCISE 10.3.11

Show that if Γ is as in Exercise 10.3.10, then the function

$$v(\mathbf{y})=\int_{\Omega}\int_0^{\infty}\Gamma(\mathbf{x},\mathbf{y},t)\,dt\,d\mathbf{x}$$

is the solution of the boundary value problem

$$Lv(\mathbf{y})=-1 \qquad \text{in } \Omega$$

$$v(\mathbf{y})=0 \qquad \text{on } \partial\Omega.$$

EXERCISE 10.3.12

Show that if

$$\int_{-\infty}^{\infty} e^{-\phi(x)}\,dx<\infty,$$

then the fundamental solution Γ of the equation

$$\frac{\partial u}{\partial t}=\frac{\frac{1}{2}\partial^2 u}{\partial x^2}+\frac{\partial(u\,\partial\phi/\partial x)}{\partial x}$$

satisfies

$$\Gamma(x,y,t)\rightarrow\frac{e^{-\phi(y)}}{\int_{-\infty}^{\infty} e^{-\phi(x)}\,dx}$$

as $t\rightarrow\infty$ (Courant and Hilbert 1962; Mandl 1968).

If the coefficients of L are bounded functions of $-t$, $-\infty<t<\infty$, then the problem (10.3.10), (10.3.11) has a unique solution for bounded $f(\mathbf{x},t)$ and $g(\mathbf{x},t)$. In particular, if L is independent of t and so are f and g, then the solution is independent of t (Friedman and Schuss 1971; Schuss 1972, 1977).

EXERCISE 10.3.13

Solve the backward boundary value problem

$$\frac{\partial^2 u}{\partial x^2}+xb(t)\frac{\partial u}{\partial x}=\frac{\partial u}{\partial t}, \qquad -\infty<t\leqslant 0$$

$$u(x,t)\to c_1 \quad \text{as} \quad x\to\infty$$

$$u(x,t)\to c_2 \quad \text{as} \quad x\to-\infty,$$

where $b(t)$ is a bounded smooth function. Try a solution of the form

$$u(x,t)=K_1\int_0^{xa(t)}e^{-s^2/2}\,ds+K_2,$$

where K_1 and K_2 are constants and $a(t)$ a function to be determined.

Appendix

Elements of Electrical Circuits

The basic elements of an electrical circuits are resistors, coils (inductance and self-inductance), capacitors, and power sources (direct and alternating). The resistance, current, and voltage between two points a and b in a circuit are denoted by R_{ab}, I_{ab}, and V_{ab}, respectively, and they may be functions of time t. Ohm's law states that

$$V_{ab}(t)=R_{ab}(t)I_{ab}(t). \tag{A.1}$$

If a coil is placed between a and b, whose self-inductance coefficient is L_{ab}, then

$$V_{ab}(t)=L_{ab}\frac{d}{dt}I_{ab}(t). \tag{A.2}$$

If a capacitor is placed between a and b, then

$$I_{ab}(t)=C_{ab}\frac{d}{dt}V_{ab}(t), \tag{A.3}$$

where C_{ab} is the capacitance of the capacitor.

The quantity $Q_{ab}(t)=C_{ab}V_{ab}(t)$ represents the charge on the capacitor. Two inductance elements can interact by inducing electromotive forces in each other according to the following inductance law. Let L_1 and L_2 be two coils whose inductance coefficient equals M. Then

$$\begin{aligned} V_1(t)&=L_1\dot{I}_1(t)+M\dot{I}_2(t)\\ V_2(t)&=L_2\dot{I}_2(t)+M\dot{I}_1(t), \end{aligned} \tag{A.4}$$

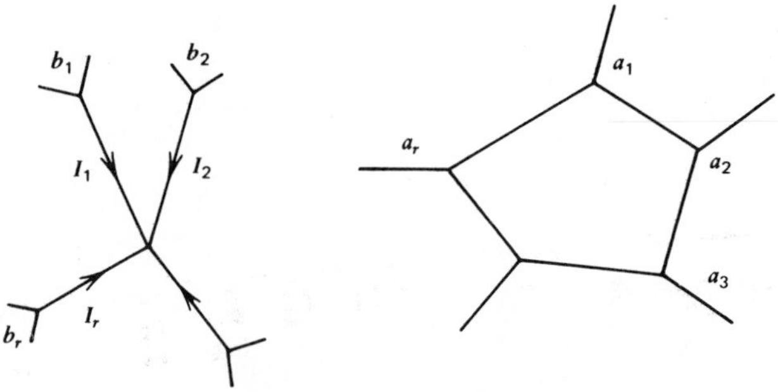

Figure A.1 Kirchhoff's laws.

where I_k and $V_k (k=1,2)$ are the current and voltage on the coils. The following two laws, due to Kirchhoff, hold (see Fig. A.1):

(i) The sum of all currents entering any node in a circuit is zero.

(ii) The sum of all voltage drops along any closed loop in a circuit equals zero.

$$\sum_{j=1}^{n} I_j = 0, \qquad \sum_{j=1}^{n} V_{a_j, a_{j+1}} = 0 \qquad (a_{n+1} = a_1).$$

Using Laplace transform language, we denote

$$\frac{1}{s} f = \int_0^\infty f(t)\, dt.$$

Equations (A.1)–(A.3) can be stated now in the form

$$V(t) = Z(s) I(t),$$

where $Z(s) = R$, $Z(s) = Ls$, and $Z(s) = 1/Cs$, respectively. The function $Z(s)$ is called the impedance operator between a and b.

The function $G(s) = 1/Z(s)$ is called the conductance operator. Thus we have

$$I(t) = G(s) V(t)$$

in (A.1)–(A.3). In (A.4) we have

$$V_1(t) = L_1 s I_1(t) + M s I_2(t)$$

$$V_2(t) = L_2 s I_2(t) + M s I_1(t).$$

The following symbols are used in sketching electrical circuits:

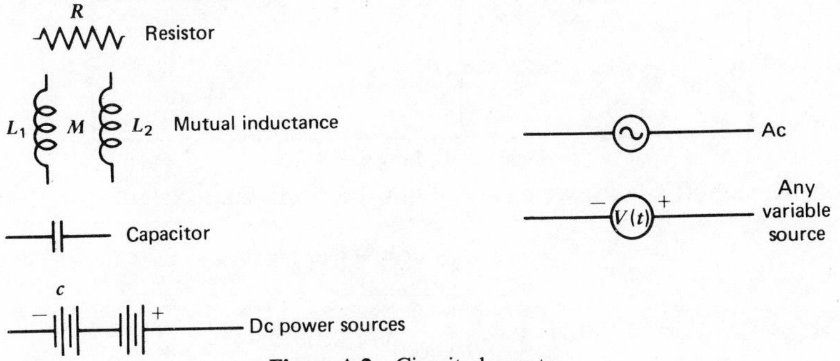

Figure A.2 Circuit elements.

An *oscillator* consists of the following elements:

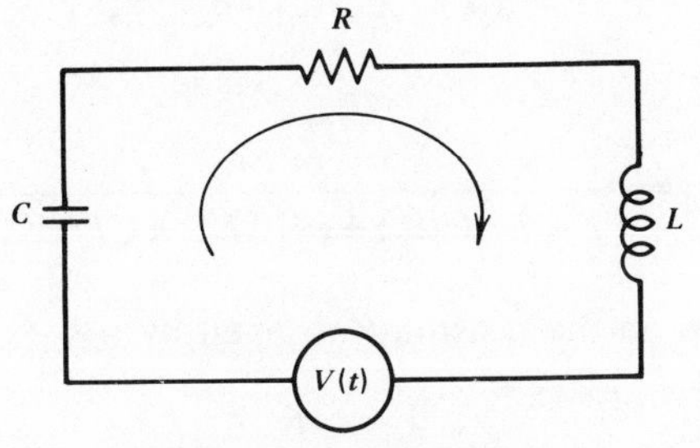

Figure A.3 An oscillator.

Using Kirchhoff's laws, one easily gets

$$\left(Ls+R+\frac{1}{Cs}\right)I(t)=V(t); \tag{A.5}$$

hence

$$G(s)=\frac{Cs}{LCs^2+RCs+1}.$$

If $V(t)=0$ and $R=0$, then (A.5) yields

$$I(t)=k\cos\left(\frac{t}{\sqrt{LC}}+\delta\right),$$

where δ depends on initial conditions. Thus $\omega=1/\sqrt{LC}$ is the frequency of oscillations in the circuit.

A *filter* is a circuit consisting of the following elements:

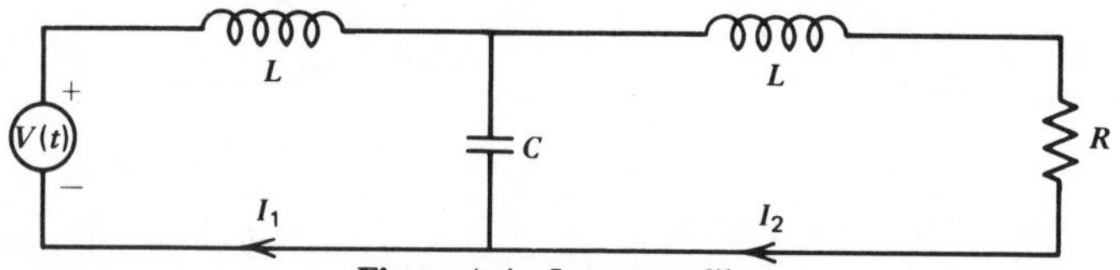

Figure A.4 Low pass filter.

From Krichhoff's laws, we easily obtain the following system

$$LsI_1+LsI_2+RI_2-V(t)=0$$

$$LsI_2+RI_2+\frac{1}{Cs}(I_2-I_1)=0.$$

Given $V(t)=be^{i\omega t}$ the solutions are given by

$$I_1=a_1e^{i\omega t}, \qquad I_2=a_2e^{i\omega t},$$

where

$$a_2=\frac{b}{(R-LRC\omega^2)+i\omega(2LC-CL^2\omega^2)}.$$

The voltage drop on the resistor R is given by

$$V=I_2R$$

Hence

$$|V|=|a_2|R\equiv a=\frac{bR/C}{\sqrt{[(R/C)-LR\omega^2]^2+\omega^2[(2L/C)-L^2\omega^2]^2}}.$$

If ω is small, we have

$$\frac{a}{b}\approx 1,$$

so that small frequencies do not affect the amplitude of the voltage oscillations on the resistor (i.e., low frequencies pass through the filter with small changes only). For high frequencies ω, we have

$$\frac{a}{b}\approx\frac{R}{cL^2\omega^2}\ll 1;$$

that is, high frequencies virtually do not propagate through the filter.

References

1. Anderson, R. F. and Orey, S. "Small random perturbations of dynamical systems with reflecting boundary," *Nagoya Math. J.*, **60**(1976), 189–216.
2. Aronson, D. G. "Linear parabolic differential equations containing a small parameter," *J. Rat. Mech. Anal.*, **5**(1956), 6, 1003–1014.
3. Aronson, D. G. "The fundamental solution of a linear parabolic equation containing a small parameter," *Ill. J. Math.* **3**(1959), 580–619.
4. Benson, S. W. *The Foundations of Chemical Kinetics*. McGraw-Hill, New York, 1960.
5. Benssousan, A., Lions, J. L., and Papanicolaou, G. *Asymptotic Methods for Media with Periodic Structure*. North-Holland, Amsterdam, 1978.
6. Bobrovsky, B. Z. and Schuss, Z. "Multiple cycle slips in a PLL," to appear.
7. Bobrovsky, B. Z. and Schuss, Z. "Singular perturbation method for the computation of the mean fir first passage time in a non-linear filter," to appear.
8. Bobrovsky, B. Z., Emanuel, A., and Schuss, Z. "Performance of a quasi-optimum FM demodulator," to appear.
9. Chandrasekhar, S. "Stochastic problems in physics and astronomy," *Noise and Stochastic Processes* (N. Wax, Ed.). Dover, New York, 1954.
10. Coddington, E. and Levinson, N. *Theory of Ordinary Differential Equations*. McGraw-Hill, New York, 1955.
11. Cohen, J. K. and Lewis, R. M. "A ray method for the asymptotic solution of the diffusion equation," *J. Inst. Math. Appl.*, **3**(1967), 266–290.
12. Courant, R. and Hilbert, D. *Methods of Mathematical Physics*, Vol. 2. Interscience, New York, 1962.
13. Cramér, H. "Über eine Eigenschaft der normaler Verteilungsfunktion," *Math. Z.*, **41**(1936), 405.
14. Crow, J. F. and Kimura, M. *An Introduction to Population Genetics Theory*. Harper & Row, New York, 1970.
15. Dyhkin, E.B. *Markov Processes* I, II Springer Verlag, N.Y. (1965).
16. Einstein, A. *Investigations on the Theory of the Brownian Movement*. Dover, New York, 1956.
17. Eliezer, S. and Schuss, Z. "Stochastic (white noise) analysis of resonant absorption in laser generated plasma," *Phys. Lett. A* Vol. 70 # **4**(1979), p. 307–310.
18. Ewens, W. J. "Conditional diffusion process in population genetics," *Theor. Popul. Biol.* **4**(1973), 21–34.
19. Feller, W. Diffusion processes in genetics. *Proc. 2nd Berkeley Symp.* (1951), 227–246.
20. Feller, W. "The parabolic differential equation and the associated semi-groups of transformations," *Ann. Math.*, **55**(1952), 468–519.

21. Feller, W. "Diffusion processes in one dimension," *Trans. AMS*, **97**(1954), 1–31.

22. Feller, W. "Generalized second order differential operators and their lateral conditions," *Ill. J. Math.*, **1**(1957), 459–504.

23. Feller, W. *An Introduction to Probability Theory and Its Applications*, Vols. 1 and 2. Wiley, New York, 1971.

24. Fisz, M. *Probability Theory and Mathematical Statistics*. Wiley, New York, 1963.

25. Friedman, A. *Partial Differential Equations of Parabolic Type*. Prentice-Hall, Englewood Cliffs, N.J., 1964.

26. Friedman, A. "The asymptotic behavior of the first real eigenvalue of a second order elliptic operator with a small parameter in the highest derivatives," *Indiana Univ. Math. J.*, **22**(1973), 1005–1015.

27. Friedman, A. *Stochastic Differential Equations*. Academic Press, New York, 1976.

28. Friedman, A. and Schuss, Z. "Degenerate evolution equations in Hilbert space," *Trans. AMS*, **161**(1971), 401–427.

29. Fürth, R. "Statistik und Wahrscheinlichkeits-nachwirkung," *Phys. Z.*, **19**(1918), 421–426.

30. Gihman, I. I. and Skorokhod, A. V. *Stochastic Differential Equations*. Springer-Verlag, Berlin, 1972.

31. Girifalco, L. *Atomic Migration in Crystals*. Blaisdell, Waltham, Mass., 1964.

32. Glasstone, S., Laidler, J. J., and Eyring, H. *The Theory of Rate Processes*. McGraw-Hill, New York, 1941.

33. Glyde, H. R. "Rate processes in solids," *Rev. Mod. Phys.*, **39**(1967), 2.

34. Hale, J. *Ordinary Differential Equation*. Wiley-Interscience, New York, 1969.

35. Halmos, P. R. *Measure Theory*. D. Van Nostrand CO., Princeton, N.J., 1959.

36. Holland, C. J. "The regular expansion in the Neuman problems for elliptic equations," *Commun. P. D. E.*, **1**(1976), 3, 191–213.

37. Itô, K. and McKean, H. *Diffusion Processes and Their Sample Path*. Springer-Verlag, Berlin, 1965.

38. Jazwinski, A. H. *Stochastic Processes and Filtering Theory*. Academic Press, New York, 1970.

39. Kac, M. *Probability and Related Topics in Physical Sciences*. Interscience, New York, 1959.

40. Kamienomostskaya, S. "On equations of elliptic type and parabolic type with a small parameter in the highest derivative," *Math. Sch.*, **31**(73), 3(1952), 703–708.

41. Kamienomostskaya, S. "The first boundary value problem for elliptic equations containing a small parameter," *Jzv. Akad. Nauk. U.S.S.R.*, **19**(1955), 345–360.

42. Kamin, S. "On elliptic perturbation of a first order operator with a singular point of attracting type." *Indiana Univ. Math. J.*, **27**(1978), 6, 935–951.

43. Kamke, E. *Theory of Sets*, Dover, New York, 1950.

44. Khasminskii, R. Z. "On diffusion processes with a small parameter," *Jzv. Akad. Nauk. U.S.S.R. Sev. Mat.*, **27**(1963), 1280–1300.

45. Khasminskii, R. Z. *Stability of Systems of Differential Equations with Random Parameters*. Nauk, Moscow, 1969.

46. Khinchine, A. I. *Asymptotische Gesetze der Wahrscheinlichkeitsrechnung*. Springer-Verlag Berlin, 1933.

47. Kramers, H. A. "Brownian motion in a field of force and the diffusion model of chemical reactions," *Physica*, **7**(1940), 284–304.

48. Kushner, H. J. "On the dynamical equations of conditional probability density functions with applications to optimal stochastic control theory," *J. Math. Anal. Appl.*, **8**(1964), 332–344.

49. Kushner, H. J. "On the weak convergence of interpolated Markov chains to a diffusion," *Ann. Probab.*, **2**(1974), 1, 40–50.

50. Ladyzhenskaya, O. A. "Linear partial differential equations containing a small parameter multiplying derivatives of highest order," *Vestn. Leningr. Univ.*, **7**(1957), 104–120.

51. Lamb, Sir H. *Hydrodynamics*. Dover, New York, 1945.

52. Landau, L. and Lifshitz, E. *Mechanics*. Nauka, Moscow, 1954.

53. Larsen, E. and D'Arruda, J. "Asymptotic theory of the linear transport equation for small mean free paths. I," *Phys. Rev. A.*, **13**(1976), 5, 1932–1939.

54. Larsen, E. W. and Keller, J. B. "Asymptotic solution of neutron transport problems for small mean free paths," *J. Math. Phys.* **15**(1974), 75–81.

55. Larsen, E. and Schuss, Z. "Diffusion tensor for atomic migration in crystals," *Phys. Rev. B.*, **18**(1978), 5, 2050–2058.

56. Levikson, B. "The effect of random environments on the evolutionary process of gene frequencies—a mathematical analysis." Thesis, Tel-Aviv University, 1974.

57. Levikson, B. and Schuss, Z. "Nonhomogeneous diffusion approximation to a genetic model," *J. Math. Pures Appl.* **56**(1977), 55–65.

58. Levinson, N. "The first boundary value problem for $\varepsilon\Delta u + Au_x + Bu_y + Cu = D$ for small ε," *Ann. Math.*, **51**(1950), 2, 428–445.

59. Lighthill, M. J. *Introduction to Fourier Analysis and Generalized Functions*. Cambridge University Press, New York, 1958.

60. Lindsey, W. C. *Synchronization Systems in Communication and Control*. Prentice-Hall, Englewood Cliffs, N.J., 1972.

61. Lions, J. L. *Perturbations Singulières dans les Problèmes aux Limites et en Contrôle Optimal* Springer Verlag, New York (1973).

62. Livne, A. and Schuss, Z. "Singular perturbations for degenerate elliptic equations of second order," *Arch. Rat. Mech. Anal.*, **52**(1973), 3, 233–243.

63. Ludwig, D. *Stochastic Population Theories*. In *Lecture Notes in Biomathematics*, Vol. 3. Springer-Verlag, New York, 1974.

64. Ludwig, D. "Persistence of dynamical systems under random perturbations," *SIAM Rev.*, **17**(1975), 4, 605–640.

65. McKean, H. P., Jr *Stochastic Integrals*. Academic Press, New York, 1969.

66. Mandl, P. *Analytical Treatment of One Dimensional Markov Processes*. Springer-Verlag, New York, 1968.

67. Mangel, M. and Ludwig, D. "Probability of extinction in a stochastic competition," *SIAM J. Appl. Math.*, **33**(1977), 92, 256–266.

68. Matkowsky, B. "On boundary layer problems exhibiting resonance," *SIAM Rev.*, **17**(1975), 1, 82–100.

69. Matkowsky, B. and Schuss, Z. "The exit problem for randomly perturbed dynamical systems," *SIAM J. Appl. Math.*, **33**(1977), 12, 365–382.

70. Matkowsky, B. and Schuss, Z. "The eigenvalues of the Fokker-Planck operator and the approach to equilibrium for diffusions in potential fields," *SIAM J. Appl. Math.*, to appear.

71. Miller, G. F. "The evolution of eigenvalues of a differential equation arising in a problem in genetics," *Proc. Camb. Phil. Soc.*, **58**(1962), 588–593.

72. Mizel, V. J. "Boundary layer problems for an elliptic equation in a neighborhood of a singular point," *Proc. AMS*, **8**(1957), 62–67.

73. Natanson, I. P. *Theory of Functions of a Real Variable*, Vol. 7 Unger, New York, 1961.

74. Oleinik, O. A. "Equations of elliptic type containing a small parameter," *Mat. Sb.* (73):1, **31**(1952), 104–118.

75. Olver, F. W. J. "Asymptotic methods and singular perturbations," *SIAM-AMS Proc.*, **10**(1976), 105–117.

76. O'Malley, R. E., Jr. *Introduction to Singular Perturbations*. Academic Press, New York, 1974.

77. Ornstein, L. S. and Uhlenbeck, G. E. "On the theory of the Brownian motion," *Phys. Rev.*, **36**(1930), 1, 823–841.

78. Papanicolaou, G. "Introduction to asymptotic analysis of stochastic equations," Lecture Notes, AMS Seminar, Renselaer Polytechnic Institute, Troy, N.Y., 1975.

79. Schuss, Z. "Regularity theorems for solutions of a degenerate evolution equation," *Arch. Rat. Mech. Anal.*, **46**(1972), 3, 200–211.

80. Schuss, Z. "Degenerate and backward parabolic equations," *J. Appl. Anal.*, **7**(1977–78), 111–119.

81. Schuss, Z. "Singular perturbation methods in stochastic differential equations of mathematical physics," *SIAM Rev.*, **22**(1980), 2, 119–155.

82. Schuss, Z. and Matkowsky, B. "The exit problem: a new approach to diffusion across potential barriers," *SIAM J. Appl. Math.*, **36**(1979), 43, 604–623.

83. Smoluchowski, M. "Drei Vortrage uber Diffusion, Brownsche Beregung und Koagulation von Kolloidteilchen," *Phys. Z.*, **17**(1916), 557–585.

84. Snyder, D. L. *The State-Variable Approach to Continuous Estimation with Applications to Analog Communication Theory*. M.I.T. Press, Cambridge, Mass., 1969.

85. Stratonovich, R. L. *Conditional Markov Processes and Their Application to the Theory of Optimal Control*. Elsevier, New York, 1965.

86. Svedberg, Th. *Z. Phys. Chem.*, **77**(1911), 177.

87. Tausworth, R. "Simplified formula for mean cycle-slip time for phase locked loops with steady state error," *IEEE. Trans. Commun.* (1972), 331–337.

88. Van Trees, H. L. *Detection, Estimation, and Modulation Theory I, II*. Wiley, New York, 1970.

89. Ventzel, A. D. and Freidlin, M. I. "On small random perturbations of dynamical systems," *Uspekhi Mat. Nauk.*, **25**(1970), 3–55.

90. Vineyard, G. H. "Frequency factors and isotope effects in solid state rate processes," *J. Phys. Chem. Solids*, **3**(1957), 121–127.

91. Vishik, M. I. and Liusternik, L. A. "Regular degeneration and boundary layer for linear differential equations with a small parameter," *A.M.S. Transl*, **20**(1962), 2, 239–364.

92. Viterbi, A. J. *Principles of Coherent Communications*. McGraw-Hill, New York, 1966.

93. Watson, G. N. *Theory of Bessel Functions* (Cambridge 1922).

94. Westgren, A. *Ark. Mat. Astron. Fys.* **11**(1916), 8, 14; and **13**(1918), 14.

95. Wong, E. and Zakai, M. "On convergence of ordinary integral to stochastic integrals," *Ann. Math. Stat.*, **36**(1965), 1560–1564.

96. Zadeh, L. A. and Desoer, C. A. *Linear System Theory, The State Space Approach*. McGraw-Hill, New York, 1963.

Index

Applied Probability and Statistics (Continued)

FLEISS • Statistical Methods for Rates and Proportions, *Second Edition*
GALAMBOS • The Asymptotic Theory of Extreme Order Statistics
GIBBONS, OLKIN, and SOBEL • Selecting and Ordering Populations: A New Statistical Methodology
GNANADESIKAN • Methods for Statistical Data Analysis of Multivariate Observations
GOLDBERGER • Econometric Theory
GOLDSTEIN and DILLON • Discrete Discriminant Analysis
GROSS and CLARK • Survival Distributions: Reliability Applications in the Biomedical Sciences
GROSS and HARRIS • Fundamentals of Queueing Theory
GUPTA and PANCHAPAKESAN • Multiple Decision Procedures: Theory and Methodology of Selecting and Ranking Populations
GUTTMAN, WILKS, and HUNTER • Introductory Engineering Statistics, *Second Edition*
HAHN and SHAPIRO • Statistical Models in Engineering
HALD • Statistical Tables and Formulas
HALD • Statistical Theory with Engineering Applications
HARTIGAN • Clustering Algorithms
HILDEBRAND, LAING, and ROSENTHAL • Prediction Analysis of Cross Classifications
HOEL • Elementary Statistics, *Fourth Edition*
HOLLANDER and WOLFE • Nonparametric Statistical Methods
JAGERS • Branching Processes with Biological Applications
JESSEN • Statistical Survey Techniques
JOHNSON and KOTZ • Distributions in Statistics
Discrete Distributions
Continuous Univariate Distributions—1
Continuous Univariate Distributions—2
Continuous Multivariate Distributions
JOHNSON and KOTZ • Urn Models and Their Application: An Approach to Modern Discrete Probability Theory
JOHNSON and LEONE • Statistics and Experimental Design in Engineering and the Physical Sciences, Volumes I and II, *Second Edition*
JUDGE, GRIFFITHS, HILL and LEE • The Theory and Practice of Econometrics
KALBFLEISCH and PRENTICE • The Statistical Analysis of Failure Time Data
KEENEY and RAIFFA • Decisions with Multiple Objectives
LANCASTER • An Introduction to Medical Statistics
LEAMER • Specification Searches: Ad Hoc Inference with Nonexperimental Data
McNEIL • Interactive Data Analysis
MANN, SCHAFER and SINGPURWALLA • Methods for Statistical Analysis of Reliability and Life Data
MEYER • Data Analysis for Scientists and Engineers
MILLER, EFRON, BROWN, and MOSES • Biostatistics Casebook
OTNES and ENOCHSON • Applied Time Series Analysis: Volume I, Basic Techniques
OTNES and ENOCHSON • Digital Time Series Analysis
POLLOCK • The Algebra of Econometrics
PRENTER • Splines and Variational Methods
RAO and MITRA • Generalized Inverse of Matrices and Its Applications
SCHUSS • Theory and Applications of Stochastic Differential Equations
SEAL • Survival Probabilities: The Goal of Risk Theory
SEARLE • Linear Models
SPRINGER • The Algebra of Random Variables
UPTON • The Analysis of Cross-Tabulated Data
WEISBERG • Applied Linear Regression
WHITTLE • Optimization Under Constraints